Integrated Farming
in Human Development

Proceedings of a Workshop

March 25 - 29, 1996
Tune Landboskole
Denmark

Organizer: **The Danish Agricultural and Rural Development Advisers Forum**

Editors: **Frands Dolberg and Poul Henning Petersen**

Integrated Farming in Human Development
Editors: Frands Dolberg and Poul Henning Petersen
1. udgave 1997

© DSR Forlag og forfatterne

ISBN 87 7432 459 4

Tryk: DSR Tryk

Ekspedition:
DSR Boghandel
Thorvaldsensvej 40
1871 Frederiksberg C
Tlf: 31 35 76 22
Fax 31 35 27 90

Mekanisk, fotografisk eller anden gengivelse af denne bog, eller dele deraf, er ikke tilladt ifølge gældende dansk lov om ophavsret.

Contents

Introduction

Poultry and Pigs

Recycling and Energy

Low Rainfall Agriculture

Goats

Methodology, Impact Assessment, Credit, etc.

Conclusions and Recommendations

☆☆☆☆☆☆☆☆

Foreword

The background for the meeting leading to the present set of proceedings is that since 1989 the Department of Animal Science and Animal Health, The Royal Veterinary and Agricultural University, Copenhagen, Denmark has conducted an annual course on a topic of current interest. However, from 1994 the course has been organised by the Danish Agricultural and Rural Development Adviser's Forum.

The participants have been technical experts, scientists, project personnel and administrators working with or interested in agricultural research and development in Third World countries. Although the majority have come from Denmark a range of countries have been represented such as the Scandinavian countries, France, the UK, US, China, Colombia, Nicaragua, Cambodia, the Gambia, Vietnam, Bangladesh, India, Zimbabwe and Nigeria. There have been representatives from international organisations like FAO and the International Fund for Agricultural Development (IFAD).

While during the course papers and other information have been exchanged among the participants, a formal set of proceedings was first time published after the 1995 meeting as the idea of the first workshops - it still is - was to allow the course to act as a breathing space for busy people to exchange ideas.

The rationale behind the workshop leading to the present volume was that within livestock development work, the emphasis has been overwhelmingly on large ruminants, especially cattle. However, small farmers typically keep a few of several types of animals in close integration with other farm resources. The purpose of this workshop was therefore to focus on all components of the integrated farming system. The emphasis was on the social and ecological dimensions of rural livestock and poultry production, recycling of nutrients and on appropriate procedures for learning, notably training of students at M.Sc. and Ph.D. level and several of the papers are the results of students' research works.

These proceedings are also available in electronic format, either on diskette from Dr. Poul Henning Petersen, Department of Animal Science and Animal Health, Royal Veterinary and Agricultural University, Bulowsvej 13, 1870 Frederiksberg C., Denmark, or via Internet from http:/ifs.plants.ox.ac.uk/tune/tune96.html by FTP from ifs.plants.ox.ac.uk (login anonymous). They are published in Adobe Acrobat format. To read them you will need the Acrobat Reader (available for DOS, Windows, MAC and Unix systems) which may be obtained on disk from the above address or from Adobe at http://www.adobe.com . The Acroread software is free.

Frands Dolberg   Poul Henning Petersen

Acknowledgements

We would like to acknowledge the contribution of all the participants and the staff of Tune Course Center. However, special mention must be made of the contribution from Sweden to this workshop. Through its Course Director Dr. Brian Ogle, The Department of Animal Nutrition and Management of the Swedish University of Agricultural Sciences sponsored the participation of eleven students from its M.Sc. course "Livestock Based Systems for the Sustainable use of Renewable Natural Resources". The quality of the appearance of the proceedings would not have been what it is without the work of Mrs. Jonna Kjær, Department of Political Science, University of Aarhus. However, we alone remain responsible for any omissions or any other mistakes.

Every year the courses have taken place at Course Centre Tune Landboskole, Greve, near Copenhagen in Denmark. The relationship with the management of the Centre has always been cordial. In recent years the Centre has paid for the tickets of some of the foreign speakers. Danida of the Danish Ministry of Foreign Affairs has every year supported with some travel funds and this time also contributed funds to the publication of these proceedings. Travel funds and course fees to individual participants were contributed by SIDA, SAREC, and Danida ENRECA projects.

Introduction

Introduction to the Concept of Human Development

John Martinussen

International Development Studies, Roskilde University,
P.O. Box 260, Dk-4000 Roskilde, Denmark
E-mail: johnm@iu.ruc.dk

Summary

The paper starts by clarifying the terms: development concept, theory and strategy before it goes on to review the origin and elaboration of the concept of human development. This is followed by a brief discussion of the different concepts of human development and some strategies proposed for promoting human development. In that context the role of donor agencies is discussed. It is concluded that the concepts provide a framework for analyses and development cooperation that are different from the 'paradigm' associated with conceptions of development as growth. It is noted that attempts to promote human development may be divided into two approaches of which one is top-down aiming at improving management of public resources while another is bottom-up aiming at empowering poor and resource-weak segments of society. It is finally mentioned that a more detailed discussion of human development and its promotion will have to be set in a specific context.

Key words: human development, origin, dimensions, strategies, top-down, bottom-up, economic growth.

Introduction

The concept of human development was originally introduced as an alternative to conceptions of development that focused on economic growth - with or without equity considerations. Therefore, it may be useful to begin by briefly referring to the definition of development as growth in one sense or another. Even prior to that, however, it may be expedient to clarify the dimensions and aspects involved in discussing development issues. Much of the debate, in my opinion, has been confused because a necessary distinction has not been applied between development concept (or development objective); development theory; and development strategy. Let me briefly try to outline these three dimensions and their interlinkages.

A *development concept* contains the answer to what development is. This answer can never be value-free. It will always reflect notions of what ought to be understood by development. These notions can be formulated as development objectives either in terms of particular conditions which must be achieved, or in terms of a certain direction of change.

To illustrate, a development concept, like the one embodied in modernization theory, may claim that the large industrialized countries, e.g. USA, are developed, i.e. they have achieved certain - positively evaluated - conditions. According to this conception, changes in Third World countries towards increasing similarity with these industrialized countries are regarded as development. Other changes are not regarded as such. The dynamic change processes through which a country moves towards greater resemblance with the developed countries is called the *development process*, according to this notion.

Other concepts of development focus more on the given conditions in Third World societies and define development in terms of bringing out, unfolding, what is potentially contained in these societies. Often, emphasis is given here to increasing the capacities for taking and implementing decisions in accordance with nationally or locally perceived priorities.

Development theory seeks to answer questions such as the following: How can chosen and specified development objectives be promoted? What conditions will possibly obstruct, delay, or detract progress towards the objectives? What causal relationships and laws of motion apply to the societal change processes? What actors play dominant roles, and what interests do they have? How do the changes affect various social groups and various geographical regions?

Questions like these are not value-neutral, but they set the stage for expounding, unlike a development concept, how social reality is actually structured - as opposed to how it ought to be structured. Theories thus contain significant normative elements, but can nonetheless be subjected to validating or invalidating tests through empirical analyses of the actual conditions and historical experiences.

Development strategy as an abstract notion refers essentially to the actions and interventions that can be appropriately used to promote strictly defined development objectives. Once again the basis is heavily value-loaded in that there are 'chosen' development goals. But there is - at least in principle - the possibility of a matter-of-fact weighing of which strategies are the most effective and least costly to promote the established objectives. In practice, though, decision makers as well as researcher often have had too little insight into the relevant contexts and causal relationships to ensure indisputable strategy choices. These are, therefore, in many cases more reflections of prejudices, ideologies and personal preferences.

The abstract interrelationship between concept, development process, theory and strategy may be depicted as shown in figure 1.

In the terms thus outlined what we shall attempt to do in the present article is to review the origin and elaboration of the concept of human development. This will be followed by a brief discussion of the various aspects of human development. Space does not allow us to go much into

Figure 1: *Development Objective, Theory, and Strategy*

Development theory
-hypotheses about promoting and obstructing conditions

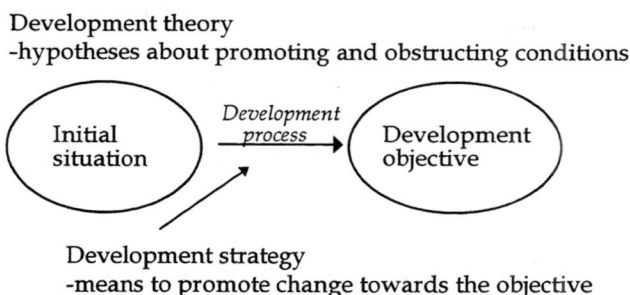

Development strategy
-means to promote change towards the objective

theories about conditions that promote or impede progress on the various dimensions, but we will touch on some of the strategies proposed for promoting human development and briefly refer to the role of donor agencies in this context. It may be added that I have dealt with these issues more extensively in recent textbooks (Martinussen, 1994; 1996b).

Origin and elaboration of the concept

As indicated above, the concept of human development emerged as an alternative to definitions of development which focused on growth. There has never been general agreement on how to define economic growth; nor on how best to measure growth in developing societies. Yet, wide approval has been gained today for a notion which defines economic development as *a process whereby the real per capita income of a country increases over a long period of time while simultaneously poverty is reduced and the inequality in society is generally diminished* - or at least not increased.

Conceptions of this kind have also been adopted in World Bank analyses. Further, they have informed Bank strategies since the early 1970s. However, considerable fluctuations over time can be observed. Up till around 1980, the World Bank was mainly interested in combining growth in per capita income with special assistance to the poor. One of the strategies was described as redistribution with growth; another went under the name 'the basic needs strategy'. In the 1980s, the focus shifted towards aggregate growth in conjunction with restoration of macro-economic balances, structural adjustment, and increased foreign-exchange earnings. Since 1990, the Bank has again emphasized growth for the poor and resource-weak groups - along with aggregate growth - in its overall conception of development (cf. World Bank, 1980; 1990).

The above definition of economic development embodies a wish to improve the living conditions, the welfare, for in principle all citizens of a society. However, the indicators for this remained in most of the literature and the international debate limited to income measurements of one kind or the other. As a corollary, it was asserted that growth in real incomes was the main target.

This was disputed by prominent economists such as Amartya Sen, Paul Streeten, Mahbub ul Haq, and others who believed that increased incomes should be regarded as a *means* to improve human welfare, not as an end in itself (Sen, 1988; Streeten, 1994). To these economists, human welfare was the overall objective - the essence of development. Increased incomes and national economic growth were crucial preconditions for improvements in standards of living, but not the only preconditions. This could be easily demonstrated, e.g. by comparing per capita incomes with indicators of education or health standards. Figures from the mid-1980s thus showed that the average life expectancy in many countries was considerably lower than one would expect from the income figures. Sri Lanka, with an average income of US$ 360, had an average life expectancy of 70 years, whereas Brazil, with over US$ 1700, had an average life expectancy of no more than 64 years (Sen, 1988, p. 12).

With the first *Human Development Report* from 1990, prepared under the leadership of Mahbub ul Haq, UNDP adopted this basic criticism of income measurements and presented a more comprehensive concept of human development (UNDP, 1990). The report defined human development as *a process of enlarging people's choices*. In the following section we shall see how proponents of human development elaborated this conception into a multi-dimensional approach.

Dimensions of human development

According to Mahbub ul Haq, the defining difference between the economic growth and the human development schools is that the first focuses exclusively on the expansion of only one choice - income - while the second embraces the enlargement of all human choices - whether economic, social, cultural, or political (Haq, 1995, Ch. 2).

It could be argued that an increase in income would enlarge all other choices as well. But this is exactly what Haq and others have questioned by asserting that the causal link between expanding income and expanding human choices depend on the quality and distribution of economic growth, not only on the quantity of such growth. They have argued that a link between income growth and human welfare has to be created consciously through public policies which aim at providing services and opportunities as equitably as possible to all citizens. This cannot be left to the market mechanisms, because these are essentially very unfriendly to the poor, to the weak, and to the vulnerable (Haq, 1995, Ch. 12).

Rejecting the automatic link, however, should not be taken to imply any rejection of the importance of economic growth, according to Haq. He very carefully tries to balance the argument by pointing, on the one hand, to the need for growth in poor societies for reducing mass poverty, and on the other, to the fact that the distribution of growth and the manner in which available resources are being utilized often matter more to the poor than aggregate growth of national income and production.

The human development school at first drew attention primarily to the choices in three essential areas: the opportunity to lead a long and healthy life; the opportunity to acquire knowledge; and the opportunity to have access to resources needed for a decent standard of living. To this was later added several other dimensions and aspects, and the name of the concept itself was changed from 'human development' to 'sustainable human development' in order to highlight the importance of sustaining all forms of capital and resources - physical, human, financial, and environmental - as a precondition for meeting the needs also of future generations.

The UNDP reports, which have been published annually since 1990, present accounts of human development in both developing countries and industrialized countries. These accounts are based on an index with three central components: (a) The average real income per capita, adjusted downwards for the rich countries by using the purchasing power of a country's currency (i.e. the number of units of a particular currency required to purchase the same representative basket of goods and services that a US dollar would buy in the USA); (b) the average life expectancy; and (c) adult literacy combined with real access to education at various levels (UNDP, 1995, p 134ff.).

The concept of human development has gradually been extended into basically all areas of societal development. To the original focus on the missing link between income and welfare has been added particular concern for the provision of social infrastructure and services that are made available on an equal basis to all citizens; special emphasis on gender equality; and equal opportunities for participation in political and economic decision-making. The latter requires both an enabling legal and institutional framework and empowerment of citizens and civil society organizations so that they become capable of reaching up to the authorities. Some of the adherents to the concept have furthermore put special emphasis on the environmental and natural resources aspects of sustainability.

As a reflection of the emphasis on gender, recent UNDP reports have included indicators for measuring gender equality. The 1995 report, in particular, focused on this issue (UNDP, 1995).

UNDP's work on human development contains some attempts at identifying causal relationships and obstacles to the enhancement of welfare and the enlargement of opportunities and choices on an equitable basis. Strategies for overcoming these obstacles have also been evolved. Based partly on UNDP's contributions in this area, the next section discusses briefly some of the main strategies for promoting human development. I have added to this a few observations on the possible roles that donor agencies may play in relation to these strategies.

Strategies for promoting human development

Based on the understanding that the core of human development comprises the enlargement of people's choices, welfare and equitable access to opportunities, it is deemed expedient to make a distinction between, on the one hand, management of public resources for human development, and on the other hand, the creation of an enabling environment for people's participation and the exercising of choices.

Management of public resources for human development involves mobilization, allocation, and utilization of resources so as to meet as equitably as possible the needs of all citizens. Taking into consideration the sustainability perspective this further implies meeting the needs of current generations without compromising the needs of future ones.

In more specific terms, management of public resources must ensure more effective and efficient provision of social infrastructure and delivery of basic social services like education and health. In order to achieve that most governments in the Third World need to be strengthened considerably. They need to have their capacities for *reaching down* and out to the citizens significantly enhanced. This emphasis on government capacity does not imply a return to notions of the state as the chief engine of growth, nor does it imply that the capacities of non-government institutions and organizations are seen as unimportant. Rather, the point is that the emerging division of labor between state and market and between state and civil society requires governments that are capable of performing a range of key functions effectively. That is a very basic precondition for achieving a higher degree of human development.

Creation of an enabling environment relates to institutional arrangements and procedures that promote people's participation in decision making, human security, and a genuine dialogue between government and civil society. In this area the shortcomings are even more evident in most Third World societies than with respect to the capacities of the state. Very few societies have a proper legal and institutional framework for interaction between civil society organizations and government agencies. Essentially, the focus of a human development strategy therefore should be on establishing a framework that will enable the citizens and their organizations to *reach up* to the authorities and to do so on a more equal basis.

Official multilateral and bilateral aid agencies, including UNDP, may assist in improving the management of public resources for human development. Even the World Bank and the regional development banks can play major roles in this area, and they sometimes have done so in the past. With respect to improving the environment for people's participation, however, official donor agencies are likely to play more limited roles. Organizations like the World Bank or UNDP are not in a position to directly empower the resource-poor segments of the population. What these official donor agencies can and should do is to propose and assist the

government in creating the necessary institutional framework and procedures and, possibly, under certain conditions agencies like the UNDP can
provide direct services to facilitate dialogue and conciliation, as has happened in some countries during transition from civil war to political stability, or from centrally planned economies to market economies.

References above to government and institutional frameworks should
be taken to mean both central and local level. As increasingly acknowledged by both scholars and practitioners there is considerable scope for
promoting human development more effectively and efficiently by transferring authority and responsibility to regional and local governments.
These sub-national authorities may be in better positions than the central
government to establish genuine partnerships with citizens and communities which, in turn, will help the authorities to become better at providing
basic services and performing key functions, while at the same time generating an enabling environment for people's participation. I have elaborated
extensively on this point in a recent study of local authorities in Nepal,
which further comprises a number of recommendations on how Danida
and other official donor agencies may assist in building capacities at the
local level within a democratic framework (Martinussen, 1995).

In more specific terms, at least four areas of intervention may be identified where donor agencies may assist Third World authorities in improving management of public resources in support of human development,
namely by strengthening their capacities to: (1) work out strategies in
accordance with national needs and priorities - and in accordance with the
objectives embodied in the human development concept; (2) reform public
administration with a further view to enhancing government capacity to
implement policies and strategies aimed at poverty eradication and promoting welfare and equitable access to services; (3) mobilize resources for
human development - from domestic and international sources; and (4)
manage aid in accordance with national priorities and procedures.

In relation to creating an enabling environment two aspects of capacity
development appear of special importance and within the capabilities of
multilateral agencies like UNDP as well as bilateral agencies like Danida:
(1) Capacities to bring about devolution and other forms of decentraliz-
ation aimed at improving resource mobilization and utilization and with a
further view to involving the citizens more in decision making; and (2)
capacities to involve NGOs and other civil society organizations in human
development efforts. A particular area of intervention in certain countries,
partly associated with the last point, would be consensus building and the
strengthening of institutional capacities for dialogue between government
and civil society organizations.

Most of the above remarks on the possible roles of donor agencies in
support of human development strategies have focused on the reaching-
down approaches. Moreover, they have been based on the assumption that
we have to do with political authorities who genuinely pursue the goal of

human development for all. As is evident from experience, this is rarely the case. Therefore, the emphasis that UNDP and some other official aid agencies put on human development may be seen primarily as an appeal to governments in both recipient countries and donor countries to engage in new forms of development co-operation that are more conducive to promoting this kind of development and with particular emphasis on equity.

Other international organizations and a particular group of development researchers who favor people-managed development, base their strategies on a different assumption. They rather take for granted that no government in the Third World will use the necessary resources on mass development and mass welfare, unless the poor population majority is sufficiently powerful to force upon the authorities such a policy. Their strategies, therefore, above all aim at empowering the poor and deprived social groups so that they become able to effectively reach up to the authorities (Gran, 1983; Chambers, 1983). Only a few official aid agencies, like the ILO and UNICEF, are seen to pursue policies which to some extent correspond to this emphasis. The assistance provided by the ILO Workers' Education Branch to strengthening trade unions in the Third World is a particularly interesting case here (cf. Martinussen, 1996a). But otherwise I would argue that genuine empowerment strategies are difficult to implement for official donor agencies because of the opposition they are likely to provoke from the authorities in most developing countries. This is an area best left to non-government organizations, particularly those that organize citizens and communities of the Third World countries themselves.

Concluding remarks

In this brief introduction to the concept of human development I have confined myself to outlining the abstract notion and some overall strategies which may be pursued in support of human development. From this it is clear that the concept provides a framework for analyses and development co-operation which is basically different from the 'paradigm' associated with conceptions of development as growth. It is further noted that efforts aimed at promoting human development may essentially be divided into two different approaches: One top-down with a focus on improving management of public resources; another bottom-up with a focus on empowering poor and resource-weak segments of society. In a more detailed discussion of human development and how to promote it general and abstract observations like the above will have to be related to a specific context, to particular interests of different social groups and to the special economic, political, and cultural obstacles that prevent millions of people from living decent lives.

References

Chambers, Robert, 1983, *Rural Development. Putting the Last First*, New York, John Wiley.

Gran, Guy, 1983, *Development by People. Citizen Construction of a Just World*, New York, Praeger.

Haq, Mahbub ul, 1995, *Reflections on Human Development*, New York, Oxford University Press.

Martinussen, John Degnbol, 1994, *Samfund, stat og marked. En kritisk gennemgang af teorier om udvikling i den 3. verden*, København, Mellemfolkeligt Samvirke.

Martinussen, John, 1995, *Democracy, Competition and Choice. Emerging Local Self-Government in Nepal*, New Delhi, SAGE.

Martinussen, John Degnbol, 1996a "Empowerment of Labor. A Study of ILO-assisted Activities in Support of Third World Trade Unions", in: Lars Rudebeck and Olle Törnquist (Eds.), *Democratization in the Third World. Concrete Cases in Comparative and Theoretical Perspective*, Uppsala, Uppsala University, 1996.

Martinussen, John, 1996b, *Society, State and Market. A Guide to Competing Theories of Development*, London, ZED Press (forthcoming; revised and extended version of the above Danish textbook).

Sen, Amartya, 1988, "The Concept of Development", in: Hollis Chenery and T. N. Srinivasan (Eds.), *Handbook of Development Economics*, Amsterdam, North Holland, 1989.

Streeten, Paul P., 1994, *Strategies for Human Development*, Copenhagen, Handelshøjskolens Forlag/ Munksgaard International Publishers.

UNDP, 1990, *Human Development Report 1990*, New York, Oxford University Press.

UNDP, 1991, *Human Development Report 1991*, New York, Oxford University Press.

UNDP, 1992, *Human Development Report 1992*, New York, Oxford University Press.

UNDP, 1993, *Human Development Report 1993*, New York, Oxford University Press.

UNDP, 1994, *Human Development Report 1994*, New York, Oxford University Press.

UNDP, 1995, *Human Development Report 1995*, New York, Oxford University Press.

World Bank, 1980, *World Development Report 1980*, Oxford, Oxford University Press for the World Bank.

World Bank, 1990, *World Development Report 1990*, Oxford, Oxford University Press for the World Bank.

Participatory Learning for Integrated Farming

Jules N. Pretty
Sustainable Agriculture Programme,
International Institute for Environment and Development, 3 Endsleigh Street,
London WC1H ODD. E-mail: iiedagri@gn.apc.org

Summary

Increasing human population will require substantial increases in food production, which is well understood. However, it is less appreciated that to deliver such increases is not only a matter of availability of physical resources as inputs in the process of production, but - as important - of conducive schools of thought in science and approaches to learning. The paper is based on the assumption that agricultural development will have to be sustainable (in the terms defined in the paper) and provides recent evidence of impact achieved. Experience demonstrates participation by people is a critical condition for success in sustainable agricultural development and interest in application of participatory approaches is growing. Assuming the meaning of sustainability encompasses activities spread beyond a project in space and time, participation requires collective analyses by interdisciplinary and intersectoral teams, and even a researcher working alone must cooperate closely with local people. The paper goes on to discuss conditions for scaling-up of sustainable agriculture. Principal among these are farmers' capacity to innovate and it is the process, which sustains this capacity which is important and much less specific technologies. A new professional able to select methodologies according to needs and work in multidisciplinary teams and not afraid of interaction with non-scientific people will facilitate spread of sustainable agriculture and so will an institutionalisation of these approaches.

Key words: Learning, agriculture, sustainable, participatory, professionalism.

Challenges for agricultural development

Agricultural development now faces some unprecedented challenges. By the year 2020, the world will have 2.5 billion more people than today. Even though enough food is produced in aggregate to feed everyone, and that world prices have been falling in recent years, some 700-800 million people still do not have access to sufficient food. This includes 180 million children underweight and suffering from malnutrition.

It is now widely accepted that over the next quarter to half century food production will have to increase substantially.

But the views on how to proceed vary hugely. Some are optimistic, even complacent; others are darkly pessimistic. Some indicate that not much needs to change; others argue for fundamental reforms to agricultural and food systems. Some indicate that a significant growth in food production will only occur if new lands are taken under the plough; others suggest that there are feasible social and technical solutions for increasing yields on existing farmland.

There are five distinct schools of thought over these future options in agricultural development.

Schools of thought

There are "*optimists*", who say supply will always meet increasing demand, and so recent growth in aggregate food production will continue alongside reductions in population growth (Rosegrant and Agcaolli, 1994; Mitchell and Ingco, 1993; FAO, 1993). As food prices are falling (down 50% in the past decade for most commodities), this indicates that there is no current crunch over demand. Food production is expected to grow for two reasons: i) the fruits of biotechnology research will soon ripen, so boosting plant and animal productivity; ii) the area under cultivation will expand, probably by some 20-40% by 2020 (by an extra 79 million ha in Sub-Saharan Africa alone). It is also expected that developing countries will substantially increase food imports from industrialised countries (perhaps by as much as 5 fold by 2050).

There are "*environmental pessimists*", who suggest that ecological limits to growth are being approached); are soon to be passed; or have already been reached (Harris, 1995; Brown, 1994; CGIAR, 1995; Kendall and Pimentel, 1994; Brown and Kane, 1994; Ehrlich, 1968). It is said that populations are too great; yield growth has slowed, and will slow more, stop or even fall; no new technological breakthroughs are likely; and that environments have been too thoroughly degraded for recovery. Solving these problems means putting population control as the first priority

The "*industrialised world to the rescue*" group believes that Third World countries will never be able to feed themselves, for all sorts of ecological, institutional and infrastructural reasons, and so the looming food gap will have to be filled by modernized agriculture in the North (Avery, 1995; Wirth, 1995; DowElanco, 1994; Carruthers, 1993; Knutson et al, 1990). Increasing production in large, mechanised operations will allow smaller and more 'marginal' farmers to go out of business, so taking the pressure off natural resources, which can then be conserved in protected areas and wildernesses. These large producers will then be able to trade their food with those who need it, or have it distributed by famine relief or food aid. It is also vigorously argued that any adverse health and environmental consequences of modern agricultural systems are minor in comparison with those wrought by the expansion of agriculture into new lands.

One group, what we might call the `new modernists', argues that biological yield increases are possible on existing lands, and that this food growth can only come from high-external input farming (Borlaug, 1992, 1994a, b; Sasakawa Global 2000, 1993, 1994; World Bank, 1993; Paarlberg and Breth, 1994; Winrock International, 1994; Crosson and Anderson, 1995). The target is both the existing Green Revolution lands, and the 'high-potential' lands that have been missed by the past 30 years of agricultural development. This group argues that farmers simply use too few fertilizers and pesticides, which are said to be the only way to improve yields and so keep the pressure off natural habitats. This repeat of the green revolution model is called 'science-based' agriculture, the objective being to increase farmers' use of fertilizers and pesticides. It is also argued that high-input agriculture is more environmentally sustainable than low-input, as low-input agriculture can only ever be low output.

The case is also being made for the benefits of `sustainable intensification', on the grounds that substantial growth is possible in currently unimproved or degraded areas whilst at the same time protecting or even regenerating natural resources (Pretty, 1995a, b; Hazell, 1995; McCalla, 1994, 1995; Scoones and Thompson, 1994; NAF, 1994; Hewitt and Smith, 1995; Röling and Wagemakers, 1996).

It is argued that empirical evidence now indicates that regenerative and low-input (but not necessarily zero-input) agriculture can be highly productive, provided farmers participate fully in all stages of technology development and extension. This evidence also suggests that agricultural and pastoral lands productivity is as much a function of human capacity and ingenuity as it is of biological and physical processes. Sustainable agriculture seeks the integrated use of a wide range of pest, nutrient, soil and water management technologies.

What is and what is not sustainable agriculture

A great deal of effort has gone into trying to define sustainability in absolute terms. Since the Brundtland Commission's definition of sustainable development in 1987, there have been at least 100 more definitions constructed, each emphasising different values, priorities and goals. But precise and absolute definitions of sustainability, and therefore of sustainable agriculture, are impossible. Sustainability itself is a complex and contested concept. To some it implies persistence and the capacity of something to continue for a long time. To others, it implies not damaging or degrading natural resources.

In any discussion of sustainability, it is important to clarify what is being sustained, for how long, for whose benefit and at whose cost, over what area and measured by what criteria. Answering these questions is difficult, as it means assessing and trading off values and beliefs.

It is critical, therefore, that sustainable agriculture does not prescribe a concretely defined set of technologies, practices or policies. This would only serve to restrict the future options of farmers. As conditions change and as knowledge changes, so must farmers and communities be encouraged and allowed to change and adapt too. Sustainable agriculture is, therefore, not a simple model or package to be imposed. It is more a process for learning (Pretty, 1995b; Röling, 1994).

The basic challenge for sustainable agriculture is to make better use of available physical and human resources. This can be done by minimizing the use of external inputs, by regenerating internal resources more effectively, or by combinations of both. This ensures the efficient and effective use of what is available, and ensures that any changes will persist as dependencies on external systems are kept to a reasonable minimum.

A sustainable agriculture, therefore, is any system of food or fibre production that systematically pursues the following farming objectives:

- A thorough incorporation of natural processes such as nutrient cycling, nitrogen fixation, and pest-predator relationships into agricultural production processes, so ensuring profitable and efficient food production;

- A minimisation in the use of those external and non-renewable inputs with the greatest potential to damage the environment or harm the health of farmers and consumers, and a more targeted use of the remaining inputs used with a view to minimising costs;

- The full participation of farmers and rural people in all processes of problem analysis, and technology development, adaptation and extension;

- A more equitable access to productive resources and opportunities, and progress towards more socially-just forms of agriculture;

- A greater productive use of local knowledge and practices, including innovative approaches not yet fully understood by scientists or widely adopted by farmers;

- An increase in self-reliance amongst farmers and rural people;

- An improvement in the match between cropping patterns and the productive potential and environmental constraints of climate and landscape to ensure long-term sustainability of current production levels.

Sustainable agriculture seeks the integrated use of a wide range of pest, nutrient, soil and water management technologies. It aims for an increased diversity of enterprises within farms combined with increased linkages and flows between them. By-products or wastes from one component or enterprise become inputs to another. As natural processes increasingly substitute for external inputs, so the impact on the environment is reduced.

New evidence on impacts

There is now emerging evidence that regenerative and resource-conserving technologies and practices can bring both environmental and economic benefits for farmers, communities and nations. The best evidence comes from countries of Africa, Asia and Latin America, where the concern is to increase food production in the areas where farming has been largely untouched by the modern packages of externally-supplied technologies. In these lands, farming communities adopting regenerative technologies have substantially improved agricultural yields, often only using few or no external inputs (Bunch, 1990, 1993; GTZ, 1992; UNDP, 1992; Krishna, 1994; Shah, 1994; SWCB, 1994; Balbarino and Alcober, 1994; Pretty, 1995a)

A recent study of 86 projects in 14 countries of East and Southern Africa discovered that improvements are now occurring for at least 230,000 farming families (Hinchcliffe et al, 1996). Over 6 million hectares are being farmed with sustainable agriculture, and on average crop yields have more than doubled. All the projects are using resource-conserving technologies and are working in a participatory fashion with local people.

But these are not the only sites for successful sustainable agriculture. In the high-input and generally irrigated lands, farmers adopting regenerative technologies have maintained yields whilst substantially reducing their use of inputs (Bagadion and Korten, 1991; Kenmore, 1991; van der Werf and de Jager, 1992; UNDP, 1992; Kamp et al, 1993; Pretty, 1995a). And in the industrialised countries, farmers have been able to maintain profitability, even though input use has been cut dramatically, such as in the USA (NRC, 1989; NAF, 1994; Hewitt and Smith, 1995); and in Europe (Pretty and Howes, 1993; Reus et al, 1994; Somers, 1996).

But this empirical evidence is still contested. In the USA, for example, some 82% of conventional US farmers believe that low input agriculture will always be low output (Hewitt and Smith, 1995). Two influential politicians have recently emphasised these beliefs. In 1991, the Secretary of Agriculture, Earl Butz, said

> "we can go back to organic agriculture in this country if we must - we once farmers that way 75 years ago. However, before we move in that direction, someone must decide which 50 million of our people will starve. We simply cannot feed, even at subsistence levels, our 250 million Americans without a large production input of chemicals, antibiotics and growth hormones".

In 1996, Under-Secretary for Agriculture, Eugene Moos, said:

> "The prospective increase in world population will double food aid needs in the next decade... and it will be necessary for agricultural producing nations to use biotechnology and hormones to meet growing demand."

Yet a selection of recent evidence (Box 1) shows quite the opposite. In the USA, some 40,000 farmers in 32 states are using sustainable agriculture technologies and have cut their use of external inputs substantially. This includes 2800 sustainable agriculture farmers in the North Western States, who grow twice as many crops compared with conventional farmers, use 60-70% less fertilizer, pesticide and energy, and their yields are roughly comparable; they also spend more money on local goods and services - each farm contributed more than £13,500 to its local economy.

Box 1. Successful sustainable and integrated agriculture in selected countries

---

Bangladesh: 5000 farming families
Fish in rice fields combined with integrated pest management technologies and participatory action learning approach; rice yields up 12% and pesticide use cut to zero.

Brazil: 38,000 farming families
Community-based microwatershed programme, with farmers growing more than 60 species of green manures and cover crops; yields more than doubling and farmers needing less labour for weeding and ploughing.

Germany: 55,400 farmers in Baden Würtemburg
Following the establishment of the MEKA scheme of incentives, farmers have cut pesticide and fertilizer use; extensified grassland systems; increased the use of cover crops and legumes; protected rare breeds; and increased undersowing.

Guatemala and Honduras: 8000 farming families
Regenerative agriculture based on soil conservation, green manures, farmer experimentation and farmer extensionists; yield increases of 2-3 fold, continued beyond the projects and spreading independently, with sustainable agriculture now the motor for local economic growth.

Indonesia: 400,000 farmers
Integrated pest management for rice programme with farmer field schools as the mechanism to enhance farmers' capacity to learn about their farming environment and innovate; rice yields have stabilised or slightly increased even though all farmers have substantially cut pesticide use (25% no longer use pesticides at all).

East and Southern Africa: 250,000 farming families (in Angola, Botswana, Ethiopia, Kenya, Lesotho, Malawi, Mozambique, South Africa, Tanzania, Uganda, Zambia, Zimbabwe):
80 community-based projects with 6 million hectares farmed with sustainable agriculture technologies; with participatory learning methods being used to develop farmers' capacity to experiment and develop their own solutions, resulting in an average doubling of crop yields using only regenerative technologies.

USA: 40,000 farmers
Farmers using sustainable agriculture technologies grow twice as many crops compared with conventional farmers, use 60-70% less fertilizer, pesticide and energy, and their yields are roughly comparable; they also spend more money on local goods and services.

Recent growth in interest in participation

There is a long history of participation in agricultural development, and a wide range of development agencies, both national and international, have attempted to involve people in some aspect of planning and implementation. Two overlapping schools of thought and practice have evolved. One views participation as a means to increase efficiency, the central notion being that if people are involved, then they are more likely to agree with and support the new development or service. The other sees participation as a fundamental right, in which the main aim is to initiate mobilization for collective action, empowerment and institution building.

In recent years, there have been an increasing number of comparative studies of development projects showing that 'participation' is one of the critical components of success. It has been associated with increased mobilization of stakeholder ownership of policies and projects; greater efficiency, understanding and social cohesion; more cost-effective services; greater transparency and accountability; increased empowering of the poor and disadvantaged; and strengthened capacity of people to learn and act (Cernea, 1991; Pretty and Sandbrook, 1991; Uphoff, 1992; Narayan, 1993, 1995; World Bank, 1994; Pretty, 1995a, b; Thompson, 1995).

As a result, the terms 'people's participation' and 'popular participation' have become part of the normal language of many development agencies, including NGOs, government departments and banks (Adnan et al, 1992). It is such a fashion that almost everyone says that participation is part of their work. This has created many paradoxes. The term `participation' has been used to justify the extension of control of the state as well as to build local capacity and self-reliance; it has been used to justify external decisions as well as to devolve power and decision-making away from external agencies; it has been used for data collection as well as for interactive analysis.

In conventional rural development, participation has commonly centred on encouraging local people to sell their labour in return for food, cash or materials. Yet these material incentives distort perceptions, create dependencies, and give the misleading impression that local people are supportive of externally-driven initiatives. This paternalism undermines sustainability goals and produces impacts which rarely persist once the project ceases (Bunch, 1983; Reij, 1988; Pretty and Shah, 1994; Kerr, 1994). Despite this, development programmes continue to justify subsidies and incentives, on the grounds that they are faster, that they can win over more people, or they provide a mechanism for disbursing food to poor people. When little effort is made to build local skills, interests and capacity, then local people have no stake in maintaining structures or practices once the flow of incentives stops.

The many ways that development organisations interpret and use the term participation can be resolved into seven clear types. These range from manipulative and passive participation, where people are told what is to happen and act out predetermined roles, to self-mobilization, where people take initiatives largely independent of external institutions (Table 1).

Table 1. *A typology of participation: how people participate in development programmes and projects*

Typology	Characteristics of Each Type
1. *Manipulative Participation*	Participation is simply a pretence, with 'people's' representatives on official boards but who are unelected and have no power.
2. *Passive Participation*	People participate by being told what has been decided or has already happened. It involves unilateral announcements by an administration or project management without any listening to people's responses. The information being shared belongs only to external professionals.
3. *Participation by Consultation*	People participate by being consulted or by answering questions. External agents define problems and information gathering processes, and so control analysis. Such a consultative process does not concede any share in decision-making, and professionals are under no obligation to take on board people's views.
4. *Participation for Material Incentives*	People participate by contributing resources, for example labour, in return for food, cash or other material incentives. Farmers may provide the fields and labour, but are involved in neither experimentation nor the process of learning. It is very common to see this called participation, yet people have no stake in prolonging technologies or practices when the incentives end.
5. *Functional Participation*	Participation seen by external agencies as a means to achieve project goals, especially reduced costs. People may participate by forming groups to meet predetermined objectives related to the project. Such involvement may be interactive and involve shared decision making, but tends to arise only after major decisions have already been made by external agents. At worst, local people may still only be coopted to serve external goals.
6. *Interactive Participation*	People participate in joint analysis, development of action plans and formation or strengthening of local institutions. Participation is seen as a right, not just the means to achieve project goals. The process involves interdisciplinary methodologies that seek multiple perspectives and make use of systemic and structured learning processes. As groups take control over local decisions and determine how available resources are used, so they have a stake in maintaining structures or practices.
7. *Self-Mobilization*	People participate by taking initiatives independently of external institutions to change systems. They develop contacts with external institutions for resources and technical advice they need, but retain control over how resources are used. Self-mobilization can spread if governments and NGOs provide an enabling framework of support. Such self-initiated mobilization may or may not challenge existing distributions of wealth and power.

Source: Pretty, 1995a

This typology suggests that the term 'participation' should not be accepted without appropriate clarification. The problem with participation as used in types one to four is that any achievements are likely to have no positive lasting effect on people's lives (Rahnema, 1992). The term participation can be used, knowing it will not lead to action. Indeed, some suggest that the manipulation that is often central to types one to four mean they should be seen as types of non-participation (Hart, 1992).

The World Bank's internal 'Learning Group on Participatory Development', in seeking to clarify the benefits and costs of participation, distinguished between different types of participation: "many Bank activities which are termed 'participatory' do not conform to [our] definition, because they provide stakeholders with little or no influence, such as when [they] are involved simply as passive recipients, informants or labourers in a development effort" (World Bank, 1994).

Another study of 121 rural water supply projects in 49 countries of Africa, Asia and Latin America found that participation was the most significant factor contributing to project effectiveness and maintenance of water systems (Narayan, 1993). Most of the projects referred to community participation or made it a specific project component, but only 21% scored high on interactive participation. Clearly, intentions did not translate into practice. It was when people were involved in decision-making during all stages of the project, from design to maintenance, that the best results occurred. If they were just involved in information sharing and consultations, then results were much poorer. According to the analysis, it was clear that moving down the typology moved a project from a medium to highly effective category.

Great care must, therefore, be taken over both using and interpreting the term participation. It should always be qualified by reference to the type of participation, as most types will threaten rather than support the goals of sustainable agriculture. What will be important is for institutions and individuals to define better ways of shifting from the more common passive, consultative and incentive-driven participation towards the interactive end of the spectrum.

Alternative systems of participatory learning and action

Recent years have seen a rapid expansion in new participatory methods and approaches to learning in the context of agricultural development (see *PLA Notes (formerly RRA Notes)*, 1988-present; *PALM Series*, 1991-present; Pretty et al, 1995; IDS/IIED, 1994; Chambers, 1994a, b, c; Mascarenhas et al, 1991[1]). Many have been drawn from a wide range of non-agricultural con-

---

[1] This list of references cannot possibly be comprehensive, as the antecedents and actors involved are too numerous to mention. The informal journal PLA Notes (formerly RRA

cont.

texts, and were adapted to new needs. Others are innovations arising out of situations where practitioners have applied the methods in a new setting, the context and people themselves giving rise to the novelty.

There are now more than 30 different terms for these systems of learning and action, some more widely used than others. Participatory Rural Appraisal (PRA), for example is now practised in at least 130 countries. This diversity and complexity is a strength, as it is a sign of both innovation and ownership.

There are six common principles of these systems of learning:

- *A defined methodology and systemic learning process* - the focus is on cumulative learning by all the participants and, given the nature of these approaches as systems of inquiry and interaction, their use has to be participative. The emphasis on visualisations democratises and deepens analysis.

- *Multiple perspectives* - a central objective is to seek diversity, rather than characterise complexity in terms of average values. The assumption is that different individuals and groups make different evaluations of situations, which lead to different actions. All views of activity or purpose are heavy with interpretation, bias and prejudice, and this implies that there are multiple possible descriptions of any real-world activity.

- *Group learning process* - all involve the recognition that the complexity of the world will only be revealed through group inquiry and interaction. This implies three possible mixes of investigators, namely those from different disciplines, from different sectors, and from outsiders (professionals) and insiders (local people).

- *Context specific* - the approaches are flexible enough to be adapted to suit each new set of conditions and actors, and so there are multiple variants.

- *Facilitating experts and stakeholders* - the methodology is concerned with the transformation of existing activities to try to bring about changes which people in the situation regard as improvements. The role of the 'expert' is best thought of as helping people in their situation carry out their own study and so achieve something.

- *Leading to sustained action* - the learning process leads to debate about change, and debate changes the perceptions of the actors and their readiness to contemplate action. Action is agreed, and implementable

---

cont.

Notes) (in issues 1 to 25) has alone published some 280 articles since 1988 based on field experiences in rural and urban communities in some 55 countries; and the IDS/IIED (1994) annotated bibliography contains a listing of some 600 references relating to PRA and RRA.

changes will therefore represent an accommodation between the different conflicting views. The debate and/or analysis both defines changes which would bring about improvement and seeks to motivate people to take action to implement the defined changes. This action includes local institution building or strengthening, so increasing the capacity of people to initiate action on their own.

Participatory learning methods

The participatory methods (sometimes called tools, techniques or instruments) used in these systems of learning and action can be structured into four classes: methods for group and team dynamics, for sampling, for interviewing and dialogue, and for visualisation and diagramming. It is the collection of these methods into unique approaches, or assemblages of methods, that constitute different systems of learning and action.

Participation calls for collective analysis. Even a sole researcher must work closely with local people (often called 'beneficiaries', 'subjects', 'respondents' or 'informants'). Ideally, though, teams of investigators work together in interdisciplinary and intersectoral teams. By working as a group, the investigators can approach a situation from different perspectives, carefully monitor one another's work, and carry out a variety of tasks simultaneously. Groups can be powerful when they function well, as performance and output is likely to be greater than the sum of its individual members. Many assume that simply putting together a group of people in the same place is enough to make an effective team. This is not the case. Shared perceptions, essential for group or community action, have to be negotiated and tested. Yet, the complexity of multidisciplinary team work is generally poorly understood. A range of workshop and field methods can be used to facilitate this process of group formation.

In order to ensure that multiple perspectives are both investigated and represented, practitioners must be clear about who is participating in the data-gathering, analysis and construction of these perspectives. Communities are not homogenous entities, and there is always the danger of assuming that those participating are representative of all views. There are always differences between women and men, between poor and wealthy, between young and old. Those missing, though, are usually the socially marginalised (Guijt and Kaul Shah, 1996). Rigorous sampling is, therefore, an essential part of these participatory approaches, and a range of field methods is available.

Sensitive interviewing and dialogue are a third element of these systems of participatory learning. For the reconstructions of reality to be revealed, the conventional dichotomy between the interviewer and respondent should not be permitted to develop. Interviewing is, therefore, structured around a series of methods that promote a sensitive dialogue. This should appear more like a structured conversation than an interview.

The fourth element is the emphasis on diagramming and visual construction. In formal surveys, information is taken by interviewers, who transform what people say into their own language. By contrast, diagramming can give local people a share in the creation and analysis of knowledge, providing a focus for dialogue which can be sequentially modified and extended. Local categories, criteria and symbols are used during diagramming, which include mapping and modelling, comparative analyses of seasonal, daily and historical trends, ranking and scoring methods to understand decision-making, and diagrammatic representations of household and livelihood systems. Rather than answering questions which are directed by the values of the researcher, local people are encouraged to explore their own versions of their worlds. Visualisations, therefore, help to balance dialogue, establish rapport and increase the depth and intensity of discussion.

These alternative methodologies imply a process of learning leading to action. A more sustainable agriculture, with all its uncertainties and complexities, cannot be envisaged without a wide range of actors being involved in continuing processes of learning.

The spread and scaling up of sustainable agriculture

Despite the increasing number of successful sustainable agriculture initiatives in different parts of the world, it is clear that most of these are still only 'islands of success'. There remains a huge challenge to find ways to spread or 'scale up' the processes which have brought about these transitions.

Sustainability ought to mean, therefore, more than just agricultural activities that are environmentally neutral or positive; it implies the capacity for activities to spread beyond the project in both space and time. A 'successful' project that leads to improvements that neither persist nor spread beyond the project boundary should not be considered sustainable.

When the recent record of development assistance is considered, it is clear that sustainability has been poor. There is a widespread perception amongst both multilaterals and bilaterals that agricultural development is difficult, that agricultural projects perform badly, and that resources may best be spent in other sectors. Reviews by the World Bank, the EC, Danida and ODA have all shown that agricultural and natural resource projects both performed worse in the 1990s that in the 1970s-1980s and worse than projects from other sectors (World Bank, 1993; Pohl and Mihaljek, 1992; EC, 1994; Danida, 1994; Dyer and Bartholomew, 1995). They are also less likely to continue achievements beyond the provision of aid inputs.

A recent analysis of 95 agricultural project evaluations logged on the DAC-OECD database shows a disturbing rate of failure, with at least 27%

of projects having non-sustainable structures, practices or institutions, and 10% causing significant negative environmental impact (Pretty and Thompson, 1996).

This empirical evidence of completed agricultural development projects suggest four important principles for sustainability and spread:

1. *Imposed technologies do not persist*: if coercion or financial incentives are used to encourage people to adopt sustainable agriculture technologies (such as soil conservation, alley cropping, IPM), then these are not likely to persist.

2. *Imposed institutions do not persist*: if new institutional structures are imposed, such as cooperatives or other groups at local level, or Project Management Units and other institutions at project level, then these rarely persist beyond the project.

3. *Expensive technologies do not persist*: if expensive external inputs, including subsidised inputs, machinery or high technology hardware are introduced with no thought to how they will be paid for, they too will not persist beyond the project.

4. *Sustainability does not equal fossilisation or continuation of a thing or practice forever*: rather it implies an enhanced capacity to adapt in the face of unexpected changes and emerging uncertainties.

Most agricultural development programmes have begun with the notion that there are technologies that work, and it is just a matter of inducing or persuading farmers to adopt them. Yet few farmers are able to adopt whole packages of conservation technologies without considerable adjustments in their own practices and livelihood systems. To some, this may not be a problem; to the majority, it is a major impediment to adopting conservation technologies and practices.

The problem is that the imposed models look good at first, and then fade away. Alley cropping, an agroforestry system comprising rows of nitrogen-fixing trees or bushes separated by rows of cereals, has long been the focus of research (Kang et al, 1984; Attah-Krah and Francis, 1987; Lal, 1989). Many productive and sustainable systems, needing few or no external inputs, have been developed. They stop erosion, produce food and wood, and can be cropped over long periods. But the problem is that very few, if any, farmers have adopted these alley cropping systems as designed. Despite millions of dollars of research expenditure over many years, systems have been produced suitable only for research stations (Carter, 1995).

There has been some success, however, where farmers have been able to take one or two components of alley cropping, and then adapt them to their own farms. In Kenya, for example, farmers planted rows of leguminous trees next to field boundaries, or single rows through their fields; and in Rwanda, alleys planted by extension workers soon became dispersed through fields (Kerkof, 1990).

But the prevailing view tends to be that farmers should adapt to the technology. Of the Agroforestry Outreach Project in Haiti, it was said that

> Farmer management of hedgerows does not conform to the extension program .... Some farmers prune the hedgerows too early, others too late. Some hedges are not yet pruned by two years of age, when they have already reached heights of 4-5 metres. Other hedges are pruned too early, mainly because animals are let in or the tops are cut and carried to animals ... Finally, it is very common for farmers to allow some of the trees in the hedgerow to grow to pole size (Bannister and Nair, 1991).

Farmers were clearly adapting the technology to suit their own needs.

Enhancing farmers' capacity to innovate

Important evidence comes from a variety of soil conservation and agricultural regeneration programmes in Central America (Bunch and López, 1994). The Guinope (1981-89) and Cantarranas (1987-1991) programmes in Honduras and the San Martin Jilotepeque programme in Guatemala (1972-1979) were collaborative efforts between World Neighbours and other local agencies. All began with a focus on soil conservation in areas where maize yields were very low (400 to 660 kg/ha), and where shifting cultivation, malnutrition, and outmigration prevailed. All show the importance of developing resource-conserving practices in partnership with local people.

There were several common elements. All forms of paternalism were avoided, including giving things away, subsidising farmer activities or inputs, or doing anything for local people. Each started slowly and on a small scale, so that local people could meaningfully participate in planning and implementation. They used technologies, such as green manures, cover crops, contour grass strips, in-row tillage, rock bunds and animal manures, that were appropriate to the local area, and which were finely-tuned through experimentation by and with farmers. Extension and training was done largely by villager farmers who had already experienced success with the technologies on their own farms.

Each programme substantially improved agricultural yields, increasing output per area of land from some 400-600 kg/ha to 2000-2500 kg/ha. Altogether improvements have been made in some 120 villages. Over time, soils were not simply conserved but regenerated, with depth increases from 0.1 metres to 0.4 - 1.3 metres not uncommon.

These programmes have also helped to regenerate local economies. Land prices and labour rates are higher inside the project areas compared with outside. There are housing booms, and families have moved back from capital cities. There are also benefits to the forests. Farmers say they no longer need to cut the forests, as they have the technologies to farm permanently the same piece of land. Before the programmes, national park authorities sought to keep villagers out of the forests; now there is no such concern since the forests are no longer threatened.

There are few published studies that give evidence of impacts years after outside interventions have ended. In 1994, however, staff of the Honduran organisation COSECHA (Associación de Consejeros una Agricultura Sostenible, Ecológica y Humana) returned to the three programme areas, and used participatory methods with local communities to evaluate subsequent changes (Bunch and López, 1994).

They first divided all 121 villages into three categories, according to where they felt there had been good, moderate and poor impact. Twelve villages were sampled from these - 4 from each programme comprising one of the best, two of the moderate and one poor. These villages had some 1000 families (with a range of 30 to 180 per village). The first major finding was that crop yields and adoption of conserving technologies had continued to grow since project termination (Table 2).

Table 2. Changes in adoption of resource-conserving technologies, maize yields, and migration patterns in three programmes in Central America during and after projects

	At Initiation	At Termination[1]	In 1994
No. of farmers with technologies			
Contour grass barriers	1	192	280
Contour drainage ditches	1	253	239
Contour rows	0	100	245
Green manures	0	35	52
Crop rotations	12	209	254
No burning fields or forests	2	160	235
Organic matter as fertilizer	44	195	397
Yields of maize (kg/ha)			
1. San Martin, Guatemala (1972-79)	400	2500	4500
2. Guinope, Honduras (1981-89)	600	2400	2730
3. Cantarranas, Honduras (1987-91)	660	2000	2050
Migration (no. of households)			
1. San Martin			
San Antonio Correjo	65	nd	4
Las Venturas	85	nd	4
2. Guinope: 3 villages	38	0	(2)2
3. Cantarranas: 3 villages	nd	10	(6)2

[1] Termination dates were: San Martin 1979; Guinope 1989; Cantarranas 1991.
[2] (2) and (6) refer to negative outmigration, ie families returning to their villages.
nd = no data

Source: Bunch and López, 1994

Surprisingly, though, many of the technologies known to be 'successful' during the project had been superseded by new practices. Had the original technologies been poorly selected? It would appear not, as many that had been dropped by farmers are still very successful elsewhere. The explanation would appear to be that changing external and internal circumstances had reduced or eliminated their usefulness, such as changing markets, droughts, diseases, insect pests, land tenure, labour availability, and political disruptions.

Altogether, some 80-90 successful innovations were documented in these 12 villages. In one Honduran village, Pacayas, there had been 16 innovations, including 4 new crops, 2 new green manures, 2 new species of grass for contour barriers in vegetables, chicken pens made of king grass, marigolds for nematode control, use of lablab and velvet bean as cattle and chicken feed, nutrient recycling into fishponds, human wastes in composting latrines, napier grass to stabilise cliffs, and home made sprinklers for irrigation.

Technologies had been developed, adopted, adapted and dropped. The study concluded that the half-life of a successful technology in these project areas is 6 years. Quite clearly the technologies themselves are not sustainable. As Bunch and López have put it *"what needs to be made sustainable is the process of innovation itself"*.

Towards a new professionalism

The central concept of sustainable agriculture is that it must enshrine new ways of learning about the world. Such learning should not be confused with 'teaching'. Teaching implies the transfer of knowledge from someone who knows to someone who does not know, and is the normal mode of educational curricula (Argyris et al, 1985; Bawden, 1992; Pretty and Chambers, 1993). Universities and other professional institutions reinforce the teaching paradigm by giving the impression that they are custodians of knowledge which can be dispensed or given (usually by lecture) to a recipient (a student). Where teaching does not include a focus on self-development and enhancing the ability to learn, then *"teaching threatens sustainable agriculture"* (Ison, 1990).

A move from a teaching to a learning style has profound implications for agricultural development institutions. The focus is less on *what* we learn, and more on *how* we learn and *with whom*. This implies new roles for development professionals, leading to a whole new professionalism with new concepts, values, methods and behaviour (Table 3).

Typically, normal professionals are single-disciplinary, work largely in ways remote from people, are insensitive to diversity of context, and are concerned with themselves generating and transferring technologies. Their beliefs about people's conditions and priorities often differ from people's own views. The new professionals, by contrast, make explicit their underlying values, select methodologies to suit needs, are more multidisciplinary and work closely with other disciplines, and are not intimidated by the complexities and uncertainties of dialogue and action with a wide range of non-scientific people (Pretty and Chambers, 1993; Pimbert and Pretty, 1995).

Table 3. *Towards a new professionalism for sustainable agriculture*

Elements	Components of the new professionalism
Assumptions about reality	The assumption is that realities are socially constructed, and so participatory methodologies are required to relate these many and varied perspectives one to another.
Underlying values	Underlying values are not presupposed, but are made explicit; old dichotomies of facts and values, and knowledge and ignorance, are transcended.
Scientific method(s)	The many scientific methods are accepted as complementary; with reductionist science for well-defined problems and when system uncertainties are low; and holistic and constructivist science when problem situations are complex and uncertain.
Who sets priorities and whose criteria count?	A wide range of stakeholders and professionals set priorities together; local people's criteria and perceptions are emphasised.
Context of researching process	Investigators accept that they do not know where research will lead; it has to be an open-ended learning process; historical and spatial context of inquiry is fundamentally important.
Relationship between actors and groups in the process	Professionals shift from controlling to enabling mode; they attempt to build trust through joint analyses and negotiation; understanding arises through this interaction, resulting in deeper relationships between investigator(s), the 'objects' of research, and the wider communities of interest.
Mode of professional working	More multidisciplinary than single disciplinary when problems difficult to define; so attention is needed on the interactions between members of groups working together.
Institutional involvement	No longer just scientific or higher-level institutions involved; process inevitably comprises a broad range of societal and cultural institutions and movements at all levels.
Quality assurance and evaluation	There are no simple, objective criteria for quality assurance: criteria for trustworthiness replace internal validity, external validity, objectivity, and reliability when methods is non-reductionist; evaluation is no longer by professionals or scientists alone, but by a wide range of affected and interested parties (the extended peer community).

Source: Pretty, 1995b, adapted from Pretty and Chambers, 1993

But it would be wrong to characterise this as a simple polarisation between old and new professionalism, implying in some way the bad and the good. True sensibility lies in the way opposites are synthesised. It is clearly time to add to the paradigm of positivism for science, and embrace the new alternatives. This will not be easy. Professionals will need to be able to select appropriate methodologies for particular tasks (Funtowicz and Ravetz, 1993).

Where the problem situation is well defined, system uncertainties are low, and decision stakes are low, then positivist and reductionist science will work well. But where the problems are poorly defined and there are great uncertainties potentially involving many actors and interests, then the methodology will have to comprise these alternative methods of learning. Many existing agricultural professionals will resist such para-digmatic changes, as they will see this as a deprofessionalisation of re-search. But Hart (1992) has put it differently: "*I see it as a 're-professionalisa-tion', with new roles for the researcher as a democratic participant.*"

A systematic challenge for agricultural and rural institutions, whether government or non-government, is to institutionalise these approaches and structures that encourage learning. Most organisations have mechanisms for identifying departures from normal operating procedures. This is what Argyris et al (1985) call single loop learning. But most institutions are very resistant to double-loop learning, as this involves the questioning of, and possible changes in, the wider values and procedures under which they operate. For organisations to become learning organisations, they must ensure that people become aware of the way they learn, both from mis-takes and from successes.

Institutions can, therefore, improve learning by encouraging systems that develop a better awareness of information. The best way to do this is to be in close touch with external environments, and to have a genuine commitment to participative decision-making, combined with participa-tory analysis of performance.

Learning organisations concerned with integrated farming will, there-fore, have to be more decentralised, with an open multidisciplinarity, and capable of responding to the demands and needs of farmers. These multi-ple realities and complexities will have to be understood through multiple linkages and alliances, with regular participation between professional and public actors. It is only when some of these new professional norms and practices are in place that widespread changes in the livelihoods of farmers and their natural environments are likely to be achieved.

References

Adnan S, Barrett A, Nurul Alam S M, and Brustinow A. 1992. *People's Participation. NGOs and the Flood Action Plan.* Research and Advisory Services, Dhaka
Argyris, C., Putnam, R. and Smith, D.M. 1985. *Action Science.* Jossey-Bass Publish-ers, San Francisco and London.

Attah-Krah, A.N. and Francis, P.A. 1987. The role of on-farm trials in the evaluation of composite technologies: the case of alley farming in Southern Nigeria. *Agric. Systems* 23: 133-152

Avery D. 1995. *Saving the Planet with Pesticides and Plastic.* The Hudson Institute, Indianapolis

Bagadion B U. and Korten F F. 1991. Developing irrigators' organisations; a learning process approach. In: Cernea, M. M. (ed). *Putting People First.* Oxford University Press, Oxford. 2nd Edition.

Balbarino E A and Alcober D L. 1994. Participatory watershed management in Leyte, Philippines: experience and impacts after three years. Paper for IIED/ActionAid Conference New Horizons: The Social, Economic and Environmental Impacts of Participatory Watershed Development. Bangalore, India: November 28 to December 2

Bannister, M.E. and Nair, P.K.R. 1991. Alley cropping as a sustainable agricultural technology for the hillsides of Haiti: experience of an agroforestry outreach project. *Amer. J. Altern. Agric.* 5, No 2, 51-59

Bawden R. 1992. Creating learning systems: a metaphor for institutional reform for development. Paper for joint *IIED/IDS Beyond Farmer First: Rural People's Knowledge, Agricultural Research and Extension Practice Conference, 27-29 October, Institute of Development Studies, University of Sussex, UK.* IIED, London.

Borlaug, N. 1992. Small-scale agriculture in Africa: the myths and realities. *Feeding the Future* (Newsletter of the Sasakawa Africa Association) 4:2.

Borlaug N. 1994a. Agricultural research for sustainable development. Testimony before US House of Representatives Committee on Agriculture, March 1, 1994

Borlaug N E. 1994b. Chemical fertilizer 'essential'. Letter to *International Agricultural Development* (Nov-Dec), p23

Brown L. 1994. *The world food prospect: entering a new era.* In: Winrock International. 1994. *Assisting Sustainable Food Production: Apathy or Action?* Winrock International, Arlington VA

Brown L R and Kane H. 1994. *Full House: Reassessing the Earth's Population Carrying Capacity.* W W Norton and Co, New York

Bunch, R. 1983. *Two Ears of Corn.* World Neighbours, Oklahoma City.

Bunch, R. 1990. *Low Input Soil Restoration in Honduras: The Cantarranas Farmer-to-Farmer Extension Programme.* Sustainable Agriculture Programme Gatekeeper Series SA23. IIED, London.

Bunch R. 1993. EPAGRI's work in the State of Santa Catarina, Brazil: Major New Possibilities for Resource-Poor Farmers. COSECHA, Tegucigalpa, Honduras

Bunch R. and López G V. 1994. Soil recuperation in Central America: measuring the impact four and forty years after intervention. Paper for IIED Conference *New Horizons:The Social, Economic and Environmental Impacts of Participatory Watershed Development,* November, Bangalore, India

Carruthers I. 1993. Going, going, gone! Tropical agriculture as we knew it. *Tropical Agriculture Association Newsletter,* 13 (3), 1-5

Carter J. 1995. *Alley Cropping: Have Resource Poor Farmers Benefited?* ODI Natural Resource Perspectives No 3, London

Cernea M M. 1991. *Putting People First.* Oxford University Press, Oxford. 2nd Edition.

CGIAR. 1995. *Sustainable Agriculture for a Food Secure World: A Vision for International Agricultural Research.* Expert Panel of the CGIAR, Washington DC and SAREC, Stockholm

Chambers R. 1994a. The origins and practice of participatory rural appraisal. *World Development* 22, No 7, 953-969

Chambers R. 1994b. Participatory rural appraisal (PRA): analysis of experience. *World Development* 22, No 9, 1253-1268

Chambers R. 1994c. Participatory rural appraisal (PRA): challenges, potentials and paradigm. *World Development* 22, No 10, 437-454

Crosson P and Anderson J R. 1995. *Achieving a Sustainable Agricultural System in Sub-Saharan Africa.* Building Block for Africa Paper No 2, AFTES, The World Bank, Washington DC

Danida. 1994. *Agricultural Sector Evaluation. Lessons Learned.* Ministry of Foreign Affairs, Copenhagen

DowElanco. 1994. What makes agriculture sustainable. *The Bottom Line*, Indianapolis, USA

Dyer N and Bartholomew A. 1995. *Project Completion Reports: Evaluation Synthesis Study.* Evaluation Report Ev583. ODA, London

EC. 1994. *Evaluation des Projets de Developpement Rural Finances Durant les Conventions de Lomé I, II, et III.* European Commission, Brussels

Ehrlich P. 1968. *The Population Bomb.* Ballantine, New York

FAO. 1993. *Strategies for Sustainable Agriculture and Rural Development (SARD): The Role of Agriculture, Forestry and Fisheries.* FAO, Rome

Funtowicz S O and Ravetz J R. 1993. Science for the post-normal age. *Futures* 25, No 7, 739-755

GTZ. 1992. *The Spark Has Jumped the Gap.* Deutsche Gessellschaft für Technische Zusammenarbeit (GTZ), Eschborn

Guijt I and Kaul Shah M. 1996. *The Myth of Community.* IT Publications, London (in press).

Harris J M. 1995. World Agriculture: Regional Sustainability and Ecological Limits. Discussion Paper No 1. Center for Agriculture, Food and Environment, Tufts University, MA

Hart R.A. 1992. *Children's Participation: From Tokenism to Citizenship,* UNICEF Innocenti Essays No 4. Florence: UNICEF

Hazell P. 1995. Managing Agricultural Intensification. IFPRI 2020 Brief 11. IFPRI, Washington DC

Hewitt T I and Smith K R. 1995. *Intensive Agriculture and Environmental Quality: Examining the Newest Agricultural Myth.* Henry Wallace Institute for Alternative Agriculture, Greenbelt MD

Hinchcliffe F, Thompson J and Pretty J N. 1996. *Sustainable Agriculture and Food Security in East and Southern Africa.* A report for The Committe on Food Security in East and Southern Africa, Swedish International Development Coperation Agency, Stockholm. IIED, London, 133 pp

IDS/IIED. 1994. *PRA and PM&E Annotated Bibliography.* IDS, Sussex and IIED, London.

Ison R. 1990. *Teaching Threatens Sustainable Agriculture.* Gatekeeper Series SA21, IIED, London

Kamp K, Gregory R and Chowhan G. 1993. Fish cutting pesticide use. *ILEIA Newsletter* 2/93, 22-23

Kang B T, Wilson G F and Lawson T L. 1984. *Alley Cropping: A Stable Alternative to Shifting Agriculture.* IITA, Ibadan.

Kendall H W and Pimentel D. 1994. Constraints on the expansion of the global food supply. *Ambio* 23, 198-205

Kenmore P. 1991. *How Rice Farmers Clean up the Environment, Conserve Biodiversity, Raise More Food, Make Higher Profits. Indonesia's IPM - A Model for Asia.* FAO, Manila, Philippines

Kerkhof, P. 1990. *Agroforestry in Africa. A Survey of Project Experience.* Panos Institute, London.

Kerr J. 1994. How subsidies distort incentives and undermine watershed development projects in India. Paper for IIED Conference *New Horizons: The Social, Economic and Environmental Impacts of Participatory Watershed Development,* November, Bangalore, India

Knutson R D, Taylor J B, Penson J B and Smith E G. 1990. *Economic Impacts of Reduced Chemical Use.* Texas A&M University

Krishna, A. 1994. Large-scale government programmes: watershed development in Rajasthan, India. Paper for *IIED New Horizons* conference, Bangalore, India. November 1994. IIED, London

Lal, R. 1989. Agroforestry systems and soil surface management of a Tropical Alfisol. I: Soil moisture and crop yields. *Agroforestry Systems* 8: 7-29.

Mascarenhas J, Shah P, Joseph S, Jayakaran R, Devavaram J, Ramachandran V, Fernandez A, Chambers R, and Pretty J N. (eds) 1991. Participatory Rural Appraisal. *RRA Notes* 13. IIED, London.

McCalla A. 1994. *Agriculture and Food Needs to 2025: Why We Should be Concerned.* Sir John Crawford Memorial Lecture, October 27. CGIAR Secretariat, The World Bank, Washington DC

McCalla A. 1995. Towards a strategic vision for the rural/agricultural/natural resource sector activities of the World Bank. World Bank 15th Annual Agricultural Symposium, January 5-6th, Washington DC

Mitchell D O and Ingco M D. 1993. The World Food Outlook. International Economics Dept. World Bank, Washington DC

N.A.F. 1994. *A Better Row to Hoe: The Economic, Environmental and Social Impact of Sustainable Agriculture.* St Paul, Minnesota: Northwest Area Foundation

Narayan D. 1993. *Focus on Participation: Evidence from 121 Rural Water Supply Projects.* UNDP-World Bank Water Supply and Sanitation Program, World Bank, Washington DC

Narayan D. 1995. *Designing Community Development.* Environment Department Papers, Participation Series No 7. The World Bank, Washington DC

NRC. 1989. *Alternative Agriculture.* National Research Council. National Academy Press, Washington DC

Paarlberg R L and Breth S A. 1994. *Assisting Sustainable Food Production: Apathy or Action?* Winrock International Institute for Agricultural Development, Arlington VA

PALM Series. 1991-present. Participatory Learning Methods Series, Myrada, Bangalore

Pimbert M and Pretty J N. 1995. *Parks, People and Professionals: Putting 'Participation' into Protected Area Management.* UNRISD Discussion Paper DP 57, United Nations Research Institute for Social Development and WWF International, Geneva

PLA Notes (formerly RRA Notes).1988-present. Issues 1-25, cont. Sustainable Agriculture Programme, IIED, London.

Pohl G and Mihaljek D. 1992. Project evaluation and uncertainty in practice: a statistical analysis of rate-of-return divergences of 1015 World Bank projects. *World Bank Economic Review* 6 (2), 255-277

Pretty J N. 1995a. *Regenerating Agriculture: Policies and Practice for Sustainability and Self-Reliance*. Earthscan Publications, London; National Academy Press, Washington DC; ActionAid, Bangalore

Pretty J N. 1995b. Participatory learning for sustainable agriculture. *World Development* 23 (8), 1247-1263

Pretty, J N and Sandbrook R. 1991. Operationalising sustainable development at the community level: primary environmental care. Presented to the DAC Working Party on Development Assistance and the Environment, OECD, Paris, October 1991

Pretty, J.N. and Chambers, R. 1993. *Towards a learning paradigm: new professionalism and institutions for sustainable agriculture. IDS Discussion Paper* DP 334. IDS, Brighton

Pretty J N and Howes R. 1993. *Sustainable Agriculture in Britain: Recent Achievements and New Policy Challenges*. IIED Research Series Vol 3, No 1. IIED, London

Pretty J N and Shah P. 1994. *Soil and Water Conservation in the 20th Century: A History of Coercion and Control*. Rural History Centre Research Series No.1. University of Reading, Reading.

Pretty J N and Thompson J. 1996. *Sustainable Agriculture at the Overseas Development Administration*. Report for NRPAD, ODA, London

Pretty J N, Guijt I, Scoones I and Thompson J. 1995. *A Trainers' Guide to Participatory Learning and Interaction*. IIED Training Materials Series No. 2. IIED, London.

Rahnema, M. 1992. Participation. In: Sachs, W. (ed) *The Development Dictionary*. Zed Books Ltd, London.

Reij, C. 1988. The agroforestry project in Burkina Faso: an analysis of popular participation in soil & water conservation. In: Conroy C and Litvinoff M (eds). *The Greening of Aid*. Earthscan Publications Ltd, London.

Reus J.A.W.A., Weckseler H.J. and Pak G.A. 1994. *Towards A Future EC Pesticide Policy*. Utrecht: Centre for Agriculture and Environment (CLM)

Röling N. 1994. Platforms for decision making about ecosystems. In: Fresco L (ed). *The Future of the Land*. John Wiley and Sons, Chichester

Röling N R and Wagemakers M A (eds). 1996. *Sustainable Agriculture and PArticiaptory Learning*. Cambridge University Press, Cambridge (in press).

Rosegrant M W and Agcaolli M. 1994. *Global and Regional Food Demand, Supply and Trade Prospects to 2010*. IFPRI, Washington DC

Sasakawa Global 2000. 1994. *Developing African Agriculture: New Initiatives for Institutional Cooperation*. SAA, Tokyo

Sasakawa Global 2000. 1995. *Feeding the Future*. Newsletter of the Sasakawa Africa Association, Tokyo

Scoones I and Thompson J. 1994. *Beyond Farmer First: Rural People's Knowledge, Agricultural Research and Extension Practice*. IT Publications, London

Shah P. 1994. Village-managed extension systems in India: implications for policy and practice. In: Scoones I and Thompson J (eds). *Beyond Farmer First*. Intermediate Technology Publications Ltd, London

Somers B. 1996. Learning about sustainable agriculture: the case of Dutch arable farmers. In Röling N G and Wagemakers M A (eds). 1996. *Sustainable Agriculture: Participatory Learning and Action.* Cambridge University Press, Cambridge (in press).

SWCB 1994. *The Impact of the Catchment Approach to Soil and Water Conservation: A Study of Six Catchments in Western, Rift Valley and Central Provinces, Kenya.* Ministry of Agriculture, Nairobi

Thompson J. 1995. Participatory approaches in government bureaucracies: facilitating the process of institutional change. *World Development* 23 (9), 1521-1554

UNDP. 1992. *The Benefits of Diversity. An Incentive Toward Sustainable Agriculture.* United Nations Development Program, New York

Uphoff N. 1992. *Learning from Gal Oya: Possibilities for Participatory Development and Post-Newtonian Science* Cornell University Press, Ithaca

van der Werf E and de Jager A. 1992. *Ecological Agriculture in South India: An Agro-Economic Comparison and Study of Transition.* Landbouw-Economisch Instituut, The Hague and ETC-Foundation, Leusden

Winrock International. 1994. *Assisting Sustainable Food Production: Apathy or Action?* Winrock International, Arlington VA

Wirth T E. 1995. *US Policy, Food Security and Developing Countries.* Undersecretary of State for Global Affairs, presentation to Committee on Agricultural Sustainability for Developing Countries, Washington DC

World Bank. 1993. *Agricultural Sector Review.* Agriculture and Natural Resources Department, Washington DC

World Bank. 1994. The World Bank and Participation. Report of the Learning Group on Participatory Development. April 1994. World Bank, Washington, DC

Poultry and Pigs

Landless Women and Poultry: The BRAC Model in Bangladesh

Md.A. Saleque and Shams Mustafa

Bangladesh Rural Advancement Committee (BRAC),
66, Mohakhali, Dhaka 1212, Bangladesh
E-mail: brac@drik.bgd.toolnet.org

Summary

On the background of the extreme poverty, most women of rural, landless households are subjected to in Bangladesh, an outline is provided of Bangladesh Rural Advancement Committee and the evolution, which led to its present poultry development model. The model is exclusively targeted at landless women and builds on GO-NGO collaboration. It involves women in a chain of activities as vaccinators, hatchery operators, chicken rearers, feed sellers, producers of hatching eggs and as producers of eggs for the market. Credit as well as marketing are integrated into the model. A recent survey has reported considerable positive impact both in terms of income and producer household egg and meat consumption. It is concluded that poor rural women can contribute to economic development as buyers and sellers of goods and services, by contributing to improved household income, and - as important - in the process their own self esteem is heightened.

Key words: Landless women, rural poultry, poultry development, BRAC, GO-NGO collaboration, impact.

Introduction: Poverty in Bangladesh

That poverty is acute in Bangladesh does not need to be told: However, a few words on poverty will be useful to set the scene in which Bangladesh Rural Advancement Committee (BRAC) operates. Bangladesh has the unfortunate label of a case of endless poverty and deprivation. The proportion of the rural population living below the poverty line in the early 1990s has been estimated to be between 38% (Sen, 1995a), 48% (BBS, 1995) and 55% (Hossain, 1995). These single index measures hide a wide range of variation among the poor: the household food intake is gender biased with the females' consumption levels being between 71 and 90% of the males' (and Chowdhury, 1995). The per capita food intake (1980 k.cal. in 1990-91) remains below the requirement (2273 k.cal.). The literacy rate is still low at 35% (for females it is 29% and for males 45%) (Hamid, 1995). Another indication of poverty is the real wage rate, i.e. kg. of course rice a day's money wage will purchase in rural areas, which has declined from 3.84 in 1987 to 3.24 in 1990 (Hossain, 1995a). Land, which is the single most important resource in the rural areas, is distributed very uneven with 50% of the households owning none or less than 0.50 acre. At the same time, for around 60% of the rural households agriculture (cultivation and source of employment) is the primary source of income (Sen, 1995).

Access to credit has been identified as a major mechanism with which a household can improve its economic condition (Rahman, 1989; Khondakar and Chowdhury, 1995). The rural households in general and the landless in particular have very little access to institutional credit. In the late 1980s less than seven% of the landless and 14% of all rural households had access to institutional credit including NGO programmes (Rahman, 1989). Given this, it is not surprising that the women in rural areas virtually had no access to institutional credit until the 1980s.

Since the beginning of 1980s some specialised programmes were launched to provide financial support on credit basis to women, who in their turn have proved themselves to be "bankable" (Rahman 1989; Hossain and Afsar, 1989). Along with the expansion of credit availability in rural areas and for women in particular through expansion of commercial banks and non-governmental programmes, other positive changes are worth noting. One of these is the expansion of safe drinking water to 87% of all households (Sen, 1995b). An important change is the improvement in the nutritional status of children under five years of age : the figures for stunting and wasting have declined from 74% and 22%, respectively in 1975 to 43% and 13% in 1991 (Khondakar and Chowdhury, 1995).

BRAC: The organisation

In 1972, following Bangladesh's War of Independence, BRAC worked on the resettlement of refuges in the Sulla area of Sylhet district1 in the northeastern part of the country. It organized relief and rehabilitation for war victims whose homes, cattle, fishing boats and other means of livelihood had been destroyed. What made BRAC set out on its remarkable journey was the realization that relief-oriented activities could only serve as a stop-gap measure. From then on the new pledge was to provide sustainable measures to improve the conditions of the rural poor by developing their ability to mobilize, manage and control local and external resources by themselves. BRAC's programmes have never been determined by a rigid set of strategies. The organization's success is attributable largely to its flexibility in responding to the needs of the people (Lovell, 1992).

Another factor that has contributed to BRAC's transformation is its capacity to learn through trials and errors. In 1973, BRAC adapted in its work

---

1. The administrative units in Bangladesh in descending order are country, division, district, sub-district or Thana and Union Porishad.

the basic rural development community strategy, focusing on entire village communities. It was at this point that BRAC realized that in fact, there was a community within the larger village community comprising the poor. By 1976, it therefore became apparent that the community approach would not work, as the poor who outnumbered the others in the community benefited very little from the interventions. This was because those who owned land and other productive assets were able to secure for themselves the larger share of the benefits. From here on began BRAC's involvement with the poorest of the poor - the landless, small farmers, artisans, and vulnerable women. The time had also come to fix the organization's goals which were identified as:

A. Poverty Alleviation
B. Empowerment of the Poor

BRAC's programmes

BRAC's definition of the poor refers to those people who own less than half an acre of land (including the homestead) and to those who earn their living by selling manual labour. Efforts to empower this group have been evaluated and adjusted many times over the years in the light of BRAC's growing capacity and the needs of the programme participants. Today , working as a development organization in the private sector, BRAC strives to attain its two goals by implementing such programmes as:

1. *Rural Development* which involves development of village organizations of the poor, credit operation, and facilitation of savings' habits. The village organizations are designed to mobilize the collective strength of the poor with a view to empowering them to be self-reliant. BRAC has a Human Rights and Legal Education Programme to further the initiatives aimed at empowering the Village Organization members. BRAC's Rural Development Programme implements these initiatives along with several income and employment generating programmes, designed particularly for the women village organization members. These women are provided with credit and training to carry out their activities. There are also some special programmes that have been introduced under the Income Generation for Vulnerable Group Development and Small Holder Livestock Development Programme initiatives. These too are implemented through Rural Development Programme.

2. *Education* initiatives in the form of the Non Formal Primary Education Programme for the children of the disadvantaged rural people.

3. *Health programmes* addressing the health and nutritional status of women and children. These initiatives seek to develop and strengthen the capacity of communities to sustain health related activities.

4. *Administrative and technical support services* that facilitate BRAC's programme activities, e.g., training, research, monitoring, the "Aarong" marketing outlet, publications, public affairs & communication, accounts & audit, logistics, computer service, and construction service.

5. Furthermore, in order to attain budgetary self-reliance the organization has set up its own revenue generating enterprises i.e., the BRAC Printing Press, BRAC Cold Storage and the BRAC Garments Factories.

Of the 86,038 villages of the country, BRAC's Rural Development Programme covers 35,961 with the Education Programme coverage in 16,946 and the Health and Population Programme in 12,056 villages. In certain areas these three programmes overlap.

The management system within BRAC is participatory and decentralized, and programme planning draws upon the experience and expertise of workers at all levels. Women comprise 23% of the staff. Founded by Mr. Fazle Hasan Abed in 1972, BRAC has now grown into an organization in which capacity building of the individual worker and the programme participants is given the topmost priority. BRAC also attaches the utmost importance to the institutionalization of the organization so that it may withstand and overcome the challenges of the future.

The state of the poultry sector in Bangladesh

The poultry and livestock sectors are integral parts of the farming systems in Bangladesh. There are about 90 million chicken and 12 million ducks in the country. About 89% of rural livestock households rear poultry and the average number per household is 6.8. It is an important source of cash income for the poor rural families, particularly for women. Most birds are kept in a scavenging system and are fed on household waste and crop residues. The predominant poultry breed in Bangladesh is the local. The productivity of the hen is about 40-60 eggs per year. Some other exotic breeds such as Rhode Island Red, White Leghorn, Barred Plymouth Rock, Australorp and Fyaumi are now available in the government poultry farms. There are six government hatcheries in Bangladesh which produce day old chicks, but there is no distribution system in rural areas. These exotic hatching eggs and day old chicks are now distributed to BRAC project areas to develop the local breed. There are some commercial farms in Bangladesh, where the production cost of eggs and meat is comparatively higher than of eggs and meat produced in scavenging system.

The annual growth rate in the chicken population was 6.5% between 1990-94 (Alam, 1996). The annual per capita egg consumption was only 23 although it should be 100 from a nutritional point of view. For optimal productivity, the high yielding varieties (HYV) of poultry requires improved feeding, but presently the feed which is prepared in the government farms is far less than the need and consequently balanced feed is not available in rural areas.

The prevailing poultry diseases in Bangladesh are Newcastle, Fowl Pox, Fowl Cholera, Fowl Typhoid, Coccidiosis, deficiency diseases and worm infestation, etc. Without interventions, the mortality rate of the poultry in the scavenging systems is high (35% to 80%) due to diseases and predators. In spite of 4 types of important vaccines are produced in Bangladesh, remote rural areas are not served due to lack of service delivery mechanisms. There are only four field staff and one livestock officer at sub-district level and they are responsible for about 200,000 poultry, 50,000 cattle and 20,000 sheep and goats.

Government institutions that are responsible for the delivery of support services in the rural areas are not geared to assist BRAC's target group. There is thus a need to assist particularly the landless in their efforts to earn an income and to the extent possible, improve their long term potential for deriving income from sustainable agricultural practices. In remote areas where government services are not operative or inadequate, BRAC collaborate with the government machinery to extend the service delivery system by developing and using local manpower.

Summary of the constraints and major issues in the poultry sector:

- High mortality of the scavenging bird.
- Low productivity of the local hen.
- Unavailability of cheap sources of HYV birds at village level.
- Supply of vaccines in remote rural areas.
- Health and veterinary care is inaccessible for the village women.
- Poor poultry rearing and management system.
- Unavailability of some feed ingredients.
- Lack of organized marketing mechanism.
- The government livestock service delivery system is inadequate and inefficient.
- The vast majority of women are left out of the formal credit system.

The scope of poultry development

The possibilities for women's participation in poultry development are as follows:

1. About 70% of the rural, landless women are directly or indirectly involved in poultry rearing activities. Traditionally these women have some experience in poultry rearing, which therefore represent skills known to them.

2. BRAC has proved that homestead poultry rearing is economically viable. If the landless women are properly trained, supported with credit

and other necessary inputs and made to operate under supervision of extension workers of both Government and BRAC and the Government machinery are activated to provide for the delivery of services, the poultry sector could be one of the most productive sectors.

3. Poultry rearing is suitable for widespread implementation as it is of low cost, requires little skills, is highly productive and can be incorporated into the household work.

4. There are few or no job opportunities for the landless, disadvantaged women. Poultry is the only activity in which a large number of landless women can participate.

5. In the small scale poultry units which support the landless, production per bird may be low, but distribution of benefits will be more equal and have great human development impact.

6. Poultry rearing is culturally acceptable, technically and economically viable. Moreover, the ownership of poultry is entirely in the hands of women. This is an asset over which the poor women actually have control. This activity can therefore play an important role in poverty alleviation which is the main goal of BRAC.

Development of BRAC's poultry programme

The history of BRAC's efforts to develop a poultry programme design can be divided into three phases, i.e. formative, development and replication (see for details in Mustafa et al., 1993). These relate roughly to an eleven year timeframe during which the programme continuously underwent changes and fine tuning.

Formative Phase. In the late 1970s BRAC identified poultry rearing as a source of income for the landless, particularly destitute women. A high mortality rate for poultry in Bangladesh, combined with its relevance as an income generating activity for poor women, led BRAC to carry out participatory 'action research' aimed at increasing productivity.

Initially, efforts were made to increase the productivity of local poultry by cockerel exchange, but this system with improved cockerels for cross-breeding failed since the improved birds tended to be sold and mortality remained high. In order to reduce bird mortality BRAC initiated an action research in its Manikganj project area. BRAC staff regularly vaccinated poultry birds in the five intervention villages for one year. The positive results in terms of reduction in mortality and increase in bird population led BRAC to realise that vaccination must be an integral part of any intervention to promote poultry rearing as an income earning source.

It was decided to involve women group members in vaccination work and allow them to vaccinate for a fee, using vaccines supplied free of cost from the Government.

It was observed that the pullets supplied by the government and other farms, like the cockerels also suffered high mortality in the scavenging system. It was therefore decided to buy day-old chicks from Government farms. Selected, trained and supervised by BRAC, rural women were to rear the day old chicks for two month on their homestead plots and thereafter sell them to key rearers. The advantage was that the two months old chicks were released into the scavenging system survived to a much higher degree as mortality in chicks is particularly high in the two first months after hatching in the scavenging system. From BRAC's point of view it was an advantage that this arrangement did not require more BRAC staff.

Between *1978 and 1982* the poultry programme of BRAC had no model or design, it was done on an ad-hoc basis. The focus changed from *1983* to supply of improved chicks, prevention of common diseases and training in improved scavenging based rearing. The following model was developed:

- One poultry worker (vaccinator) for every 1000 birds. The poultry worker is given a five day training on preventive and curative aspects as well as rearing management.
- Vaccinations to take place at dawn.
- key rearers in each village keeping one HYV cock and ten hens, trained in the improved scavenging rearing management.
- Model rearers with three cocks and 20 hens.
- Chick rearing units (CRUs) at the level of a cluster of villages, each with a capacity to nurse 200 day-old chicks for two months.
- Feed production centres at the level of a cluster of villages to supply feed to the programme participants who are all women.
- In late 1980s, two more components were added to this early model: egg merchants to protect the producers from middlemen, and experiments are underway, from 1993, to create the supply sources of day old chicks through hatchery operators.

Development phase

Having developed a model for rural poultry development, the District Livestock Office in Manikganj (60km from the capital Dhaka) was approached for cooperation by BRAC's project staff. Between *1983 and 1985* an informal collaboration developed in Manikganj whereby the Government officers supplied vaccines and provided technical advice on the chick nursing units. After extensive evaluation by officials from the Directorate of Livestock Services in Dhaka the BRAC model was accepted as viable and replicable. Based on this experience the model was tested further.

Between *1985 and 1987*, the model was tested in 54 Area Offices of BRAC's core Rural Development Programme. The Sub-district Livestock

Officers in the respective areas ensured the supply of vaccines to the participants through the Area Offices of BRAC in 32 Thanas. This produced positive results in terms of increased income for the participating women, a reduction in mortality rates and an increase in bird population. Through the intermediation of BRAC the government structure was brought closer to the people.

In *1987* BRAC integrated the experiences of poultry development collaboration and the Government food aid for destitute women, into an independent programme. The Income Generation for Vulnerable Group Development Programme was launched in August 1987, in collaboration between BRAC, the Departments of Livestock and Relief and Rehabilitation, and the World Food Programme (WFP).

In *1988*, it was found that the income earned by the rearers was very low because the participants were able to buy only one or two HYV birds. This prevented development of crossbreeds and improved productivity, resulting in slow generation of income. In the *late 1980s* credit support for poultry rearing was introduced in BRAC.

Having incorporated the credit component, efforts were directed at *sustainability*. The range of income generation activities are being increased so that the beneficiaries can undertake additional enterprises, which need not be related to poultry. To support this the credit operation is also being scaled up. The interest earnings from financial operations and service charges to be levied for technical services, are estimated to cover the major portion of the programme costs. Furthermore, the need for technical services form BRAC is expected to decline over time as the beneficiaries become adept at using different technologies.

The programme framework

Through the process of "learning by doing" BRAC identified a framework for rural poultry development the aim of which is to enhance the income earning capability of very poor women.

BRAC's poultry development programme - Objectives

The programme aims to provide the women an entry point to diversify income earning and employment opportunities through training in poultry activities in order to improve their socio economic situation.

Specific objectives of the programme

1. Integration of poor village women into poultry rearing activities, so that they can earn a monthly income of at least Tk.200 (US$ 5).
2. To reduce poultry mortality from 40-85% to 15%
3. To increase the poultry population.

4. To introduce crossbreeds and increase the production of eggs and meat.

5. Improve the protein intake level of the rural poor.

Table 1. *Programme framework: The steps*

Subject	Method(s)	Objective
1. Check the bird mortality	1. Selection & Training of beneficiaries 2. Develop poultry worker. 3. Regular supply of vaccine & medicine. 4. Motivate the people for vaccination of their birds.	To create a favourable environment, so that the people are interested to rear poultry.
2. Upgrading of local breed.	1. Training and development of key rearers. 2. Supply of HYV birds through establishment of chick rearing unit. 3. Supplementary feed supply	To increase production and income
3. Marketing facilities	1. Develop egg collector	To ensure reasonable price of egg to key rearers.
4. Permanent network development.	1. Improvement of management system components such as housing, feeding, rearing etc. 2. Medium scale poultry unit development such as model rearer to produce hatching eggs. 3. Set up of small rice husk hatcheries. 4. Service charge.	To develop self supported programme.
5. Credit facilities.	1. Small scale credit to the rearers.	To start the poultry enterprises just after completion of training.
6. Involvement of Government.	1. Delivery of input and other support service. 2. Coordination meeting with Govt. staff.	To increase access to Government resources and services for the poor women

Methodology and development model

Considering the steps described earlier, BRAC designed a specific model for poultry development from the practical experience in 1983, which was accepted by the Government as a model for poultry development. The approach consists of an integrated package support to the rural women and includes the following:

Selection
BRAC through its Rural Development Programme organizes the landless women into groups. There are 45-55 members in each group out of which 30 are selected for poultry activities and provided different types of training on poultry rearing and management.

Training

(a) *Poultry workers*: One woman group member is selected from each village and given 5 days of training on poultry rearing, management, vaccination and treatment. The poultry workers are engaged in vaccination and treatment of birds in their respective villages. Once in a month they attend one day refresher courses and they receive poultry vaccine and medicine twice a month. The workers charge Tk. 0.25-0.50 per bird attended to as a fee.

(b) *Key rearers*: They are given three days of training on ideal methods of poultry rearing. Every key rearer must have one HYV cock and 10 hens (of which 4-5 HYVs) and a good housing system.

(c) *Chick rearers*: Chick rearers are given 7 days of training. They rear 200 chicks from day old chicks till two months of age and sell them to the key rearers. The chick rearers are supplied with chicks form the Government farms, BRAC's own farms and the poultry hatcherers (see (f) below).

(d) *Feed sellers*: One poultry feed sale's centre is established in each area. With the spread of HYV birds, people are gradually getting accustomed to buy balanced feed. Feed producers receive three days of practical training on feed preparation. She prepares poultry feed with ingredients from locally available sources under close supervision of BRAC.

(e) *Model rearers*: They are given three days of advanced training on poultry rearing and management. They rear HYV 22 hens and three cocks, and produce hatching eggs which are supplied to the poultry hatcherers.

(f) *Poultry hatcherer*: To meet the demand of day old chicken, five small hatcheries operated by the rice husk method are established in each area, the capacity is about 5000 chicken per month. The hatching eggs are purchased from model rearers through egg collectors.

Input supply

(a) After completion of training, the poultry workers are provided with vaccination kits. There are specific dates for vaccine distribution. Vaccines are supplied by the Government and distributed twice a month to the vaccinators by the Government veterinary staff. All poultry workers come to the Union Porishad to collect their vaccines. Initially the Government field staff did the vaccination work by themselves, but now the strategy is that they are responsible for distribution of vaccine instead of doing the vaccination work.

(b) Medicines are supplied by BRAC at cost price each month. Initially, medicines worth Tk. 25 are given per poultry worker as a revolving fund from where they buy and sell the medicines.

(c) Day old chicken are supplied by the Government to BRAC and BRAC distributes them as per requisition. BRAC also assists in distribution of the day old chicken produced by the rice husk hatcheries.

(d) The eight week old chickens are sold from the rearing units to the key rearers from BRAC or Union Porishad offices.

(e) To ensure complete feed, BRAC supplies ingredients like fishmeal, sesame oil cake, vitamins etc. at cost price to the feed sellers.

(f) Marketing.

(g) There are 10-15 egg collectors for each area who are responsible for buying eggs at reasonable prices from the group members and marketing of the eggs.

Credit

To ensure proper utilization of the skills imparted during training, BRAC provides credit as initial investment capital to start poultry or chick rearing, feed selling, egg collection and hatchery activities.

Coordination with government

To cooperate with Government and to ensure smooth implementation of the programme, BRAC has one Project Officer (livestock) for every 10 areas and one Project Assistant for each area office. BRAC's staff are responsible for initial surveys of participants, motivation, group formation, training and credit to the group members. The Government staff is responsible for training and input supplies.

A flow chart of the poultry programme is presented below.

Institutional arrangements

BRAC's poultry programme is also a case study in multi-agency action between the State, the Aid Community and BRAC (Mustafa, 1993). The poultry programme was developed in close cooperation with the Directorate of Livestock Services of the Ministry of Livestock and Fisheries, Government of Bangladesh, an interaction, which began in the formative phase of programme development. Thereafter the Aid Community joined hands with Directorate of Livestock Services and BRAC, firstly the World Food

Programme (WFP) and later the International Fund for Agricultural Development (IFAD) and Danida. The institutional arrangement for poultry development by BRAC has two aspects: (a) the providers' aspect, and (b) the implementation aspect.

BRAC, RDP
Poultry Programme Flow Chart

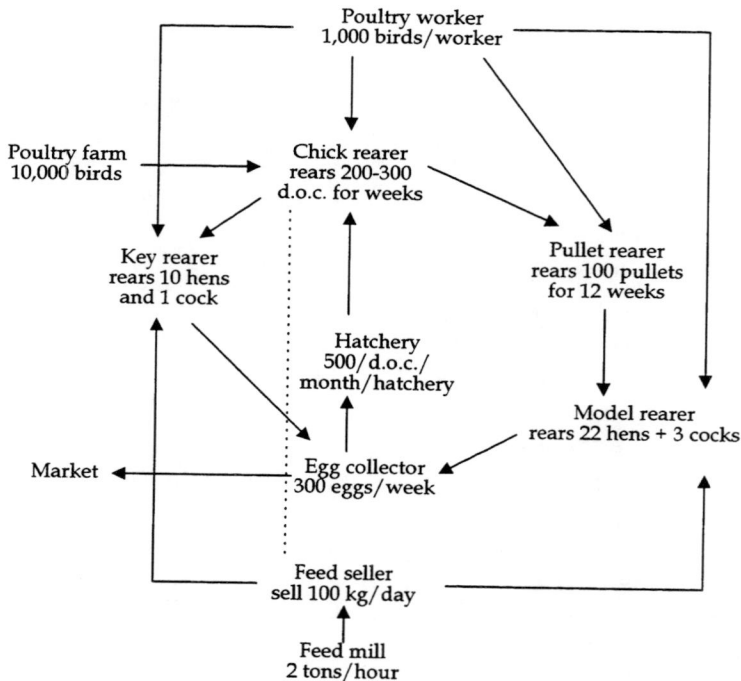

```
                        Poultry worker
                        1,000 birds/worker
                             │
                             ▼
Poultry farm            Chick rearer
10,000 birds   ───►     rears 200-300              Pullet rearer
                        d.o.c. for weeks           rears 100 pullets
                                                   for 12 weeks
         Key rearer
         rears 10 hens        Hatchery
         and 1 cock           500/d.o.c./
                             month/hatchery        Model rearer
                                                   rears 22 hens + 3 cocks
Market ◄──              Egg collector ◄──
                        300 eggs/week

                        Feed seller
                        sell 100 kg/day

                        Feed mill
                        2 tons/hour
```

BRAC and Directorate of Livestock Services work with the aid community under different arrangements: with the WFP and the Ministry of Relief and Rehabilitation in one set, and with IFAD and Danida in another set of arrangements, for the implementation of the programme. With respect to the implementation aspect the programme is executed by three organizations:

a) BRAC implements the programme through its Rural Development Programme in the latter's permanent operational areas.

b) In collaboration with WFP, the local government (Union Porishad) and the Ministry of Relief and Rehabilitation, which is targeted specifically at the destitute women who receive food aid from the Government under Income Generation for Vulnerable Group Development.

c) BRAC is a partner in the implementation of the Smallholder Livestock Development Programme.

The last two arrangements may spatially overlap BRAC's core programme for rural development.

Rationale for collaborative action

A brief discussion on the reasons why the different agencies came forward to work together will contribute to a better understanding of a phenomena that was encouraged by the donors in the late 1980s. The limitations of the government structures and the comparative advantage of NGOs like BRAC in reaching the poor provided the motivation for joint action to develop.

Limitations of the public sector

The emergence and the proliferation of non-governmental organisation (NGOs) in Bangladesh and elsewhere in South Asia, has been explained as a response to the fact that the State has had limited success in meeting the needs of the rural poor - particularly the women. A recent review of experiences of agricultural technology development in South and South-East Asia, has identified three broad trends.

- Limited public sector success in meeting the needs of the rural poor.
- The recent establishment of a large number of NGOs which claim advantages over the public sector in reaching the rural poor.
- The increasing weight attached to views that the prospects of successful change are enhanced if the poor participate in its design (Farrington and Lewis, 1993)

The Independent South Asian Commission on Poverty Alleviation (ISACPA) also pointed out the failure of the Government initiated programmes to reach the poor. The Commission identified the following among other reasons for this: "Conventional top-down development strategies, inequitable distribution of assets, inaccessibility of the poor to technological innovations and finance , the misuse of development resources and viewing the poor as a liability". These factors obviously led to the exclusion of a large number of poor from benefiting from Government initiated programmes and hence to their increasing marginalisation (SARC, 1992).

Comparative advantages of BRAC and the state

The Government structures which extend to the sub-district level and that are relevant for BRAC's poultry programme are the Ministries of Fisheries and Livestock and of Relief and Rehabilitation. The two are briefly described below with a view to identify the weaknesses in them.

The Directorate of Livestock Services, Ministry of Fisheries and Livestock, is the sectoral Government structure whose line functions extend to the Union level. The Directorate of Livestock Services is the Government agency responsible for promoting livestock and poultry development in the country.

The Directorate of Livestock Services has several constraints of its own, particularly inadequate manpower, to cope with the magnitude of the tasks involved. The activity spectrum involves providing of development inputs, training of beneficiaries, prevention and cure of diseases of the animals and birds.

The workload of the present Thana Livestock Development Complex staff appears to be quite heavy as it includes delivering services to around 40,000 households owning over 40,000 cattle, 20,000 goats and 150,000 poultry. This is excessive by any standard of user:provider ratio. As a result, the extent of coverage has obviously been limited and confined virtually to prophylactic treatment i.e., vaccination (Samdani, 1991).

The Directorate of Relief and Rehabilitation under the ministry of the same name, is responsible for providing emergency relief at times of natural disasters, for implementing the "Food for Work" programme during the slack employment season, and for the Vulnerable Group Development (formerly Vulnerable Group Feeding) Programme throughout the year. In terms of the number of beneficiaries the Vulnerable Group Development Programme administered by the Directorate of Relief and Rehabilitation is perhaps the largest. Some 450,000 families in rural Bangladesh are recipients of a monthly income transfer ration of 31.25 kg of wheat for a period of two years. However, the programme has virtually turned into a relief and welfare programme as the target women received wheat only because of lack of adequate personnel for extension services, technical ability and other resources.

The Union Porishads are one of the mechanisms for the distribution of food relief. With one chairman and nine members, directly elected by the rural population, they are close to the recipients of food aid. Village level developmental activities, such as infrastructure construction, are organised through the Union Porishads. The representatives are also involved in health and family planning activities. For all of their developmental activities and resources they are entirely dependent on the respective governmental structures at the sub-district level.

Dysfunctioning government structures

Questions of dysfunctioning arise when structures exist, but do not deliver their services to a large segment of the population. In particular, the rural poor men and women are not reached by the services provided by government structures.

In the context of agricultural technology development, a recent six-country study of NGO-State collaboration has found that NGO approaches emphasise those areas in which "Government services have either disregarded the needs of the poor or have responded to them inadequately". These areas include:

- Technologies and management practices adapted to difficult areas
- Technologies to meet the needs of the rural landless
- Technologies to meet the specific needs of women
- Approaches that "de-mystify" complex technologies and make them suitable for neglected groups
- Approaches helping to form local groups which then carry forward the technology in a sustainable fashion , linking in with input supplies and markets (Farrington and Lewis, 1993).

These general comparative advantages of NGOs in Asia, are also applicable to the concrete situation in Bangladesh and BRAC in particular. The government has traditionally concentrated on the physically favourable areas, large scale lumpy technologies, literate male farmers, emphasising the individual, and so on. Such an orientation reaches only a few and thus diverts resources away from a large number of people. On the other hand the BRAC approach considers people as active participants in development, emphasises the poorer section, particularly women and adapts existing technology to the specific condition of the poor (Mustafa, 1993).

Implementation arrangements

At the top level the poultry development programme is led by a senior manager who is reportable to the Director of Field Operations. At the implementation level three separate organisations with different funding sources, are engaged.

BRAC implements the poultry programme on its own through its Rural Development Programme which is a multisectoral intervention that comprises group formation, social development, credit and sectoral programmes such as fisheries, agriculture and livestock. Poultry development is part of BRAC's livestock sector development programme. It is implemented through the area offices of the Rural Development Programme.

The organisation of the Income Generation for Vulnerable Group Development Programme focuses primarily on the poultry sub-sector and its beneficiaries are the destitute women who receive the two-year long wheat ration from the Ministry of Relief and Rehabilitation. A part of the food aid from the WFP is used to provide the ration and another part is monetised to establish the Revolving Loan Fund. The Revolving Loan Fund is managed by BRAC staff. The management cost is primarily borne by the Rural Development Programme's budget for 'special programmes'. The officers of the Directorate of Livestock Services provide technical supervision. The role of the local government (Union Porishad) is to select the relief recipients (whose eligibility is verified by BRAC), distribute the relief, provide space for training and motivate villagers on the need for poultry health care. Coordination committees which are comprised of representatives from the Government, the WFP and BRAC, exist at four levels i.e. national, district, sub district and Union.

In 1993 a new organisation was set up to undertake poultry development on a large scale, jointly by the Government's Directorate of Livestock Services and a number of NGOs including BRAC. The funding sources of the Smallholder Livestock Development Programme are entirely different from the two above. IFAD extends soft, long term loan to the government for training and for the Revolving Loan Fund, through Bangladesh's central bank. The Bank on-lends the fund to a) the Directorate of Livestock Services which then finances the training activities undertaken by the NGOs, b) the Bangladesh Krishi Bank which on-lends to the participating NGOs who conduct the credit operations. The social development activities such as group formation, awareness building and other human resource costs are met by Danida on a grant basis. The Smallholder Livestock Development Programme is implemented in 80 sub-districts (out of a national total of 460) of which BRAC is responsible for 66. After the expiry of the Smallholder Livestock Development Programme the groups which are formed will become part of BRAC's core programme for rural development.

Impact of the poultry programme

A recent evaluation of the Smallholder Livestock Development Programme (Alam, 1996) reports positive improvements for the beneficiaries. The findings of the assessment are summarised below:

Poultry and breed

The number of poultry reared per farm in 1995 was 17 for key rearers, 11 for chick rearers, 32 for model rearers, 10 for poultry workers, 10 for feed sellers and 11 for mini hatcherers. The average number of poultry reared per farm was 16 which was much higher than the national average (of 6.8 in 1988-89).

All birds reared by different categories of beneficiary households were classified by type of breed. It was observed that 47.4% of all birds was of the improved type while 52.6% was local. The %age of improved breed was highest for model rearers (79.2%), followed by mini hatcheries (57.0%), key rearers (44.8%), feed sellers (42.7%) and poultry workers (39.7%). In the case of chick rearers, all chicks were identified to be of improved breed.

Considering the national average figure for improved breeds of chicken is at around 5%, one can safely conclude that the Smallholder Livestock Development Programme has contributed substantially to breed improvement in the areas it covers.

One of the objectives of the Smallholder Livestock Development Programme was to reduce the mortality rate of chicken in rural areas. The mortality rate of adult chicken was less than 3% for each category of farms and one can thus conclude that the Smallholder Livestock Development Programme has had a significant, positive impact on the mortality rate of chicken.

Income generation

The repayment behaviour of group members suggests that the loans were properly used and that investment in Smallholder Livestock Development Programme activities was profitable. It was noted that the average net income per household from the Smallholder Livestock Development Programme activities was Taka 427 per month. The amount of income was highest (Taka 1047) for mini hatcherers followed by Taka 761 for chick rearers, Taka 757 for feed sellers, Taka 500 for model rearers, Taka 394 for key Rearers and Taka 265 for poultry workers. The average monthly benefit/cost ratio was 1.5 : 1 for all farms. The benefit/cost ratio was highest, 3.86 : 1 for key rearers and lowest 1.06 : 1 for feed sellers.

Consumption

As the economic condition of the beneficiary households improved, one would expect that the intake of food by household members would increase after the intervention of the Smallholder Livestock Development Programme. The proposition was thoroughly investigated and analysed. It appears that the consumption of all food items increased after membership. The increase in consumption was substantial in the case of eggs by 159.6% for per week and of chicken by 137% per annum. With regard to consumption of eggs within the households, children, and especially boys, were given priority.

Decision making

All the beneficiaries of the Programme are women. The Smallholder Livestock Development Programme has ensured employment and income for them and thereby enhanced their status in the family. Their relationships with their husbands have improved and their participation in decision making has increased. Sole decision making by men has declined sharply from 21% in the pre-project period to 2% in the project period. The evidence suggests that the socio-economic status of women within the household has increased after the intervention made by the Smallholder Livestock Development Programme, whereas no change has yet been registered with regard to the beneficiaries' status in the village society.

Conclusion

The poultry programme has made a significant contribution in raising the income level of disadvantaged women, who would otherwise be left without work. They are now an active work force and even if their income is not much, it helps to augment the meagre earnings of the family as well as improving their quality of life. For many it is the sole source of income.

What is noteworthy about this programme is that poor, rural women can actively participate in the rural economy both as buyers and sellers of goods and services. Moreover, a strong linkage is developed with the Government services, which are now accessed. Another, more important aspect of the programme is the feeling of dignity, the women develop as a result of their participation.

References

Alam, I (1996). Socio-economic Impact of the Smallholder Livestock Development Programme. MFL (GOB) and Danida, Dhaka

BBS (1995). Summary Report of Household Expenditure survey 1991-1992. Bangladesh Bureau of statistics, GOB. Dhaka

BRAC (1995). The BRAC Annual Report -1994. BRAC, Dhaka.

Khondakar and Chowdhury (1995). Nutritional Dimension of Poverty. In: Rahman and Hossain (eds)(1995).

Farrington and Lewis (eds)(1993). Non-Governmental Organisations and the State in Asia: Rethinking Roles in Sustainable Agriculture Development. Overseas Development Institute (UK). Routledge, London.

Hamid, S (1995). Gender Dimension of Poverty. In Rahman and Hossain (eds)(1995).

Hossain, M (1995a). Structure and Distribution of Household Income and Income Dimension of Poverty. In: Rahman and Hossain (eds) (1995).

Hossain, M (1995b). Socioeconomic Characteristics of the poor. In: Rahman and Hossain(eds)(1995).

Hossain, M and Afsar, R (1989). Credit for Women's Involvement in Economic Activities in Rural Bangladesh. Research Report no. 105, Bangladesh Institute of Development Studies, Dhaka.

Lovell, K (1992). Breaking the Cycle of Poverty - The story of BRAC. Kumarian Press, New York.

Mustafa, S (1993). The State and BRAC: a case study of Joint Public Action in Bangladesh. Paper for Workshop on NGO-Local Government Collaboration in Management of Local Development in South Asian Countries. Organised by the UN Centre for Human Settlement, Rajendrapur, (Bangladesh).

Mustafa, S, Rahman, S and Sattar, G (1993). BRAC: Backyard Poultry Development and Landless Irrigators' Programmes, in Farrington and Lewis (eds) (1993).

Rahman, A (1989). Credit for the Rural Poor, in: Institutions (compendium Volume V), Bangladesh Agriculture Sector Review, UNDP, Dhaka.

Rahman, H Z and Hossain, M (eds) (19995). Rethinking Rural Poverty: Bangladesh as a case study. UPL, Dhaka.

Samdani, G. (1991). Poultry Development in Rural Bangladesh - A case of GO & NGO collaboration. Paper presented at National Workshop on Livestock Development. Centre for Development Management, Rajendrapur.

SARC (1993). Meeting the Challenge: an Overview of the Report of the Independent South Asian Commission on Poverty Alleviation. ISACPA secretariat, Colombo.

Sen, B (1995a). Rural Poverty Trends, 1989-90. In Rahman and Hossain (eds) (1995).

Sen, B (1995b). Selected Living Standard Indicators. In Rahman and Hossain (eds) (1995).

Semi-Scavenging Poultry Flock

Hans Askov Jensen
Team, Bangladesh Smallholder Livestock Development Project,
Strindbergsvej 104, 2500 Valby, Denmark
Fax: +45 36 161406
E-mail: askov@ibm.net

Summary

Defining semi-scavenging as a system in which poultry flocks are under partly controlled management and where the scavenged feed account for a significant part of the total feed eaten, the goes paper goes on to describe the Bangladesh Rural Advancement Committee (BRAC) model as it is organised in production, supply and service lines. The technology transfer approach of the model is briefly discussed. Data of recent field surveys and on-farm studies are used to undertake a financial evaluation. Factors affecting feed supply in a scavenging system are discussed using data from different locations. It is concluded the model is the most structured and most carefully designed available for smallholder poultry development, it has the potential to open new grounds in smallholder and scavenger poultry production. While the institutional set-up, structure and implementation procedure are well developed and documented, the technical part of the model needs still further documentation. However, there is no doubt that the model is viable, but also, that there is a great potential for technical improvements.

Key words: semi-scavenging, poultry, organisation, financial evaluation, location effect, technical improvement.

Introduction

The terminology semi-scavenging is used for small poultry flocks under partly controlled management and where the scavenged feed account for a substantial part of the total feed consumed. A Semi-scavenging Poultry Model (Saleque and Mustafa, 1996) is an integrated system to provide supplies and services to establish and to maintain a semi-scavenging poultry sector.

Scavenging poultry account for far the largest number of domesticated animals in the developing countries. Scavenging hens are, however, more or less neglected as an income generating activity by institutions as well as by the poultry holders themselves. The main activities to improve scavenging poultry holdings have been introduction of cockerel exchange programs and vaccination campaigns. However, the effect has been rather small because they have not been followed by other management activities.

In Bangladesh there has, during the past decade, been developed a successful model for semi-scavenging poultry holding. In 1996 it was established more than 800,000 semi-scavenging smallholders and the number is now increasing with more than 100,000 per year.

The smallholder, the producer of the end product, constitute around 95% of the total number of small entrepreneurs involved in the model.

The Bangladesh model

Organization

The Model is a three pronged organization where each prong has its specialized functions. The institutional structure behind the Model is the Government through the Department of Livestock Services, DLS and NGO's, mainly Bangladesh Rural Advancement Committee (BRAC). In figure 1 is shown the three lines involved in the Model and the tasks of each line.

Figure 1.
Integrated semi-scavenging production model

PRODUCTION	SUPPLY	SERVICE
Breeders	Parent stock	Village groups
Hatcheries	Feed	Training
Chicken rearers	Vaccine/medicine	Credit/saving
Smallholders	Marketing	Extension

Production line

1. *Breeders (model rearers).* Small low cost parent farms with 25 parent hens and cocks per farm. The hens are kept in confinement and fed with balanced feed. The Parent Stock are of improved breed such as White Leghorn, Rhode Island Red and Fayoumi and the males and females are of different breeds.

2. *Mini hatcheries.* Small low cost hatcheries operated with close to 100% solar energy. Black pillows filled with rice husk are heated in the sun

and the eggs are placed in a cylinder between 2 pillows for hatching. Each hatchery has a capacity to hatch 1,000 chicken per month.

3. *Chicken rearers.* Small rearing farms, each with a capacity of 200-300 chicken. The chicken are reared in a low cost house from day old to 8 weeks of age. The chicken are fed with balanced feed supplied by the local feed seller.

4. *Smallholders (key rearers).* Small farms with only 10 hens, mainly improved breeds supplied by the chicken rearers and a few Desi (local) hens. The hens are kept under semi-scavenging conditions and fed 30-70% supplemented feed and scavenge for the remaining part.

Supply line

1. *Parent stock.* The parent stock are supplied by the Directorate of Livestock Services to market price for day old chicken. The breeds are mainly Fayoumi, White Leghorn and Rode Island Red.

2. *Feed.* The feed is supplied by a number of small feed sellers located in the villages. The sellers purchase local by-products from the milling industry and mix it with fishmeal, vitamins and minerals. A feed seller sells about 1 ton of feed per month.

3. *Vaccine/medicine.* A number of poultry workers are trained to vaccinate the birds. The vaccine is supplied free of charge by the Government but the poultry workers charge a vaccination fee.

4. *Marketing.* The eggs are collected by egg collectors and marketed in the nearby towns or the poultry holders sell the eggs and chickens themselves in the village.

Service line

1. *Group formation.* The involved NGOs form small village groups with some 30 members. The groups hold weekly meetings to discuss relevant subjects and new poultry holders are selected from the groups.

2. *Training.* Before a poultry holder is established she has been through a 4 days training programme followed by refreshment courses.

3. *Credit.* Depending on the activities each member is provided with a small loan ranging from USD 25 to USD 200. The repayment period is 1 year.

4. *Extension.* Extension services are provided as a cooperation between the Government and the involved NGOs.

The organization of the Model is well developed and well functioning. There is, however, a big gap on the technical side. The scientific resource

base for semi-scavenging poultry holdings is rather weak and a professional network for this discipline is not established yet (Dolberg, 1996).

Smallholders

Smallholder structure
Smallholders constitute 95% of the units in the integrated model shown in
figure 1. The sustainability of the model relay fully on the viability of the
smallholders because all other links are established to serve the smallholders.

Figure 2.
Smallholder flow

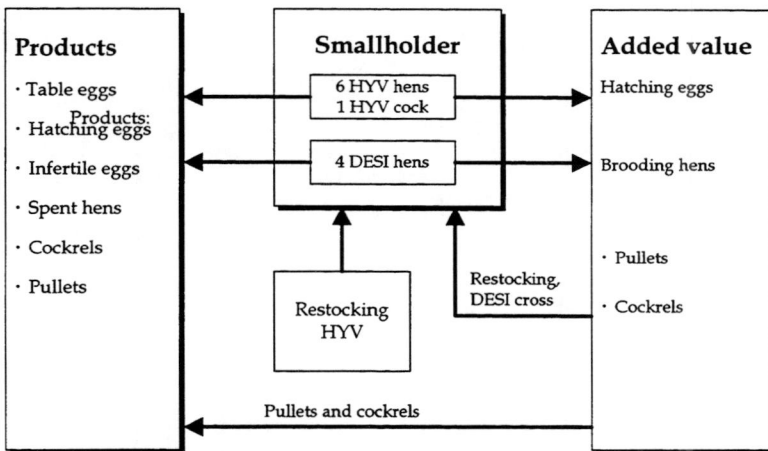

A smallholder unit is a rather complex operation and, even small in size,
it comprises several activities as shown in figure 2. This complexity of activities makes it possible for the individual smallholder to adapt her operation to the prevailing market conditions and demand.

The use of Desi (local) hens as a value adding element for the eggs produced by the exotic hens is an essential activity in the smallholder set-up.

Transfer of technology

The structure of a smallholder unit does in many ways mirror an entire
poultry sector with parent stock, hatchery, rearers and broiler and egg producers. The infrastructure is further supported by the supply and service
functions which are an integral part of the model.

The concept behind the Bangladesh Smallholder Livestock Project implies that 10% of the population in the project area are directly involved as
smallholders or in one of the supply and service activities. There is in this

way an environment established for others to establish themselves within the poultry business. The smallholders' increasing standards of living serve as examples of the viability of poultry holdings and thereby establish the awareness of using poultry as an income generating activity.

Financial evaluation

The investment to establish a smallholder unit amounts to some US$ 30 allocated as 50% to pullets and Desi (local) hens, 25% to housing facilities and 25% as working capital. The cash flow is positive from the first year of operation and the average annual profit is between Taka 3.000 and 6.000 or US$ 75 to 150 (Alam, 1996).

In table 1 is shown the distribution of income and operation cost for a smallholder belonging to the Smallholder Livestock Development Project. The figures are from a survey planned by Hanne Nielsen (Nielsen, 1996) and conducted by Jahangir Alam (Alam, 1996).

Table 1.   *Distribution of income and expenditure per week*

Income	Amount (Taka)	Percentages
Sale of egg	94.7	76.42
Sale of chicken	10.29	8.30
Home consumption of eggs	7.07	5.70
Home consumption of chicken	1.12	0.90
Other poultry income	10.74	8.66
Total income	123.95	100.00
Expenditure		
Feed	19.68	61.26
Transport	0.31	0.95
Medicine/vaccine	1.27	3.97
Labour	9.11	28.38
Others	0.09	0.28
Total expenditure	32.12	100.00

Source: Alam, 1996.

Another survey conducted in 1994 and mainly in Tangail District showed that income from sale of chicken is higher than indicated above (Jensen, 1995).

Sensitivity analyses are presented in tables 2 and 3 of a smallholder with both exotic and Desi (local) hens. The parameters used in the analyses originate partly from surveys and partly from information provided by Bangladesh Rural Advancement Committee and cover a normal year of production.

Table 2. *Sensitivity analysis for 6 HYV hens*

Parameter	standard	variation	Gross profit (Taka/year) standard	+variation
Egg yield, hd, %	50	+5	1066	+226
Age at lay, month	7	-1	1066	+202
Feed sup. rearing, kg	6	-1	1066	+56
Feed sup. lay. g/h/day	70	-10	1066	+184
Mortality rate, %	25	-5	1066	+68

Table 3. *Sensitivity for 4 Desi hens*

Parameter	standard	variation	Gross profit (Taka/year) standard	+variation
Clutches per year	3	+1	2170	+703
Hatchability, %	67	+10	2170	+267
Mortality, chickens, %	50	-10	2170	+299
Mortality, adult, %	25	-5	2170	+30
Feed sup. g/hen/day	0	+10	2170	-158
Feed sup. kg/chicken	0.5	+0.25	2170	-116

Even though the main income seems to come from the Desi (local) hens it is stressed that income is based on hatching eggs from the exotic hens.

Location effect

The cornerstone in the semi-scavenging system is that scavenged feed constitute a substantial part of the total feed consumed. As such, there are two prerequisites to the system:

1. Scavenged feed shall be available in sufficient amounts.
2. Birds must have safe access to scavenge the feed available.

Sufficient feed available for scavenging depends on an area's carrying capacity influenced by factors such as cropping patterns and on the density of birds. Regarding the problem with density it can be solved in a manner seen in an location in Orissa, India, where each household maximum is allowed to keep 4 hens. Another method is to increase the amount of feed available for scavenging.

Predators are one of the main constraints met by scavenging poultry holdings. Especially young chicken below 8 weeks are vulnerable to predators which can account for more than 80% of the mortality.

Even though, the birds are kept in confinement as in the semi-scavenging model, predators are still a threat.

Performance data from three locations in Bangladesh are shown in table 4. A detailed description of the scavenging conditions at the three locations is under preparation. However, Manikganj is the area where the model has been developed and Bangladesh Rural Advancement Committee as well as the villages have experiences in this type of poultry holding, which may be the reason for the good results. Rajshahi is a typical sugar cane area with many predators and with scarcity of feed to scavenge.

Table 4. *Location effect on performance, 6 to 11 months of age*

Traits/location	Jessore	Manikganj	Rajshahi
Number of egg/hen	48	740	33
Mortality, diseases, %	19	11	16
Mortality, predators, %	1	0	8
Sup. feed g/bird/day	59	41	52

Conclusions

The Smallholder Model developed in Bangladesh is the most structured and the most carefully designed smallholder poultry programme in any developing countries. Chicken mortality has been brought down to an acceptable level and the resource consumption, mainly feed, seems even to be competitive with the intensive poultry production.

The model has a potential, apart from improving the living conditions of the smallholders, to open new grounds in smallholder and scavenger poultry production.

The institutional set-up, structure and implementation procedure are well developed and documented. The technical part of the model needs still more documentation of performance and structure of the operations. However, there is no doubt that the model is viable, but also, that there is a great potential for technical improvements.

Acknowledgements

The author is grateful to DANIDA and IFAD for funding the Smallholder Livestock Development Project in Bangladesh and to the staff of the Directorate of Livestock Services, Government of Bangladesh; Bangladesh Rural Advancement Committee and Bangladesh Livestock Research Institute for fruitful discussions and cooperation. Also thanks to the Technical Assistance Team: Frands Dolberg, Hanne Nielsen and Tamas Fehervari, for providing information and documentation materials.

References

Alam, I (1996). *Socio-economic Impact of the Smallholder Livestock Development Programme*. MFL (GOB) and Danida, Dhaka

Dolberg, F. (1996). Research supported by a Development Project. *These proceedings.*

Jensen, H.A (1995). *Semi-Scavenging Poultry Holding. Review of Performance at Village Level. Status Progress Report*. Bangladesh Smallholder Livestock Development Project.

Nielsen, H (1996). Socio-Economic Impact of Smallholder Livestock Development Project Bangladesh. *These proceedings.*

Saleque, Md.A. and Mustafa, S (1996). Landless Women and Poultry: The BRAC Model in Bangladesh. *These proceedings.*

Socio-Economic Impact of Smallholder Livestock Development Project, Bangladesh

Hanne Nielsen,
DARUDEC,
Stamholmen 112, DK-2650 Hvidovre, Denmark,
E-mail: darudec@inet.uni-c.dk

Summary

The Bangladesh Smallholder Livestock Development Project implemented by the Directorate of Livestock Services, Ministry of Fisheries and Livestock, Government of Bangladesh and the NGOs Bangladesh Rural Advancement Committee, Proshika and Swanirvar Bangladesh and sponsored by IFAD and Danida aims at reaching some 270 000 landless women belonging to the very poor with poultry income generating activies. The initial project period is three years from 1993 till 1996. The paper reports on an impact survey conducted by interviewing a sample of 1000 women, who had been involved in the Project for at least one year. The results show that 28 % of the households had moved above the poverty line and that the project had made a substantial contribution to human development as nearly all the women reported improvement in their economic conditions. Women are the primary beneficiaries and increased earnings are used to buy more food, send children to school and increase physical assets including purchase of land. The women's participation in household decision making is increased. Very few projects have recorded the same level of impact. It is concluded the experiences of using poultry as a tool in poverty eradication employing this particular model have so far been confined to Bangladesh, but the results clearly suggest replications should be tried in other countries.

Key words: Poultry, semi-scavenging, impact, landless, women, human development

Introduction and background

The Bangladesh Smallholder Livestock Development Project is a cooperation between the Directorate of Livestock Services, Ministry of Fisheries and Livestock and the NGOs Bangladesh Rural Advancement Committee (BRAC), Proshika and Swanirvar Bangladesh.

Project objectives are increased per capita income and increased animal protein consumption among rural poor in Bangladesh as well as increased poultry productivity for the poultry rearers involved.

It is estimated that about 89 per cent of rural households rear poultry. Ownership of backyard poultry is almost entirely in the hands of women.

The target group is defined as follows: operates less than 0.5 acres of land, has an annual income below Tk 3,740 per year sufficient to meet only 80 per cent of per capita caloric requirement and are dependent on selling their labour as the main source of income.

Implementation, supported by the International Fund for Agricultural Development (IFAD) and Danida, started mid-1993 by the Directorate of Livestock Services in cooperation with the NGO Bangladesh Rural Advancement Committee (BRAC). The NGOs Proshika and Swanirvar Bangladesh began implementation one and a half years later and it was consequently decided to include locations covered by BRAC in a socio-economic survey.

The project is based on a structured concept developed by the Director-ate of Livestock Services and BRAC (Saleque and Mustafa, 1996). The approach followed is given below:

1. A technical package comprising interrelated activities involving bene-ficiaries as poultry workers, chick rearers, key rearers, model rearers, feed sellers and mini hatcherers.

2. Support is given by the NGOs, and includes: selection of beneficiaries, establishment of village organisations, conducting of awareness educa-tion, issue based meetings as well as organisation of savings schemes, technical training for poultry rearing and credit programmes.

During the first phase from 1993 to mid-1996, the Bangladesh Small-holder Livestock Development Project involves the following numbers and types of beneficiaries:

Table 1.

Type	No.
Poultry workers	16,000
Chick rearers	2,400
Key rearers	240,000
Model rearers	1,600
Mini hatcheries	40
Feed sellers	16,000

The specific function by each type is described by Saleque and Mustafa (1996).

The objectives (Alam, 1996) of the socio-economic impact survey (Alam, 1996) has been to assess the repercussions on the following aspects:

• Income

• Nutrition/intake of animal protein

• Poultry productivity

• Gender relations and status of beneficiaries in the households and in society.

Survey methodology

Survey planning has taken place cooperatively between the involved parties, namely, the Directorate of Livestock Services, BRAC, Bangladesh Livestock Research Institute and members of the technical assistance team. The survey has been carried out under the supervision of Dr. Jahangir Alam (Alam, 1996) of the Bangladesh Livestock Research Institute.

To measure impact over the longest possible period, survey locations, where project implementation started first, were chosen in the following districts: Natore, Kustia, Chuadanga and Rajshahi. The interviewed beneficiaries have been affiliated to Bangladesh Smallholder Livestock Development Project for at most one and a half years.

Selection of interviewees was made on a random basis. A total of 1000 beneficiaries were interviewed. The number of interviewees in each beneficiary group were proportional to the size of the group. Prior to the interview phase, a structured questionnaire was pre-tested, modifications and adjustments made and the questionnaire tested again before final adjustments. Interviews were carried out during October 1995 by 12 Investigators: 8 men and 4 women. The original aim was to gender balance the investigator team so that it was composed of 50% women and 50% men, but it was not possible to identify enough women in a position to undertake field interviews. Before the survey, the investigators were trained on the purpose of the survey, interview techniques, and administration of the questionnaire. During the interview phase, the investigators were supervised in the field.

Information on the socio-economic status of beneficiaries before their membership of the Project was obtained from baseline surveys available at the local BRAC offices.

Six tabulators were selected among the investigators for tabulation and compilation of data, which took place during December 1995. A draft report was finalised by February 1996.

Summary of survey findings

1. *Project beneficiaries were all women within the target group at the time of being affiliated to the Project. 24% of the beneficiaries are heads of households.*

    All surveyed beneficiaries were at the time of affiliation to the project identified as hard core poor landless households only able to meet 80% of the per capita calorie requirement or an annual income of less than Taka 3,740,[1] operating less than 0,5 acres of land and dependent on sale of manual labour as their main source of income. This part of the population represents 22% of the population in Bangladesh.

2. *All types of beneficiaries have increased their income during the time of project affiliation. The Project related income constitutes on average 23% of*

---

[1] Tk 100 = US$ 2.44 (Oct. 1995).

total household income. This Project related income alone has ensured that 28% of the households now are above the poverty line.

99.9% of the surveyed persons report improved economic conditions.

Average monthly net income from poultry is distributed among the different types of beneficiaries as follows:

Table 2.   *Average monthly net income for each type of beneficiaries (TK)*

Mini hatchery owners	1,047
Chick rearers	761
Feed sellers	757
Model rearers	500
Key rearers	294
Poultry workers	265

Source: Alam, 1996

The largest percentage of households below the poverty line is found among key rearers, poultry workers and model rearers.

Referring to the finding mentioned above that 28% of the interviewed beneficiaries now are above the poverty line because of income from the Project activities alone, and if the poverty alleviation tendency is valid for project beneficiaries as a whole and continues, a total of 72,800 beneficiary households will by mid-1996 be above the poverty line. In the same vein, Todd (not dated) has found that after a decade of Grameen Bank membership, the women borrowers contribute 54% of the total household income.

None of the interviewed beneficiaries have reported problems in repaying their loans. According to BRAC records, no loans are reported overdue or have been written off.

In addition, average compulsory group savings per beneficiary is TaKa 413 and average personal savings is TaKa 768. Although very marginal, all beneficiaries, except Mini hatchery owners, have increased their cultivable land. Further, beneficiaries' assets comprising houses, radios, bicycles, sewing machines and rickshaws have increased. Only the number of weaving looms have decreased.

3.   *The households' intake of animal protein has increased. The average increase in egg consumption per household has increased from 2 to 5 eggs per week, the average number of chicken consumed per household has increased from 2 to 5 per year, meals with fish has on average increased from 10 to 12 per month, meals with meat have increased from 1 to 2 times per month, and milk consumed increased from 0.8 to 2.5 liters per month.*

The findings on egg consumption correspond with data from BRAC's monitoring of egg consumption[2] in Project locations outside those included in this impact survey. The intake of vegetables has remained on

---

[2] BRAC Monitoring Reports March/April to September/October 1994: Monthly Egg Consumption.

average 12 times per week and the consumption of grain (rice) increased from an average of 12 to 14 kg per week. All beneficiary types report shortage of food at times.

4. The Project has enhanced the productivity of small stock by increasing the average number of chicken reared by 270%, of which 47% is of improved type. The average number of goats has increased by 30%. The mortality rate of chicken is remarkably low with an average of 2% per month. This finding corresponds to findings in the breeding experiment (Jensen, personnel communication).

    The number of sheep has remained marginal and the number of ducks has decreased during the present project period.

    In addition to small stock, the number of cattle reared by beneficiary households has increased from 239 to 571 within the 1000 households surveyed. Assuming that the surveyed beneficiaries represent a typical income and investment pattern for all beneficiaries, the number of cattle reared by beneficiary households participating in the project will have increased from approximately 62,000 to approximately 150,000, when the targeted number of beneficiaries has been reached.

    In Todd's study on investment patterns of poor rural women in Bangladesh, for most loaners the progression has been from paddy husking and poultry, over sharecropping and cows, to leasehold - while some proportion of the loan continued to be put into building up the homestead assets in poultry and livestock as well as stocking paddy, both for consumption and resale. Further, Todd's study has found that the first loan cycles gave the women basic food security. The latter findings correspond to findings of the present impact survey, where food consumption increased considerably - although basic food security still has not been reached.

5. *Beneficiaries have gained more influence on decision-making regarding schooling of children.*

6. *The number and percentage of school age children in beneficiary households going to school has increased from 86% to 99%.*

    The increase in schooling has been in favor of girls. This indicates that with higher income and influence of women in the household, more priority is given to schooling of children in general and girls in particular.

7. *Beneficiaries have gained more influence in deciding the use of income. With the Project income husband and wife joint decision making has increased from 54% to 60% of all households.*

    Expenditure on food has on average increased by almost 32 % per year, whereas the percentage spend on food has decreased from an average of 79% to an average of 67% of total household expenditure.

The amount of money spend on all other measured items (clothing, animals, schooling, medicare, marriage gifts/dowry, housing and savings) has also increased.

8. *No change can be registered in the beneficiaries status in the village societies.*

None of the interviewed beneficiaries participate in village organisations (other than the Project's village organisations), social clubs, etc..

9. *No indication of the strength of the Project's village organisations, as a forum for non-economic activities can be established.*

The Project's village organisations are mainly regarded by the beneficiaries as forums for loan transactions, savings and poultry business. Few beneficiaries refer to supporting activities of the organisations such as issue based meetings and awareness education.

10. *All interviewed beneficiaries are interested in continued membership of BRAC.*

Particularly model rearers, key rearers and mini hatchery owners state that the reason for continued membership is that poultry rearing is profitable. As a suggestion to changes in future projects of a similar nature in other parts of Bangladesh, the majority of key and model rearers suggest larger loans.

11. 36% of the beneficiaries want larger loans and 15% state that the biggest problem was that Fowl Cholera vaccine was not supplied in time. Only a few mention problems with poultry diseases and supply of eggs and chicks or with feed.

Conclusion

Human development is defined as a process of enlarging people's choices (Martinussen, 1996). Applying these criteria, it can be seen the Project clearly makes a contribution in that direction as practically all the women reported improvement in their economic conditions. Women are the primary beneficiaries and increased earnings are used to buy more food, send children to school and increase physical assets including purchase of land. The women's participation in household decision making is increased.

Very few projects have recorded the same level of impact. The experiences of using poultry as tools in poverty eradication employing this particular model have so far been confined to Bangladesh. However, the results clearly suggest replications should be tried in other countries.

References

Alam, I (1996). Socio-economic Impact of the Smallholder Livestock Development Programme. MFL (GOB) and Danida, Dhaka.

Martinussen, J (1996). Introduction to the Concept of Human Development. These proceedings.

Saleque, Md.A. and Mustafa, S (1996). Landless Women and Poultry: The BRAC Model in Bangladesh. These proceedings.

Todd, H. (not dated). Women at the Centre, Grameen Bank Women Ten Years On. A Cashpor Pre-Publication, Malaysia. (Undated).

Research Supported by a Development Project

Frands Dolberg
Department of Political Science,
University of Aarhus, 8000 Aarhus C, Denmark,
E-mail: frands@po.ia.dk

Summary

Small livestock including poultry can play an important role in poverty eradication and several rural development projects contain components for promotion of these animals. In terms of formal knowledge generated by research these projects do, however, rest on an extremely weak foundation as demonstrated in this paper. It is argued that till rural smallholder livestock production, in casu poultry, gets the required attention from research institutions, development projects will have to undertake such work and an example is mentioned from Bangladesh.

Key words: smallholder, poverty eradication, animal production, poultry, research.

Introduction

It is common knowledge that much development work is inadequately supported by research. Seen as an aspect of the wider rich-poor countries' relationships this is not surprising. Reddy (1981) is one among many, who has described the general difference in research between rich and poor countries and pointed out that as a common rule whatever research is carried out in developing countries tends to bypass the rural poor.

Context and justification

Gupta et al. (1989) analyzed the problem within agriculture in India. They studied the abstracts of research and extension theses completed in 32 agricultural colleges and Universities between 1974-84 and found that of 1128 theses 73.5% dealt with irrigated, 22.0% with rainfed and only 4.5% with drought-prone agriculture. Out of 900 theses dealing with extension one way or another hardly any dealt with methods of science. Of 329 theses dealing with sociology and extension only 1.2% dealt with livestock. It can be seen that for the period analyzed most of the research was conducted on problems dealing with intensive agriculture. Methods of science may have to be different under small farmer conditions (Pretty, 1995), but the question was barely addressed. Very few theses dealt with livestock and it is very likely that if they had been further qualified by analyzing for - say - women's roles in poultry production, the percentage might be even lower than 1.2%. If we confine our search to rural poultry production for poor women, we will find extremely few reports through the established channels of scientific communication.

Thus a literature search conducted by the author through the library of the Royal Danish Veterinary and Agricultural University in 1995 using the international data bases produced no records from India, but several from Bangladesh. The search was defined to cover Asia and with keywords such as: village, scavenging, India, Bangladesh, Indonesia, Sri Lanka and Thailand.

Many more examples can be quoted to demonstrate the bias in research against the rural poor (Speedy, 1993 and Ørskov, 1996). However, the point is that for a project designed to develop smallholder livestock production not the least rural poultry production, the established research systems have got little to offer. Thus rural poultry was never included in the research agenda of the International Livestock Centre for Africa's (now International Livestock Research Institute) or any of the other CGIAR Centres' research agenda.

It is on this background that a budget line for research was included in the Bangladesh Smallholder Livestock Development Project. So far 5 research projects have been supported on: (1)salmonella and (2)gumboro vaccine production, (3)evaluation of HYV crossbred hens under semi-scavenging conditions (Jensen, 1996) (4)study of the comparative performance of scavenging, but feed supplemented pure, local and crossbred ducks and (5)effect of anthelminthic drug administration and supplementary feeding on the performance of Black Bengal goats. More than 500 poor women are involved as hosts in these trials, which are strongly facilitated by a NGO, Livestock Research Institute and Directorate of Livestock Services cooperation in the field.

Conclusion

The positive experience of supporting research in a development project is identical with a growing body of results accumulating across disciplines (Dolberg, 1991 and Pretty, 1995). It is a means to overcome the knowledge gap and enhance the learning process (Pretty, 1995) within projects.

References

Dolberg, F. (1991). Adding A Learning to A Blueprint Approach - or What a Small Amount of Flexible Money Can Do - Livestock Res.for Rural Dev.Vol.3 No.1 pp.1-10.

Gupta A K, Patel N T and Sanh R N (1989). Review of post-graduate research in agriculture (1973-1984): are we building appropriate skills for tomorrow? Centre for Management in Agriculture, IIM, Ahmedabad, India.

Jensen, H.A. (1996). Semi-scavenging poultry flock. These proceedings.

Ørskov, E.R. (1996). Are we pursuing the right science for rural development?. In F.Dolberg and P.H.Petersen (eds.) Agricultural Science for Biodiversity and

Sustainability in Developing Countries. Proceedings of a Workshop, April 3-7, 1995, Tune Landboskole, Denmark.

Pretty, J.N. (1995). Participatory Learning for Sustainable Agriculture. World Development, pp. 1247-1263.

Reddy, A.K.N. (1981). Technology and Society in Rural Development. Lecture notes. India, Bangalore, Indian Institute of Sciences, March 20 - April 3.

Speedy, A. (1993). Electronic Communication for Development.In F.Dolberg (ed.) Democratizing Development. Dept.of Pol. Sci., University of Aarhus, Denmark. pp.55-75.

Studies on Village Poultry Production in the Central Highlands of Ethiopia

Tadelle Dessie* and Brian Ogle**

*Debre Zeit Agriculture Research Centre,
P.O. Box 32, Alemaya University of Agriculture,
Debre Zeit, Ethiopia,
E-mail: dzarc@padis.gn.apc.org

**Department of Animal Nutrition and Management,
Box 7024, Swedish University of Agricultural Sciences,
S-75007 Uppsala, Sweden,
E-mail: brian.ogle@huv.slu.se

Abstract

This paper is based on a thesis work focusing on village poultry production in the central highlands of Ethiopia, which consisted of a survey, an investigation into the diets of scavenging hens and an on-farm feeding trial. The survey described village poultry production in three villages at three different altitudes, including high (Derek Wonz, 2, 850 m above sea level), medium (Gende Gorba, 1, 850 m above sea level), and low (Awash, 1, 500 m above sea level), and in three seasons, namely the dry (October to February), short rainy (March to May) and main rainy (June to September) seasons. The approaches used included Participatory Rural Appraisal (PRA), checklists and year round case studies. The village poultry production systems, constraints and potentials were described. Village poultry have a unique position in the rural household economy, as suppliers of high quality protein to the family food system, in addition to their social and cultural roles in the daily life of the society. The birds are owned by mainly by women and maintained under scavenging systems with no or little additional inputs for feeding (small grain supplements), housing (only 11.5% the households have small night enclosures) and health care. The objectives of keeping poultry were reported to be for sale (26.6%), sacrifice, i.e. healing ceremonies (25%), replacement (20.3%) and home consumption (19.5%) and for producing eggs for hatching (51.8%), sale (22.6%) and home consumption (20.2%). Even though the hatchability of set eggs (80.9 ±11.1%) is good, chick mortality (61%) is very high. Disease is the major problem of village poultry production, and some times it eradicates whole flocks. It was found to be important to involve women actively in poultry projects. This trial was followed by a second study, the objective of which was to determine the dietary status of local scavenging hens. A total of two hundred and seventy local hens, ninety from each of the three villages at different altitudes (high, medium and low) were purchased from the villagers and slaughtered in each of the three seasons (short rainy, rainy and dry) for physical and chemical analysis of the crop contents and carcass measurements. The physical analysis results revealed that the overall mean proportions of the materials present in the crop, as estimated by visual analysis, were: seeds 30.9%, plant material 23.3%, worms 6.7%, insects 11.1% and 23.9% unidentified materials. The overall mean liveweight and dressing percentage of the local birds were 1129.8 ± 59.9g and 65.6 ± 1.5%, respectively. There was a significant difference (P<0.05) between seasons and altitudes in liveweight and dressing percentage. There was a significant (P<0.05) difference in the chemical composition of the crop contents of birds slaughtered at different seasons and altitudes and the overall means were: DM 50.7±12.5%;% CP 8.8 ± 2.3,% CF 10.2 ± 1.6,% ash 7.8 ± 2.7,% Ca 0.9 ± 0.4,% P 0.6 ± 0.3 and ME 11.9 ± 0.9 KJ/g. The DM, CP, Ca, and ME concentrations were below the recommended requirements for egg producing strains, and protein to energy and calcium to phosphorus ratios were too low. The mean CF of the crop contents was over 10%, which would have reduced the digestibility of the other components of the diet. The third study was an on-farm trial that was conducted with 300 local laying hens over a six month period in Angu village in the central highlands, to assess egg production performance of scavenging hens with and without supplementary feeding. The dietary treatments were supplements of maize only (MS), noug cake only (NCS), maize and noug cake (MNCS) and scavenging only (SO). The% hen-day egg production was 28%, 22%, 32% and 14% for the MS, NCS, MNCS, and SO treatments, respectively (P<0.01), and there were also significant differences (P<0.05) in egg weight between treatments. Groups given the MNC, M and NC supplements were found to utilize the additional feed efficiently, requiring only 6.58, 5.19 and 7.33 kg feed/ kg eggs respectively (assuming a mean egg weight of 40 g). The best economic performance was found for the MNCS treatment due to the combined effect of both energy and protein supplementation, followed by NCS and MS. However, the NCS treatment, followed by MNCS and MS, gave the lowest cost of feed/dozen eggs.

Key words: Village poultry, altitudes, crop analysis, carcass measurement, supplementation, maize, noug cake, central highlands of Ethiopia.

Introduction

Rural poultry1 production in Ethiopia represents a significant part of the national economy in general and the rural economy in particular, and contributes 98.5 and 99.2% of the national egg and poultry meat production, respectively (AACMC, 1984), with an annual output of 72,300 metric tonnes of meat and 78,000 metric tonnes of eggs (ILCA, 1993).

Comparatively little research and development work has been carried out on village poultry, despite the fact that they are more numerous than commercial chickens2, accounting for around 99% of the total number in the country. Studies carried out at the College of Agriculture, Alemaya (Bigbee, 1965) and Wolita Agricultural Development Unit (WADU) (Kidane, 1980) and by the Ministry of Agriculture (1980) indicated that av-

---

[1] Chickens (Gallus domesticus) are the only species of domestic poultry in Ethiopia.
[2] Mature chickens will be referred to as either poultry or birds to avoid confusion with immature chicks.

erage annual egg production of the native chicken was 30 - 60 eggs under village conditions and that this could be improved to 80-100 eggs on station. A recent study at Asela Livestock Farm revealed that the average production of local birds around Arsi was 34 eggs/hen/year, with an average egg weight of 38 g (Brännäng et al, 1990). These results look unimpressive when compared with the egg type exotic breeds which can produce more than 250 eggs/hen/year, with an average egg weight of 60 g. These studies show then that local birds are poor producers of small sized eggs, but smallholder poultry production using unimproved stock can be the most appropriate system, with low input levels that makes the best use of locally available resources. Village poultry are important providers of eggs and meat as well as being valued in the religious and cultural life of society in general and by the rural people in particular. As pointed out by Sonaiya (1990), in recent years rural poultry have come to assume a much greater role as suppliers of animal protein for both rural and urban dwellers. This is because of the recurrent droughts, disease outbreaks (rinderpest and trypanosomiasis) and decreased grazing land, which have resulted in significantly reduced supplies of meat from cattle, sheep and goats. Poultry is the only affordable species to be slaughtered at home by resource poor farmers, as the prices of other species are too high, and have increased substantially in recent years. Consumption of pork is not allowed for religious reasons for most Ethiopians (Orthodox Christians and Muslims), but fortunately there are no such cultural or religious taboos in relation to the consumption of poultry and poultry products. The per capita egg and chicken meat consumption was found to be about 57 eggs and about 2.85 kg of chicken meat per annum in Ethiopia by Alemu (1987) which are very low figures by international standards. Although there is no current data on the present per capita consumption of poultry products, a similar or even declining trend is probable because the population of Ethiopia has increased by about 3% per annum over the last ten years without any marked increase in the production of poultry meat and eggs. Therefore innovative ideas and programmes are required to promote rural poultry production for the improvement of rural household incomes and nutrition, as poultry production is an effective means of transferring wealth from the high income urban consumers to the poor rural and peri-urban members of the community. Small scale poultry development should therefore concentrate on the rural and peri-urban areas of the country, and the focus of this study was on villages in the central highlands of Ethiopia.

Brief description of the central highlands of Ethiopia

Ethiopia has the largest highland (defined as areas above 1500 m elevation) area of the African continent, amounting to around 40% of the country. The highlands are home to about 90% of the total population (Brännäng, 1990) and have about 70% of the livestock population of Ethiopia (ILCA, 1983; Brännäng, 1990). Ethiopian highland topography is rugged and complex and agricultural conditions vary widely in the different ecological zones of the country (ILCA, 1983). The major crops grown are tef (Eragrostis tef), barley, horse beans, maize, sorghum, wheat, chick peas, lentils, false banana and coffee. In the central highlands livestock are a major component of the farming systems and contribute significantly to agricultural production. Their most important function is as draught animals. Traditionally paired oxen are used for cultivation and are also used for threshing. Milk, meat and hides from cattle are relatively less important by products, but manure is collected, dried and used as the major household fuel source. Sheep and goats are kept mainly as a secondary investment and a source of cash in times of need. Donkeys are widely used to transport goods from and to the market and harvests from the field, and in some areas to fetch water if the water source is distant. In addition to donkeys, mules and horses are used as pack animals and horses are also used for riding. Productivity is low for all livestock species (ILCA, 1983; Brännäng, 1990). This low productivity is attributed to various factors, in particular the progressive extension of the area under cultivation and a consequent reduction of grazing areas caused by the rising requirements for food grain for the ever increasing human population. The seasonal availability of feed resources, which are in excess and of good quality during the rainy season and in short supply during the extended dry season, and reduction of the fallow areas are also important constraints.

The agricultural extension services are limited, and the rugged terrain is a major constraint to economic development because of the inherent communication problems. Between 50 to 70% of the population live more than a day's round trip walk from an all weather road (ILCA, 1983). Even though livestock productivity is low it contributes about 35% of the national agricultural output, and additionally draught animals supply the power to cultivate the crop land and are important in rural transport (mainly donkeys). Poultry are widely kept in central highlands of Ethiopia and make up the largest population of vertebrate animals found in the area. They have a high degree of adaptation to a wide range of environments, are important parts of the farming systems, and do not contribute to land degradation and global warming. These and the other facts put poultry in the forefront for research and development work for resource poor farmers in the highlands of Ethiopia. According to Teketel (1986), the distribution of chickens in Ethiopia varies with altitude, and high concentration of village chicken is found in Gojam, Gonder, Shewa, Sidamo, Tigray and Wollo administrative regions (EMA, 1981), most of them being found in the highlands region.

Present structure of poultry production in Ethiopia

The total poultry population in Ethiopia is estimated to be 56.5 million (ILCA, 1993). Poultry production systems in Ethiopia show a clear distinction between traditional, low input systems on the one hand and modern production systems using relatively advanced technology on the other hand (Alemu, 1995). Ninety nine percent of the population consists of local breed types under individual farm household management (Alamargot, 1987), and the remaining 1% of birds are mainly in state-run modern production systems, with a very small proportion in private units. Of the total national egg and poultry meat production 98.5 and 99.2%, respectively, are contributed by local birds (AACMC, 1984), resulting in an annual output of 72,300 metric tonnes of poultry meat and 78,000 metric tonnes of eggs.

Large-scale commercial systems

Modern poultry production started in Ethiopia about 30 years ago, mainly in colleges and on research stations. The activities of these institutions mainly focused on the introduction of exotic breeds to the country and the distribution of these breeds to farmers, including management, feeding, housing and health care packages. The history of poultry production in the industrialized countries may offer some basic knowledge and guidelines for poultry development in the developing countries as a whole and in Ethiopia in particular, but in view of the particular conditions in different countries and regions, specific research and approaches are needed to determine which are the optimum production systems and development strategies. Most of the research work is still being carried out on intensive poultry production, with modern housing, and sophisticated feeding systems. However, the great majority of poultry production is based on extensive rural production systems where the results of current research are often not applicable. Today a number of large commercial state farms have been established and private poultry farms are starting to operate in the country. This would seem to be a positive trend in increasing the supply of animal protein for the Ethiopian people, whose primary source of protein is of plant origin, because poultry are efficient converters of by-products and grains into eggs and meat, and have a fast turnover rate and rapid growth rate. In spite of these advantages, including intensive poultry production in the livestock development strategy must be questioned, due to the fact that commercial poultry compete with human beings for scarce food grains. This statement is justified if we consider the composition of diets used on the industrial poultry farms, where the major ingredients are high quality cereals like maize and wheat (AACMC, 1984).

If we consider commercial poultry production under Ethiopian conditions, which implies national shortages of grain to feed an ever increasing human population and a negative trade balance, then allocating hard currency to import breeding stock, medicines, vitamin-mineral premixes and concentrates to support intensive poultry farms will involve critical political as well as economic decisions. So, in a country like Ethiopia, the outcome will be the converting of food that resource-poor people can usually afford to buy, to smaller amounts of luxury food items that only the minority wealthy members of the society can afford. No attempts have been recorded to evaluate the performance of exotic birds under local farmer conditions. The only serious on-station attempt carried out in Ethiopia was a comparative study of the performance of six different exotic breeds, namely: Brown Leghorn, White Leghorn, Rhode Island Red, New Hampshire, Light Sussex, and Barred Plymouth Rock at Debre Zeit Agricultural Research Centre. This study showed that the White Leghorn was the best performing exotic layer breed (DZARC, 1984).

Rural poultry production systems

There is no generally accepted definition of rural poultry production, and various production systems have been described by a number of authors, including Huchzermeyer (1967), Aini (1990), Cumming (1992), Alemu (1995) and (Dessie and Ogle, 1996a). The production systems are characterized as including small flocks, no or minimal inputs, with low outputs and periodic destruction of the flocks by disease. Birds are owned by individual households and are maintained under a scavenging system, with little or no inputs for housing, feeding or health care. Typically the flocks are small in number with each household flock containing birds from each age group (Dessie and Ogle, 1996a) with an average of 7-10 mature birds in each household, consisting of 2 to 4 adult hens, a male bird and a number of growers of various ages. Gunaratne et al. (1992) and Cumming (1992) also described village poultry flocks in Asia as including 10-20 birds of different ages per household. According to AACMC (1984) in Ethiopia there is an average of six indigenous birds per household, and according to Sonaiya (1990), the average flock size in Africa ranges from 5-10 birds. As described by Dessie and Ogle (1996a) the village poultry production system is characterised by minimum inputs, with birds scavenging in the backyard, and no investments beyond the cost of the foundation stock, a few handfuls of grain and possibly simple night enclosures.

Past research and development attempts

Comparatively little research and development work has been carried out on village chickens despite the fact that they are usually more numerous (table 1) than commercial chickens in most developing countries (Cumming, 1992) and they have been marginalized by planners and decision makers (Panda 1987), which is certainly true for Ethiopia.

Table 1. *Percentage contribution of local birds in selected African and Asian countries to the poultry population*

Country	% Contribution	Reference
Sri Lanka	28	Fonseka (1987)
Zimbabwe	30	Kulube (1990)
Cameroon	65	Agbede et al (1990)
Cote d'Ivoire	75	Diambra (1990)
Kenya	80	Mbugua (1990)
Gambia	90	Andrews (1990)
Malawi	90	Upindi (1990)
Nigeria	91	Adene (1990)
Ethiopia	99	Alamargot (1987)
Bangladesh	99	UNDP/FAO (1983)

Few attempts have been made to increase protein supply by improving the egg and meat production potential of local birds, and upgrading and crossbreeding with exotic germ plasm has been the main focus of the research and development organizations. For the last three decades scientists and government have promoted schemes in which cockerels from selected strains are reared up to 15 to 20 weeks of age, mainly on government poultry stations, and then exchanged for local cockerels owned by rural subsistence farmers. In our study area (Dessie and Ogle, 1996a) there had been an introduction of exotic breeds to the three villages at various times and in different forms, such as cockerels, pullets, and fertile eggs, but their impact in upgrading the village chickens has been minimal. The farmers were given advice on improved feeding and housing and were asked to remove all remaining local cockerels. In addition improved hens were introduced to boost egg production in co-operative based intensive poultry farms in rural Ethiopia, but most of these projects collapsed mainly due to inadequate feed supply, management, medicines and discontinuation of the schemes. However, these approaches led to only limited improvement, due to the high mortality rate of the modern breeds because of their lack of adaptation to the rural environment, poor management, ultimate discontinuation of the schemes and above all the farmers' lack of interest and awareness, because the programmes were usually planned without farmer participation and without parallel improvement in management and feeding. Many crossbreeding projects failed because the crosses were not accepted by local people, who feared they would be vulnerable to harsh village conditions. Above all those development strategies did not pay attention to local social and cultural aspects of poultry production. For example farmers prefer to have double-combed cocks for sacrifice purposes, in addition to their colour preferences (Dessie and Ogle, 1996a). Local scavenging chickens, in addition to providing cash income, have nutritional, cultural and social functions which require consideration from planners, professionals and farmers, which is rarely given. However, planning and execution of research and development work on local birds could result in considerable improvement in egg production performance of local hens (Dessie and Ogle, 1996c). As was emphasized by Roberts (1992) after analysing the scavenging feed resource base of local birds in three different production systems (two from Sri Lanka and one from Indonesia) strategic supplementation of birds according to age and production status can be a suitable solution. Cumming (1992) describes the feed resource base as variable, depending on the season and rainfall, which is in agreement with the results of Dessie and Ogle (1996a and 1996b). Generally, non-genetic factors such as poor nutrition, Newcastle disease and other management practices have a much greater effect on production parameters than the genetic influence under scavenging systems (Sazzed et al. 1988). Improved feeding systems for scavenging birds were suggested by Huque and Hossain (1991) as a way of achieving optimum production (Dessie and Ogle, 1996c).

Socio-economic aspects of rural poultry production

Rural poultry represent a significant part of the rural economy. This segment of production in Africa as a whole represents an estimated asset value of US $ 5.75 billion (Sonaiya, 1990). In addition to their contribution of high quality animal protein and as a source of easily disposable income for farm households, rural poultry integrate very well and in a sustainable way into other farming activities, because they require little in the way of labour and initial investment as compared with other farm activities (Dessie and Ogle, 1996a). A number of authors, including Veluw (1987), Sonaiya (1990), and Gunaratne et al (1992), have outlined that rural poultry play a significant role through their contribution to the cultural and social life of rural people. This was confirmed by Dessie and Ogle (1996a). They found that the main objectives of poultry production in the three villages were eggs for hatching (51.8%), sale (22.6%) and home consumption (20.2%), and production of birds was for sale (26.6%), sacrifice (healing ceremonies) (25%), replacement (20.3%) and home consumption (19.5%). In some cases farmers give live birds (8.6%) and eggs (5.4%) as a gift to visitors and relatives, as starting capital for youths and newly married women. They also invite special guests to partake of the popular dish "doro wat", which contains both chicken meat and eggs and is considered to be one of the most exclusive national dishes. Birds are also given as sacrificial offerings in traditional worship, and finally they perform a valuable sanitary function in the villages through eating discarded food and cockroaches for example. The feed resource base for the scavenging chicken production system described has no alternative use, and if they were not present, other scavengers, particularly dogs and crows would perform this function, with no associated benefit to the farming community.

For a better understanding of the role played by poultry in the lives of rural people, it is necessary to know exactly the purposes for which households keep poultry. The five major uses and benefits of poultry and eggs in rural societies in the central highlands of Ethiopia are reported briefly by Dessie and Ogle (1996a). It is very difficult to determine the most important purpose, because it is imposable to compare the spiritual benefit of sacrifice with the financial benefit of a sale. A ranking of purposes based on the number of birds used has very little to do with the order of importance, and understanding this is a considerable challenge for development workers, which is also confirmed by Veluw (1987) in Northern Ghana.

The existence of poultry in the household does not imply necessarily that the farmers are willing and in a position to expand poultry production. Experience has shown that intensive persuasion is needed to convince them to introduce regular watering and feeding, to clean the birds' night shelter and to take care of the young chicks, before starting any research or development programme to attain the genetic potential of the local birds. The first critical step of rural poultry development is therefore the encouragement of farmers to change their attitudes towards poultry keeping and to the traditional system.

Poultry keeping in most of the developing countries is the responsibility of women. As reported in Dessie and Ogle, 1996a, in the three study villages it was found that it is the women that look after the birds and the earnings from the sale of eggs and chickens are often their only source of cash income. It is therefore important to actively involve women in the process of poultry improvement, which actually has been neglected in the past. Most of the poultry extension workers, vaccinators and key poultry farmers are men. In some parts of Ethiopia contacts between women and male extension workers are restricted by cultural and religious factors and information has to be passed indirectly through their husbands. In Dessie and Ogle (1996a) 15 women were selected to take part in the study, all of them involved in poultry keeping and willing to take part in the trial. Based on our results it was clear that the poultry were owned and managed mainly by women, and therefore it was decided to work only with women, although the husbands also participated in the process. It was also decided to form self-sustaining women's group, whose main concern would be the economic viability of the project once implemented. It is important to plan poultry development projects in such a way that women participate actively as poultry advisers, extension workers, and vaccinators as well as poultry farmers.

Input-output relationships

As mentioned above traditional poultry production based on low input-output levels represents a part of a balanced farming system. Despite the fact that more than 70% (table 1) of the poultry population in Africa (Sonaiya, 1990) and 99% of the poultry population in Ethiopia (Alamargot, 1987), consists of local birds, their contribution to farm household and national income is not in proportion to the high numbers. Productivity is observed to increase in direct proportion to the level of confinement (Sonaiya, 1990) and other feeding and management factors, up to a certain level of production corresponding to the upper limits of the genetic potential of the local birds. As described by Smith (1990), this system of production, although appearing primitive can be economically efficient because although the output from the individual birds is low the inputs are even lower or virtually non-existent. The low output is expressed as low egg production, small sized eggs, slow growth and low survivability of chicks (Smith, 1990; Dessie and Ogle, 1996a) but small management changes, for example regular watering, night enclosures, discouraging them from getting broody, vaccination for common diseases and small energy and protein supplements can bring about significant improvements in the productivity of local birds (Dessie and Ogle, 1996c). In the Eastern hills of Nepal, indigenous birds kept under semi-intensive management conditions produced 125 eggs per annum and under this system of management eleven clutches of eggs were produced per year as compared with three to four produced under normal scavenging systems (Smith, 1990). In a rural poultry project in the Niamery region in Niger (Bessei, 1990), economic calculations based on the assumption that eggs and meat were partly consumed by the family and partly sold in the local markets showed that rural poultry keeping could be self - sustaining, since the feed for maintenance of the chickens is free. As reported in Dessie and Ogle, 1996c, with minimal additions of inputs, improving the existing management, night enclosure, regular watering and changing the attitudes of farmers, it is possible to bring about considerable improvements in terms of egg production.

Feed resources and requirements

The feed resource base for rural poultry production is scavenging, and consists of household waste, anything edible found in the immediate environment and small amounts of grain supplements provided by the women. As shown in Dessie and Ogle, 1996a and 1996b, the scavenging feed resource base is not constant. The portion that comes as a grain supplement and from the environment varies with activities such as land preparation and sowing, harvesting, grain availability in the household and season and the life cycles of insects and other invertebrates. From the results of our work (Dessie and Ogle, 1996b), it is also possible to conclude that protein supply may be critical, particularly during the drier months, whereas energy may be critical during the rainy season, this result being in agreement with the conclusions of Cumming (1992), who describes the feed resource as variable, depending on the season and rainfall. In the absence of an event which diminishes the flock biomass (number x mean liveweight), such as disease or occurrence of a major festival, the village flock will normally be at the maximum biomass that can be supported by the scavenging feed resource base. Any additions to the village flock which increases the poultry biomass will result in increased survival pressure and selection against the weakest members of the flock. According to the finding of Dessie and Ogle, 1996b, the feed resource is deficient in protein, energy and probably calcium for layer birds, and this is confirmed from the results of Dessie and Ogle, 1996c, which show that supplementation of local birds with food sources containing energy and protein and a calcium source brings about a considerable increase in egg production.

Feed requirements and supplementation of local laying hens

There is no doubt that feed supply is a major constraint in rural poultry production, and it has been calculated that scavenging birds are usually capable of finding feed for their maintenance needs and about 40 eggs per year, but higher levels of production require supplementary feed (Dessie and Ogle, 1996c). In our study reported in Dessie and Ogle (1996b), the nutritional status of local laying hens from the chemical analysis of crop contents, assuming this accurately reflects the feeds consumed, indicates that the %DM (52.3 ±12.5), CP (9.1± 2.3%), Ca (0.9± 0.4), P (0.7± 0.3) and ME (11.9± 0.9 KJ/g) were below the requirements for egg production, indicating the importance of supplementation. Compound feeds are usually not available in remote areas, or are too expensive, so it is therefore necessary to use locally available materials, such as household waste and cheap conventional and non-conventional feed resources such as brans, and oil-seed cakes. The choice of raw materials for poultry feed is limited and it is not possible to formulate balanced diets in rural Ethiopia. Suboptimal rations of supplementation may be economically justified under rural conditions and accordingly this supplementary feeding should be complementing, but not replacing the feeds scavenged by the birds. However, it must be tested and examined from an economical point of view. Special attention will need to be paid to local sources of minerals and vitamins (Dessie and Ogle, 1996b), although scavenging birds would normally find a significant proportion of their requirements of vitamins, trace minerals, although probably not in the case of Ca for laying hens (Dessie and Ogle, 1996b).

According to our work (Dessie and Ogle, 1996c) it is possible to attain hen-day production of over 30% using a supplement consisting of 30 g/day maize and 30 g/day noug cake, 28% from 30 g maize and over 20% from 30 g per day per bird of noug cake supplementation which is more than double the 13.9% from scavenging only. This result is also supported by the study of Islam et al. (1992) who showed that by giving a supplement that provided 30% of the daily energy and protein requirements of local birds it is possible to produce as many eggs as the unsupplemented Fayoumy breed in the villages, and that egg production from scavenging birds increased by a factor of three when they received a supplement covering 50% of their dietary needs.

Protein requirements of local laying hens

The protein requirements of high producing laying hens varies from 16-18% of the diet and is needed for egg production, maintenance and growth of body tissues, feather growth and also depends on the energy content of the feed. In addition to the above the feed consumption and protein requirements are influenced by a number of factors, the most important being size of the bird, stage of production and ambient temperature. It is possible to estimate the requirements for protein factorially. According to Nesheim et.al (1979) a fresh egg contains 66% water, 12% protein, 10% fat, 1% carbohydrates and 11% ash. The average weight of a local hens egg is 38 g (Sazzad, 1986; Bränäng et al., 1990 and Dessie and Ogle, 1996a). Thus a 38 g egg contains 4.56 g protein, and at an efficiency of protein utilization of 55% (Scott et al, 1982) hens must consume 8.29 g protein per egg. Harris (1966) indicated that the endogenous nitrogen excretion is estimated to be 2.55 g per day for a bird weighing 1.14 kg. According to Scott et al (1982), protein required for feather growth is 0.49g/bird/day. The sum total of calculated protein requirements for all these functions is 11.194 and 11.317g/day for birds producing a 35 g egg in phase one and a 38 g egg in phase two of lay. In this study (Dessie and Ogle, 1996b,) the mean crude protein (CP) percentage in the crop contents is 9.1 ± 2.3% which is below the above calculated requirement of the local laying hens. Protein deficiency was even more serious in the short rainy and dry seasons, when the

CP content of the crop contents was 7.6% and 8.7%, respectively. This is confirmed by the results of the supplementation trial reported by Dessie and Ogle (1996c), where provision of additional protein in the form of noug cake increased egg production by a factor of two as compared with scavenging birds not receiving a supplement.

Energy requirement of local laying hens

In moderate environmental temperatures high producing White Leghorn hens requires 300-320 kcal of metabolizable energy per hen per day. Local birds are low producers of small sized eggs and their liveweight is lower than that of the White Leghorn. According to Scott et al (1982), the net energy requirement of adult hens is NEm= 83 x BW (kg) 0.75. Thus for a local hen weighing 1.13 kg (overall mean), the NEm is 90.97 kcal/hen/day, since this figure is approximately 82% of MEm value, then 90.97/0.82 = 111 kcal/hen /day, and adding 50% of this value for activity, the total requirement for a non laying hen without travelling energy will be 166.5 kcal/hen/day. However, in addition to that local birds need more energy for travelling, and Bessie (1989) reported that a scavenging layer travelled about 4 km per day at an average environmental temperature of 200C which implies a requirement of approximately 107 Kcal per day, giving a total requirement of 273.5 kcal/day. The mean true metabolizable energy of 286 ± 23 Kcal from calculated values is sufficient to meet calculated requirements for a non laying hen only.

Production and productivity of village birds

The production level of scavenging hens is generally low, with only 40-60 small sized eggs produced per bird per year under smallholder management conditions. According to the results of Dessie and Ogle, 1996a the total output of scavenging birds is low not only due to low egg production, but also due to high chick mortality, as half of the eggs are hatched to replace birds that have died, and the brooding time of the mother bird is long in order to compensate for its unsuccessful brooding. Smith (1990) estimates that under scavenging conditions the reproductive cycle consists of a 10 day laying phase, a 21 day incubation phase and finally a 56 day brooding period. This implies a theoretical maximum number of 4.2 clutches per hen each year, although in reality the number is probably 2-3. Overall the system is quite productive in relation to the very low input levels and this is underlined by McArdle (1972), who states that the net output from poultry rearing is higher in scavenging systems as compared to commercial systems, and the scavenging flock is not in competition with humans for feed. This is true if we consider the input-output relation only.

Chick mortality represents a major loss in scavenging village chicken production systems (table 2), and reports from different countries show that 50% to 70% of chicks die between hatching and the end of brooding. Kingston (1980) and Kingston and Cresswell (1982) in Indonesia, and Roberts (1992) in Sri Lanka and Matthewman (1977) in Nigeria, calculated mortality rates of chicks as being 69%, 65% and 53% respectively, up to 6 weeks of age. Alamargot (1987) also reports on chick mortality in Ethiopia, and during some severe epidemics, rates as high as 80% have been recorded. In the study reported in Dessie and Ogle, 1996a, the overall chick mortality was 61 + 17% (n=160) in the first two months after hatching, and is higher when there was a disease outbreak in the area. Various authors attribute these losses to different causes, for example Roberts (1992) reported that in Indonesia losses were due to a combination of poor nutrition, predators and various disease factors, and although predators were blamed for the majority of the losses, other biological and environmental factors made significant contributions.

Table 2. *Reported chick mortality in rural production systems in different African and Asian countries in the first 6 to 8 weeks of age*

Country	% Mortality	References
Sri Lanka	65	Gunaratne et al (1992)
	46	Roberts (1994)
Indonesia	79	Kingston and Cresswell (1982)
	56	Hadiyanto et al (1994)
Northern Ghana	80	Veluw (1987)
Ethiopia	61	Dessie and Ogle (1996a)
Cote D'Ivoire	50	Diambra (1990)

The newly hatched chicks have access to the same feed resource base as stronger and more vigorous members of the flock, but are unable to compete. In addition the low protein and energy content of the available feed, the low hatching weight of the chicks, high ambient temperatures and other associated factors are major causes of losses, both directly, and also by increasing vulnerability to predation and susceptibility to disease.

Newcastle disease is the most important disease recognised in tropical countries in village poultry production systems (table 3).

Table 3. *Reported village bird mortality caused by Newcastle disease in selected African countries*

Country	% Mortality	References
Togo	50	Aklobessi (1990)
Sudan	50	El Zubeir (1990)
Nigeria	70	Nwosu (1990)
Comoros	80	Mohammed (1990)
Ethiopia	80	Alamargot (1987)
Morocco	100	Houadfi (1990)

In Dessie and Ogle (1996a) disease was cited as the most important problem by most of the members of the community with whom it was discussed, reducing both the number and productivity of the birds, and the problem intensified after the villagization programme (1984-86) in the country. The time of the disease outbreaks before the villagization programme was usually at the beginning of the rainy season, that is at the end of May and beginning of June, but after villagization it remains a problem throughout the year, even though it is still more serious at the beginning of the rainy season. Sonaiya (1990), after summarising the reports from six African countries, reported that the mortality caused by Newcastle disease ranges from 50-100% per annum and that severity is higher in the dry season although in our work the disease is more widespread in the rainy season in the central highlands of Ethiopia (Dessie and Ogle, 1996a). The farmers do not have any preventive medicine or practice for this fatal disease, and only after the start of an outbreak do they treat their birds with socially accepted medicines (Dessie and Ogle, 1996a). However, the effectiveness of these treatments is not satisfactory.

Although the local chickens are slow growing and poor layers of small sized eggs they are, however, ideal mothers, good sitters and hatch their own eggs, excellent foragers, are hardy and possess some degree of natural immunity against common diseases. These traits are of great importance as the farmers cannot afford to buy expensive concentrates and incubators, which at the moment are considered necessary for raising exotics. Brännäng et al. (1990) reported that 50% and 75% exotic blood birds did not show any signs of broodiness at the Asela Livestock Farm. However, as reported by Panda (1987) in India, the productivity of the Kadaknath, or indigenous fowl, can be improved without sacrificing any of the characteristics required by village fowls. Egyptian scientists, taking a different approach, achieved significant improvements in egg production of over 21% recently by simple crossbreeding between two local strains raised in the traditional way in the near-tropical conditions of upper Egypt. This success illustrates a way of stemming the genetic erosion of local poultry breeds. Although there is a lot of evidence in the literature about genetic improvement resulting from heterosis and crossbreeding techniques with regards to egg production and growth rate, so far little research effort has been directed towards these in Ethiopia. Some information is provided by Brännäng et al. (1990), who reported average yearly egg production of 129 and 114 eggs and 48 g and 53 g mean egg weight, for birds with a 50% and 75% exotic blood levels, respectively. The only other attempt to evaluate the performance of crossbreeds with different exotic blood levels was made at Debre Zeit Agricultural Research Centre, and involved crossing local birds with White Leghorns to determine the egg production performance of the crossbreeds. A preliminary analysis showed that the annual egg production of the 50% and 62.5% crosses was 146 and 193 eggs, respectively (DZARC, 1991). This shows that it is possible to improve egg number and egg weight by crossing, but the results only apply to on-station conditions, and no information is available for crossbred birds kept under local farmer management conditions. In any case it is not possible to substantially improve egg production if the hens incubate and rear their own chicks.

Conclusions

From the results of these studies it can be concluded that the scavenging system is an appropriate system for the rural areas that makes relatively good use of locally available resources. The requirement now is to improve these production systems to make the best possible use of these assets. The system is characterized by no or few inputs, and a low output level, although appearing primitive, these systems can be economically efficient because although the output from the individual birds is low the inputs are even lower or virtually non-existent.

The system is also characterised by huge chick mortality in the first two weeks of life, caused by different factors such as disease, predators, and the hostile environment for newly hatched chicks. The feed resource base for local birds in the villages is from scavenging and is inadequate for the production of more than around 40 eggs/bird/year. However the results of Dessie and Ogle, 1996c show that supplementation of energy and protein in addition to other management changes can increase egg production by more than 100%. Rural poultry production is an important part of the farming systems and needs relatively small additional resources and inputs from farmers to achieve substantial improvements in productivity and profitability by changing to semi-scavenging systems. However, because of very high mortality rates, particularly due to Newcastle disease, farmers are generally reluctant to invest in improvements in feeding, health care and housing for example. The development of a new heat tolerant vaccine that can be administered via the feed opens up the possibility of significantly reducing mortality in village poultry, which should make producers more positive towards genetically improved birds and inputs to improve feeding and housing.

Recommendations

- Village poultry production deserves greater attention from government, research and development organizations, and above all from rural farmers.

- Preferential access to feed by the newly hatched chicks should be given through some kind of creep feeding system.

- Strategic supplementation of both protein and energy, providing small night enclosures, regular water and disturbing the broody bird results in more than 100% increase in egg production of local birds.

- Vaccination against Newcastle disease with the new heat resistant vaccine administered via the feed will substantially reduce mortality.

- It is important to focus on working with women's groups, both to use their knowledge about poultry production, and to improve their incomes.

- On-farm and on-station trials on new vaccines for the prevention of Newcastle disease are needed, particularly the heat resistant vaccine which does not need a cold chain and can be administered through the feed.

- Genetic improvement should be introduced only when the current systems have been improved in terms of dietary supplementation, housing, controlling Newcastle disease and regular water and management and to change the system to semi-scavenging.

Acknowledgements

Appreciation is expressed to the Swedish Agency for Research Cooperation with Developing Countries (SAREC) for its financial and material support of this work. The study of the first author was made possible through his supervisor, Dr. Brian Ogle, and sincere thanks are extended to him for his professional guidance, continuous help, constructive criticism, and patience. I also wish to acknowledge the hospitality and dedication of farmers and development agents in the three villages in the central highlands of Ethiopia. I would like to express my sincere gratitude to Dr. Inger Ledin, Dr. Thomas R. Preston, Mr. Frands Dolberg, Dr. Luu Trong Hieu, Dr. Peter Uden, Mrs. Phoc, Dr. John Öhrvik and Mr. Howard Benson for their continued support and scientific guidance during my study. I am also indebted to the University of Agriculture and Forestry of Ho Chi Minh city (UAF), Vietnam; Department of Animal Nutrition and Management of the Swedish University of Agricultural Sciences (SLU), Debre Zeit Agricultural Research Centre (DZARC) of Alemaya University of Agriculture (AUA), Faculty of Veterinary Medicine (FVM) of Addis Abeba University (AAU) and Ministry of Agriculture Debre Brehan, Debre Zeit and Nazret regional offices for allowing me to use their facilities and their cooperation. I would like also to thank my student colleagues: Thi Mui, Thi Loc, Thi Dung, for An and Men from Vietnam, Goromela from Tanzania, Borin from Cambodia, Tania from Nicaragua, Lylian from Colombia and the last but not the least Dereje from Ethiopia. Special thanks are due to my family, to my friends Ato Eshete, D., Ato Getachew, T., W/O konjit, A., Ato Selamyhun, K., Ato Tekelu, T. and Dr. Yilma, J. for their concern and having always encouraged me to go ahead, and to my girl friend Altaye for her great love, endless patience and constant support.

References

Adene, D. F. (1990). The management and health problems of rural poultry stock in Nigeria. CTA-Seminar proceedings on Smallholder Rural Poultry Production. 9-13 October 1990, Thessloniki, Greece, 2:175-182.

Agbede, G., Demey, F., Verhuls A. T., Bell, J. G. (1990). The impact of Newcastle disease in the traditional chicken farms of Cameroon. CTA-Seminar proceedings on Smallholder Rural Poultry Production. 9-13 October 1990, Thessloniki, Greece, 2: 49-54.

Aini, I. (1990). Indigenous chicken production in South East Asia. World Poultry Sci. Journal, 46:51-56.

Aklobessi, K. K. (1990). Smallholder Rural Poultry production in Togo. CTA-Seminar proceedings on Smallholder Rural Poultry Production. 9-13 October 1990, Thessloniki, Greece, 2: 37-42.

Alamargot, J. (1987). Avian pathology of industrial poultry farms in Ethiopia. Proceedings of the First National Livestock Improvements Conference 11-13 February 1987, Addis Abeba, Ethiopia, pp 114-117.

Alemu, S. (1987). Small scale poultry production. Proceedings of the First National Livestock Improvements Conference 11-13 February 1987, Addis Abeba, Ethiopia, pp 100-101.

Alemu, Y. (1995). Poultry production in Ethiopia. World Poultry Scince Journal, 51: pp 197-201.

Andrews, P. (1990). Rural poultry development in The Gambia. CTA-Seminar proceedings on Smallholder Rural Poultry Production 9-13 October 1990, Thessaloniki, Greece, 2: pp 81-85.

Australian Agricultural Consulting and Management Company (AACMC) (1984). Livestock Subsector Review, Volume 1, Annex 3.

Bessei, W. (1990). Experiences with rural poultry development projects. CTA-Seminar proceedings on Smallholder Rural Poultry Production 9-13 October 1990, Thessaloniki, Greece, 2: pp 53-60.

Bigbee, D. G. (1965). The management of the native chickens of Ethiopia. Miscellaneous publication, No. 5. HSIU, College of Agriculture, Alemaya.

Brännäng, E. and Persson, S. (1990). Ethiopian Animal Husbandry, Uppsala, Sweden, 127 pp.

Cumming, R.B. (1992). Village chicken production: Problems and potential. In: P.B. Spradbrow (Ed.) Proceedings of an International Workshop on Newcastle disease in village chickens, control with Thermostable Oral Vaccines 6-10 October, 1991, Kuala Lumpur, Malaysia, pp 21-24.

Debre Zeit Agricultural Research Center (DZARC) (1991). Annual Research Report 1990/1991. Debre Zeit, Ethiopia.

Dessie, T. and Ogle, B. (1996a). A Survey of Village Poultry Production in the Central Highlands of Ethiopia. Part I of M.Sc. Thesis. Swedish University of Agricultural Sciences, Department of Animal Nutrition and Management.

Dessie, T. and Ogle, B. (1996b). Nutritional Status of Village Poultry in the Central Highlands of Ethiopia as assessed by analyses of Crop contents and Carcass measurements. Part II of M.Sc. Thesis. Swedish University of Agricultural Sciences, Department of Animal Nutrition and Management.

Dessie, T. and Ogle, B. (1996c). Effect of Maize (Zea mays) and noug (Guizotia abyssinica) cake supplementation on egg production performance of local birds under scavenging conditions in the Central Highlands of Ethiopia. Part III of M.Sc. Thesis. Swedish University of Agricultural Sciences, Department of Animal Nutrition and Management.

Diambra, O. H. (1990). State of Smallholder Rural Poultry production in Cote d'Ivoire. CTA-Seminar proceedings on Smallholder Rural Poultry Production 9-13 October 1990, Thessaloniki, Greece, 2: pp 107-116.

El Zubeir, A. (1990). Smallholder Rural Poultry production in the Sudan. CTA-Seminar proceedings on Smallholder Rural Poultry Production 9-13 October 1990, Thessaloniki, Greece, 2: pp 217-226.

Ethiopian Mapping Agency (EMA) (1981). National Atlas of Ethiopia, Addis Abeba, Ethiopia.

Feltwell, R. and Fox, S. (1978). Practical poultry feeding, Faber and Faber limited, Great Britain, pp 36-59.

Gunaratne, S. P., Chandrasiri, A. D., Mangalika Hemalatha W. A. P. and Roberts, J. A. (1992). The Productivity and Nutrition of Village Chickens in Sri Lanka. In: P.B. Spradbrow (Ed.) Proceedings of an International Workshop on Newcastle disease in village chickens, control with Thermostable Oral Vaccines 6-10 October, 1991, Kuala Lumpur, Malaysia.

Harris, L. E. (1966). Biological energy interrelationships and glossary of energy terms. Publ. 1411, National Academy of Sciences-National Research Council, Washington, D.C.

Houadfi, EL. M. (1990). Rapport sur la production avicole et problemes lies aux elevages traditionnels au Marco. CTA-Seminar proceedings on Smallholder Rural Poultry Production 9-13 October 1990, Thessaloniki, Greece, 2: pp 161-171.

Huchzermeyer, F. W. (1967). Some thoughts on poultry keeping in African areas of Rhodesia under subsistence economy conditions. Rhodesian Agriculture Journal, 64: 133-139.

Huque, Q.M.E. and Hossain, M.J. (1991). Production potentials of ducks under scavenging system of management. Bangladesh Journal of Animal Sci., 20(1&3):119-122.

International Livestock Center for Africa (ILCA) (1983). Research on farm and livestock productivity in the central Ethiopian highlands: Initial results, 1977-1980. ILCA research report, No. 4, Addis Abeba, Ethiopia.

International Livestock Center for Africa (ILCA) (1993). Handbook of African Livestock Statistics. ILCA, Addis Abeba, Ethiopia.

Islam, M.K., Hussain, M.A., Haque, M.F. and Paul, D.C. (1992). Performance of Fayomi Poultry Breed with Supplemental Ration Under Farmers' Management. In: R. Abidur (Ed.) Proceedings of the Fourth National Conference of Bangladesh Animal Husbandry Association, December 26 - 27, 1992.

Kassimo, M. (1990). Rapport sur L'Operation coqs raceurs a' anjouan Comores. CTA-Seminar proceedings on Smallholder Rural Poultry Production 9-13 October 1990, Thessloniki, Greece, 2: 55-65.

Kidane, H. (1980). Performance of F1 Crossbred Birds. Wollaita Agricultural Development unit. Animal Husbandry and Breeding. Wollaita Sodo, Ethiopia Bulletin No. 4: 38 pp.

Kingston, D. J. (1980). The productivity of scavenging chicken in some villages of West Java, Indonesia. Proceedings 1980 South Pacific Poultry Science Convention. 13-16 October 1980 Auckland, New Zealand, pp. 228-237.

Kingston, D. J., and Creswell, D. C. (1982). Indigenous chickens in Indonesia: population and production characteristics in five villages in West Java. Bogor, Indonesia, Research Institute for Animal Production, Report No. 2: 3-8.

Kulube, K. (1990). Small holder rural poultry production in Zimbabwe. CTA-Seminar proceedings on Smallholder Rural Poultry Production 9-13 October 1990, Thessaloniki, Greece, 2: 263-270.

Matthewman, R. (1977). A survey of small livestock production at village level in the derived savanna and lowland forest zones of South West Nigeria (Monograph, Univ. of Reading, 1977)

Mbugua, N. P. (1990). Rural smallholder poultry production in Kenya. CTA-Seminar proceedings on Smallholder Rural Poultry Production 9-13 October 1990, Thessaloniki, Greece, 2: 119-131

McArdle, A.A. (1972). Methods of poultry production in developing area. World Animal Review. 2: 28-32.

Ministry of Agriculture (MOA) (1980). Development Strategy of Animal Breeding and Improvement: 10 year Development Strategy. Animal Breeding and Improvement Team, MOA, Addis Abeba, Ethiopia.

Nesheim, M. C., 1986. Austic, R. E., and Card, L. E. (1979). Poultry production. 57 pp.

Nwosu, C. C. (1990). The state of small holder rural poultry production in Nigeria. CTA-Seminar proceedings on Smallholder Rural Poultry Production 9-13 October 1990, Thessloniki, Greece, 2:183-194.

Panda, B. (1987). Role of poultry in socio-economic development of small farmers in India. Asian Livestock, 12: pp 145-148.

Roberts, J.A. (1992). The scavenging feed resource base in assessments of the productivity of scavenging village chickens. In: P.B. Spradbrow (Ed.) Proceedings of an International Workshop on Newcastle disease in village chickens, control with Thermostable Oral Vaccines 6-10 October, 1991, Kuala Lumpur, Malaysia, pp. 29-32.

Savory, C.J. (1989. The importance of invertebrate food to chicks of gallinaceous species. Proceeding of Nutrition Society No. 48: 113-133.

Sazzad, M.H. (1986). Reproductive performance of Desi hens under scavenging and intensive systems of rearing. Proceedings of first Annual Livestock Research Workshop, Bangladesh Livestock Research, Institute, Savar, Bangladesh.

Sazzad, M.H., S.M.H. Mamotazul and Asaduzzamn, A.U. (1988). Growth pattern of Desi and Khaki Campbell ducks under rural condition. Indian Journal of Poultry Sci. 23(2):165-166.

Scott, M. L., Nesheim M. C. and Young, R. J. (1982). Nutrition of the chicken. pp 48-92.

Smith, A .J. (1990a). Poultry. Tropical Agriculturist series. CTA, Macmillan Publishers, London. pp 179-184.

Smith, A .J. (1990b). The integration of rural production into the family food supply system. CTA-Seminar proceedings on Smallholder Rural Poultry Production 9-13 October 1990, Thessloniki, Greece, 1: 15-128.

Sonaiya, E. B. (1990) The context and prospects for development of smallholder rural poultry production in Africa. CTA-Seminar proceedings on Smallholder Rural Poultry Production 9-13 October 1990, Thessaloniki, Greece, 1: 35-52.

Teketel, F. (1986). Studies on the meat production potential of some local strains of chickens in Ethiopia. PhD Thesis, J.L. University of Giessen, pp 210.

Upindi, B. G. (1990). Smallholder Rural Poultry production in Malawi. CTA-Seminar proceedings on Smallholder Rural Poultry Production 9-13 October 1990, Thessaloniki, Greece, 2: 141-146.

UNDP/FAO. (1983). Rural poultry improvement project of the Govt. of the People's Republic of Bangladesh. No. BGD/82/003. Project document, Dhaka, Bangladesh.

Veluw, V. K. (1987). Traditional poultry keeping in Northern Ghana. ILEIA, 3(4): 12-13.

Supplementing Poultry Diet with Tree Leaves or Seeds: On-Farm Research in Nicaragua[1]

Niels Kyvsgaard* and Ramiro Urbina

Proyecto de Desarrollo Rural Integral "Manuel López"
El Sauce, Nicaragua,
(*Present address: Royal Veterinary and Agricultural University,
Bülowsvej 13, 1870 Frederiksberg C, Denmark.
E-mail: Niels.C.Kyvsgaard@vetmi.Kvl.Dk)

Summary

The effect of supplementing a traditional sorghum feed for chicken was studied in farmer-managed on-farm experiments including 12 farms. The supplements were dried leaves of 3 local trees (Cordia dentata, Gliricidia sepium and Guazuma ulmifolia) or the seeds from the fruit of Crescentia alata. The seeds or leaves were grinded together with the sorghum grain on a hand-mill. The participants reported a higher daily egg production, shorter interval between clutches, increased shell thickness and improved colour of the yolk. The new technology was demonstrated to other communities on courses. In 1993, approximately two years after the initial study, 300 families (25% of the farms) were practising one or more of these methods to supplement the sorghum regularly. This number had increased to 398 by the end of 1994. The range of supplements was extended to include e.g. leaves of other tree species, legume seeds, sun-dried fish and whole pulp and seeds of Crescentia alata. Different methods to prepare the feeds were also developed by the farmers.

Key words: Poultry, supplementation, tree leaves, on-farm, Nicaragua.

Introduction

Poultry plays an important role in the production system of Nicaraguan small-holders. The peasants are mainly crop-producers, but many keep cattle or pigs, and almost all households keep poultry, mostly chicken. The poultry is scavenging and additionally fed sorghum grown on the farm. Although grain consuming, poultry also serves as an important buffer for the grain production: The flock size is increased in years of good harvest,

---

[1]. A version of this paper has been published in Livestock Research for Rural Development, Volume 8, number 1, January 1996.

where low grain prices follow, and is limited to a minimum in years of harvest failure.

Grains are however becoming more scarce due to sub-division of farms by inheritance, lower yields in the traditional slash-and-burn agriculture, and increasing drought problems. Although the use of other feed resources such as the leaves of fodder trees and the fruit of Crescentia alata are well-known practices in cattle feeding, there is no local experience of using these resources for poultry. These feed items are of interest as they compared to sorghum have a higher content of protein, some minerals and vitamins (Doggett 1988, Ravidran and Blair 1991, 1992, D'Mello 1992). Green meals are also known to be rich in lysine, the limiting amino acid of sorghum (Ravidran and Blair 1991, 1992).

On-farm research was therefore started to study the use of these under-exploited feed resources in mixtures with the traditional sorghum grain. The methods followed were mainly inspired by the review of Farrington (1988).

Materials and methods

Study area

Our study was conducted as part of the Rural Development Project "Manuel Lopez" in El Sauce, Nicaragua. The project started in 1990 in 5 pilot areas within El Sauce municipality. The project includes approximately 1200 rural families. The aim is to improve life conditions by promoting low input agriculture and reforestation.

El Sauce municipality is located at 13°N. The altitude is between 200 m in the valley and 1000 m in the mountains. Mean monthly temperatures range from 25 to 30°C in the valley. The mean yearly precipitation is 1750 mm (range: 866 to 2734 mm from 1963 to 1987). There are normally 2 crop cycles per year, the early from May to July and the late from August to November. Rainfall is irregular, especially in the early season. The main crops of the area are maize, sorghum (a drought resistant long season variety of Sorghum vulgare), beans (Phaseolus vulgaris) on the slopes and sesame in the valley.

Baseline data on poultry production

Information about production systems, productivity of the hens, possible feed alternatives and the main problems of the poultry production were gathered in two individual and four group interviews before the main study, as well as during the visits to the participants in the on-farm experiments.

Almost all rural families keep poultry, whereas only about 50% keep pigs and/or cattle. Poultry keeping is normally the responsibility of the women. The poultry is mainly chicken of a rather heavy, coloured breed, probably of mixed origin. An egg production of 10 to 15 eggs per clutch and one clutch per month was mentioned as normal. Productivity was reported to vary markedly according to season. Both feed supply and seasonal change of climate were mentioned as reasons.

The chickens are scavenging during the day. They are fed whole grain of sorghum or maize, which is spread on the ground. The daily feed ration is very variable, the highest rate being reported as 100 grams per hen per day. Sorghum was preferred to maize for poultry feed as egg production is higher on sorghum and as maize is important for human food. Maize is, however, occasionally used for poultry as two crops can be grown per year whereas sorghum is only harvested once. Maize is known to be good for fattening of poultry.

Poultry meat and eggs are consumed by the family, sold by the farmer in the town, or sold to middlemen from the regional capital. It is normally easy to sell the products and prices are relatively high. The consumers are paying 15 to 25% more for the free-ranged chicken or their eggs than for the corresponding products from the poultry industry. A hen is sold for approximately 15 cordobas, the equivalent of the salary from a days field work. (The exchange rate in 1993 was: 1 US$ = 6 cordobas.)

When asked about the main limitations to the poultry production, diseases, feeding and predators were mentioned to be important. The high mortality is mainly caused by epidemics, where the symptoms are compatible with Newcastle Disease, and endemic diseases among the young chicken, mostly Fowl Pox. (Parallel to the activities described in the present article, a vaccination program was set up to control Newcastle Disease.) The feed problem consisted both of shortage of grain in years of harvest failure and of seasonal fluctuations in grain supply.

Selection of participants

The study was carried out in four communities (two in the valley and two in the mountains), where the women had requested support for their poultry production. Initially, two lectures were given to each community about general poultry nutrition and poultry diseases. During the discussions on these courses the women expressed interest in improving the nutrition of their poultry. The use of commercial concentrates was ruled out mostly due to the cost of transportation. Each of the four groups elected three to five participants to carry out on-farm experiments with local feed resources. A total of 14 women and two men were elected. The number of participants had declined to 10 women and 2 men by the start of the experiments. The participants had small to medium scale farms with from two to 75 chicken. No economic incentives were given.

On-farm experiments

The basic design of the experiments was: The hens should be fed dried leaves of fodder trees or the seeds of C. alata mixed with sorghum by grinding. The relation between leaves and sorghum was set as one kg. of dried leaves to four kgs. of sorghum. We recommended to administer 100 grams per hen per day. The feeds should be prepared by grinding on the handmill, which is found in almost all households for the milling of boiled maize for "tortillas". It was agreed that the experiments should be carried out with own resources. The productivity before intervention was used for comparison and evaluation of the innovations. Control groups were not included, as it was considered difficult to separate groups of hens on the farms.

The participants were visited individually at least 4 times from June to September 1991 after the following scheme: During the first visit possible feed alternatives on the farm were identified and it was discussed how to dry the leaves and how to grind the ingredients on the handmill. At the second visit, which was made about one week after the first visit, practical experiences with the preparation were discussed. Demonstrations were included when necessary.

The third visit was made 3 to 4 weeks after the first visit and the fourth about one month later. Problems and solutions to the practical administration of the home-made feeds to the chicken were discussed. Palatability of the different feeds was discussed, and data on productivity were collected. Other changes, like changes in shell thickness, yolk colour or the length of the interval between clutches were registered when mentioned by the participants. (We did not put this question directly, as we wished to use it as a control for what could be called a "please the visitor" bias, which might occur when we asked if her hens had begun to lay more eggs after changing diet).

Innovator workshop

By the time of the third visit, the participants were gathered for an "innovator workshop" (Chambers and Jiggins 1987) to allow them to exchange experiences and opinions about the different feed alternatives, elaboration of the feeds, and productivity results. Practical demonstrations by the most advanced farmers were included to facilitate the discussion.

Evaluation of acceptance

As the results were considered positive, the new feed formulations were demonstrated on courses offered to other communities.

The acceptance of the new technology was later evaluated by four methods:

- house to house visits to 15 families in four communities as a follow up to lectures held in these areas (November 1991)
- follow-up interviews with farmers who had participated in the original experiment
- extensive questionnaire survey at the end of 1993 and 1994 as part of project monitoring

Results

On-farm experiments

During the initial meetings and the first visits a number of trees were identified as possible candidates for further research: the leaves of Cordia dentata (local name Tigüilote), Guazuma ulmifolia (Guácimo) and Gliricidia sepium (Madreado or Madero Negro), and the seeds of Crescentia alata (Jícaro). The latter tree is found mainly in the parts of the valley where the soil is of vertisol type. The chemical composition of these feeds is shown in table 1 together with the corresponding values for sorghum. The data for C. alata is for pulp with seeds, of which the seeds are considered to be the protein source.

Table 1. *Chemical composition of available feedstuffs*

	DM	CP	EE	CF	Ash	NFE
	%	----------------	% of DM	----------------		
Sorghum		11.0	3.0	3.0	2.0	81.0
Cordia dentata	20.0	19.6	2.9	12.8	13.7	51.0
Gliricidia sepium	29.0	22.5	7.9	14.1	9.2	46.0
Guazuma ulmifolia	25.0	13.4	9.7	24.8	10.4	41.7
Crescentia alata (pulp + seeds)	31.1	12.2	12.8	16.6	6.3	52.1

DM: Dry Matter, CP: Crude Protein, EE: Ether Extracts, CF: Crude fibre, NFE: Non Fatty Extracts.
Analyses at Universidad Nacional Agraria. Data for sorghum are from Ravidran and Blair (1991).

The participants used different methods for preparation of the feeds of which the most accepted was: The branches were cut and the leaves were dried first for a few hours in the sun whereafter the process was finished in the shade. Later, leaves and sorghum were mixed by coarse grinding on the handmill. During this process the veins were separated and discarded. The relation between sorghum and leaves had been set to 4:1 on a weight basis, but as only a few of the participants had scales they often varied the relation, mostly in direction of a lower content of leaves.

The participants normally prepared the feed twice a week. Longer storing was not possible as the grinded feed would absorb humidity especially during the rainy season. Some farmers moistened the feed immediately before use to reduce losses by wind or scraping.

Data on the productivity on different feed mixtures are given in table 2.

Table 2.   Individual results and observations from the on-farm experiments

Participant	Number of chicken	Supplement type	Productivity after supplementation	Other observations after supplementation
JA San Ramón	70 chicken 8 hens in experiment	C.den	Laying almost daily	Shell becomes thicker, yolk more yellow. The sorghum lasts longer
AG San Ramón	5 hens	C.den G.sep G.ulm	Now 5 hens lay daily, before 4-6 eggs from 9 hens	
MRV SanNic	10 hens	C.ala	Laying almost daily. The chicks grow fast	
AB Tololos	15 hens 60 chicks	C.den G.sep	The 15 hens lay daily	Daily ration: 80 g per hen
DT Tololos	25 hens	C.ala	Laying daily	The grinded mixture attracts humidity
CV Montañita	3 hens	C.den	Laying daily	The sorghum lasts longer
RMVV Montañita	2 hens	C.den G.sep	Now laying daily. Before 1 egg per 2 hens or 2 per 3	Do hardly pause between clutches
MPLV Montañita	12 hens	C.den G.sep	The chicks grow fast Now 6-8 eggs from 12 hens, before 4-5	The hens like the feed
MVV Montañ	1 hen 3 chicks	G.sep	The hen laid daily during 2 months	
IV Cig	8 hens 8 chicks	G.sep	5 hens lay 5 eggs, before 2 to 3	The shell becomes thicker
BM Cerr	2 hens 6 chicks	G.sep	Laying daily	They don't like the G.sep
SG Montaña Exp. 2:	4 hens 4 chicks 11 hens	C.ala C.ala	Laying daily 7-11 eggs daily, before 4-5	The interval between clutches becomes shorter

C.ala: Seeds of Crescentia alata, C.den: Leaves of Cordia dentata, G.sep: Leaves of Gliricidia sepium, G.ulm: Leaves of Guazuma ulmifolia.

There was a good consistency between the participants that the inclusion of the different leaves of fodder trees was beneficial. They reported of a higher daily production and a shorter pause between clutches. Apart from the improvement in production, some participants observed increased thickness of the shell and a more yellow colour of the yolk. No negative side effects with any of the tree species were reported, although rejection of mixture with G. sepium was seen in a few cases.

The highest productivity was reported with the seeds of C. alata. On a farm where both these seeds and the leaves of G. sepium were tried, the seeds were superior to the leaves. However, C. alata seeds have an alternative use in a popular soft-drink. Consequently, they have a rather high market price, which limits their incorporation in poultry feed.

The participants expressed that the work by preparing the feeds was reasonable. It was observed by one participant that 10 lbs. (4.5 kg) could be mixed and grinded in 30 minutes. The work load was mostly a concern of the women with large stocks.

A frequently expressed problem was the handmill. It is manufactured for milling boiled maize for "tortillas" and is worn by milling dry grains. This problem was overcome by some by using an old mill which was not in use for food. Even an old mill was able to grind the sorghum as it only was necessary to grind the grain coarsely.

Evaluation of acceptance

The evaluation in November 1991 of the acceptance of the techniques after they had been demonstrated on courses showed that 7 of 15 families interviewed had tried the new feed formulations. As the interviews were carried out just before the sorghum harvest, all 7 had stopped making the concentrates at the time of the interview.

Interviews with key farmers showed that some of the limitations for the acceptance of the new methods were: The wear of the mill, the work load, feeding troughs are necessary if the feed is grinded, and it was considered to be easier to distribute a limited quantity of feed between the flock if it was in the form of whole grain.

The survey carried out by the end of 1993 showed that 300 persons (of approximately 1200 families) were using the new feed formulations on a regular basis. The combinations they were using are shown in table 3. The number of feed items included in the feeds had expanded to include leaves of other tree and bush species, legume seeds, sun-dried fish and dried whole pulp of C. alata.

The prepared feeds were not used following a fixed receipt, but varied according to availability of the ingredients. The main reason given when the women had stopped producing the feeds was shortage of sorghum.

The project monitoring for 1994 showed that 398 persons in 40 communities were using the methods.

New methods for preparation of the feeds were developed by some of the farmers between 1991 and 1994. One method consisted of boiling the ingredients in a small volume of water for 30 minutes, whereafter the mixture was mashed to form a semi-solid mass, which could be fed on the

Table 3. *The number of farms which were using different supplements regularly during*
        1993

Sorghum with	leaves of Cajanus cajan 30
	leaves of Gliricidia sepium........................50
	leaves of Guazuma ulmifolia 60
	leaves of leucaena......................................20
	whole pulp of Crescentia alata 1
	seeds of Vigna unguiculata 2
	sun-dried fish..4
	seeds of Crescentia alata............................5
	seeds of Cajanus cajan 1
	leaves of Cordia dentata 117
Maize with	leaves...10
Total	300

soil or on a simple board. The ingredients were initially: C. dentata leaves,
sorghum and cowpea (Vigna unguiculata), but variations were made later
on to include other bean species, which had been introduced for soil improvement (Mucuna deeringiana, Canavalia ensiformis). The main limita-
tion of this method was mentioned to be the consumption of firewood.

Discussion

The main finding of this study was that a substantial improvement in the
productivity of the hens was possible, when the traditional sorghum diet
was replaced by a mixture made by grinding sorghum with leaves of
fodder trees or seeds of C. alata. Increased shell thickness and improved
yolk colour was also reported in some cases. These changes are believed to
be caused by:

- the high protein content of the supplements or their amino acid profile
  (table 1 and Ravidran and Blair 1992)

- the grinding of the sorghum which leads to a higher digestibility (Gohl
  1981)

- the increased shell thickness is caused by the high calcium content of
  the leaves and the improved yolk colour is caused by beta-carotene and
  xanthophyll in green plants (Ravidran and Blair 1992).

The influence of the two factors, i.e. grinding and protein supplementation from the leaves or seeds, was not separated in the present study. It has
been argued that the dominant deficiency in the diet of scavenging chicken
is energy. Huchzermeyer (1973 cited by Smith 1990) showed that maize
supplementation of scavenging hens alone could improve the productivity.
Ravidran and Blair (1991) stated that energy rather than protein was deficient in poultry feed formulations in most Asian countries. In our environment, where grains already are used at a rather high level, we expect

that protein supplementation will be beneficial. The total protein content of the leaves is generally high, but their profile of essential amino-acids is also an important factor: Sorghum is deficient of lysine, whereas green meals are good sources of this amino acid (Ravidran and Blair 1992). Green meals are, however, poor sources of sulphur containing amino acids, so methionine and cystine probably become the limiting amino acids after supplementation.

Our study did not attempt to find a substitute for the grains as carbo-hydrate source. Sorghum is grown on almost all farms, and its exploitation can be improved by grinding and supplementation. The use of a well-known feed as a basis for the experiments might also have contributed to the high rate of acceptance by the farmers (table 3). In the evaluations of acceptance, lack of grain was often mentioned as the reason why the women had stopped using these homemade feed. Therefore, the introduc-tion of a cheaper carbohydrate source such as e.g. sugar cane juice could be the next subject for investigation.

The main biological limitations to the inclusion of dried leaves in poul-try diets are their high fibre content and the occurrence of toxic or anti-nutritional factors especially in leaves of legumes (D'Mello 1987, 1992, Ravidran and Blair 1992). We did not observe any adverse effects in our study, where the participants used leaves of both legumes (G. sepium) and non-legumes (C. dentata and G. ulmifolia). Sundrying of the leaves might have helped in reducing the content of antinutritional factors (D'Mello 1992). However, rejection of the feeds with G. sepium was observed in a few cases. The evaluation of long-term acceptance showed that the most popular species were C. dentata and G. ulmifolia, both of which are non-legumes.

The main technical constraint was identified as the hand-mill, which is not designed for dry grain. Therefore the work load to grind the feeds and the wear of the mill will limit the acceptance by some farmers.

The genetic potential of the rural hens might limit the response to im-proved feeding. The breeds of our area seem to have sufficient potential to respond to better feeding. Their origin is probably mixed from previous introduction of specialized breeds. Indigenous birds, on the other hand, might not be able to respond to improved feeding. Sazzad (1992) found that the egg production of indigenous birds was almost as high under scavenging conditions as in an intensive system, whereas the exotic breeds performed much better in the intensive system.

The results of the present study have mainly been found following a participatory research methodology (see the review of Farrington 1988). Using the classification employed by Franzel and Houton (1992), our pro-cedure can be classified as jointly designed by farmers and researcher, and as farmer managed. Egg production is well suited for this type of research as the output is easy to quantify and as an improved diet will give an im-mediate response in productivity.

The conditions on the different farms were quite variable and it was not possible to introduce control groups. Furthermore there might exist a "please the visitor"-bias during the interviews, which could had led the women to exaggerate the benefit of a technology promoted by the visitor. It is therefore important to include long-term acceptance as an indicator of the benefits of a new technology (table 3).

An important advantage of the participatory research approach in this study was that it initiated a process of farmers' innovations, where a large number of farmers began to carry out experiments with different feed alternatives, including other tree species, legume seeds, sun-dried fish and dried C. alata pulp as well as with different ways of preparation. The initiation of this process might in fact prove to be the most important achievement of the work. These observations are in agreement with Sumberg and Okali (1987) who stated that the benefit of a technology should by evaluated by its acceptance and also by the variation that the farmers could give the technology.

In the present study, on-farm research was able to reveal a number of options to the farmers, whereas on-station research is needed to get more precise data on the different ingredients. The inclusion of the legumes, which presently are being widely introduced for soil improvement (Mucuna, Canavalia and Cajanus) should also receive attention.

References

Chambers R, Jiggins J 1987 Agricultural Research for Resource-Poor Farmers. Part II: A Parsimonious Paradigm. Agricultural Administration and Extension 27:109-128

D'Mello J P F 1987 Underexploited tropical feedingstuffs for poultry. World Review of Animal Production 23:37-43

D'Mello J P F 1992 Nutritional potentialities of fodder trees and fodder shrubs as protein sources in monogastric nutrition. In: Legume trees and other fodder trees as protein sources for livestock. FAO Animal Production and Health Paper 102:115-127

Doggett H 1988 Sorghum. Second Edition. Longman Scientific and Technical, pp 450-453

Farrington J 1988 Farmer participatory research: Editorial introduction. Experimental Agriculture 24:269-279

Franzel S, Houten H V 1992 Research with Farmers. Lessons from Ethiopia (Appendix 2). C.A.B. International

Gohl B 1981 Tropical Feeds. FAO Rome. Computer Version 1993

Huchzermeyer F W 1973 Free ranging hybrid chickens under tribal conditions. Rhodesian Agricultural Journal 70:73-75

Ravidran V, Blair R 1991 Feed resources for poultry production in Asia and the Pacific region. I. Energy sources. World's Poultry Science Journal 47:213-231

Ravidran V, Blair R 1992 Feed resources for poultry production in Asia and the Pacific. II. Plant protein sources. World's Poultry Science Journal 48:205-231

Sazzad H M 1992 Comparative study on egg production and feed efficiency of different breeds of poultry under intensive and rural conditions in Bangladesh. Livestock Research for Rural Development 4 (3):65-69

Smith A J 1990 The Tropical Agriculturist: Poultry. MacMillan Publishers, London

Sumberg J, Okali C 1987 Farmers, on-farm research and the development of new technology. Experimental Agriculture 24:333-342

Studies on Duck Production in the Mekong Delta, Vietnam

Bui Xuan Men, * *Brian Ogle and* ** *T.R. Preston* ***

Faculty of Agriculture, Cantho University, Vietnam

**Department of Animal Nutrition and Management,*
Swedish University of Agricultural Sciences,
P. O. Box 7024, S-75007, Uppsala, Sweden
E-mail: Brian.Ogle@huv.slu.se,

***Finca Ecologica, University of Agriculture and Forestry,*
Thuduc, Ho Chi Minh City, Vietnam
E-mail: thomas%preston%sarec%ifs.plants@ox.ac.uk

Summary

Five experiments were carried out both on station and on small farms in the Mekong Delta, Vietnam to evaluate the use of duckweed as a replacement for soyabeans in broken rice based diets for growing ducks, and to determine the effects of feed supplementation of scavenging ducks.

The two first experiments were carried out with crossbred common ducks on station, and with Muscovy ducks on a small farm. The ducks were fed broken rice diets, in which soyabeans were replaced partially or completely by fresh duckweed in the growing period. There were no significant differences in daily weight gains between the crossbred ducks fed the diet with 100% duckweed replacing soyabeans and the conventional diet. The gains were significantly higher (P<0.001) for the ducks fed 30 and 45% soyabeans replaced by duckweed. Feed conversion ratios (FCR) were significantly inferior in all diets with duckweed. There were no significant differences in the gains between female Muscovy ducks fed the diet with 56% and 100% duckweed replacing soyabeans and the conventional diet. However, the daily gains were significantly lower with 47% and complete replacement for the male Muscovy ducks. The FCRs tended to be inferior for the duckweed diets. There were no significant differences in carcass traits between treatments for the crossbred, and the female and male Muscovy ducks.

Two experiments were conducted with improved common ducks and female Muscovy ducks at the Cantho University research station. The ducks were fed levels of 80 or 60 g/duck/day of broken rice and duckweed ad-libitum from 28 to 63 days of age for the common ducks and 28-70 days for the females. Daily gains were significantly lower with the restricted broken rice diets, except for the 80 g/day broken rice diet for common ducks. The FCRs were significantly improved in the restricted broken rice diets. Complete replacement by duckweed in the diet with 60 g/day broken rice had the lowest feed cost.

An experiment was carried out using scavenging ducks on farms in the rural area. The ducks were given supplements consisting of a mixture of broken rice and dried fish meal, or broken rice and dried fish meal alone. These diets were fed to crossbred scavenging ducks from 28 to 70 days of age at night in the farmers' households. The liveweight gains were significantly higher for the fish meal and broken rice mixture than for the broken rice diets. However, the cost per kg gain for these diets were higher compared with no supplement.

Key words: Crossbred ducks, Muscovy ducks, duckweed, soyabean, daily gain, intake, local resources.

Introduction

There are around 30 million ducks raised annually in Vietnam, of which some 65% are estimated to be in the Mekong Delta. In the countryside the ducks are raised by scavenging in the rice fields especially in the harvesting season and immediately after. In other systems the ducks are raised in the backyard or in the gardens of households or kept on the canals. Ducks are easy to raise and develop in the area because they can resist diseases and consume many kinds of different feeds to produce valuable products in a short time. Duck production has contributed a considerable amount to the income of households. However, nowadays many varieties of high yielding rice are planted and harvested in a short period with only a limited time available for the duck flocks to scavenge, so this system is becoming less feasible. Also, in the dry season the farmers cannot herd their ducks in the fields as there is no feed. Another constraint is that the law of the country now prohibits the farmers from keeping their ducks on the canals and rivers so as to eliminate environmental pollution. As a result, the population of ducks has decreased and this leads to fluctuations in the supply of meat and eggs. In order to meet the increasing demand of the consumers there is increasing interest in confinement of ducks making use of locally available feeds.

The population of Muscovy ducks is about 20% of the total duck population raised in the Delta. The ducks are commonly allowed to roam around in the backyards and gardens or confined in simple enclosures around farmer households, and today raising has become especially popular around the cities. Muscovy ducks are easy to raise because they are more resistant to diseases and efficiently consume different feeds to produce valuable products that can be sold at high prices. In the integrated animals-pond-garden farming system Muscovy ducks are suitable to raised due to the fact that they are less active, not noisy and easy to breed and manage.

Duckweed, *(Lemna minor)* which is common throughout the Delta, is a tiny water plant that grows very well on stagnant pond surfaces. It can tolerate high nutrient stress, and appears to be more resistant to pests and diseases than other aquatic plants in the area. It has a high content of nutrients, especially protein (26-40% of dry matter) and carotene, that are necessary for growing animals. Duckweed has been commonly used in Vietnam to feed fish and poultry. So, as part of the overall development strategy of the integrated farming system, duckweed can be a useful candidate to be developed as a feed resource for ducks so as to improve production all the year round.

The main objectives of these experiments therefore were to determine the optimum levels of duckweed as replacement for soyabeans in diets for fattening crossbred meat-type and Muscovy ducks based on broken rice fed *ad-libitum,* and evaluate effects on carcass quality. In addition, the economic benefits obtained from the use of duckweed were calculated. Feed supplementation for scavenging ducks was investigated in order to bring about improvements in the traditional duck-rice integrated farming system.

Trials with duckweed in diets based on broken rice offered *ad-libitum*

Methods and materials

Experiment 1 was carried out at the Experimental Duck Farm, Cantho University in the Mekong Delta between March and May 1995, in the late dry and early rainy season. A total of two hundred four-week old crossbred ducklings [female Pekin (imported a long time ago) X male Cherry Valley (Czechoslovakia)] with initial average liveweights of between 830 and 860 g were used in the trial. The one day old ducklings were selected from scavenging breeding flocks and incubated at the traditional duck hatchery of Omon district, Cantho province, then brooded and fed a conventional diet ad-libitum from 1 to 28 days of age at the experimental farm. The birds were identified and then individually weighed initially, weekly and at slaughter. They were allocated at random to the five treatments with four replicates and ten birds balanced for sex per replicate.

Experiment 2 was carried out on a small farm in the suburbs of Cantho town where Muscovy ducks are commonly raised. The experiment was conducted between May and July 1995, in the early rainy season. One hundred and twenty 4-week old local Muscovy ducklings (sixty females and sixty males) were used, with six treatments (three for each sex) and four replicates of five ducks per replicate. The one day old Muscovy ducklings were selected from clusters of small breeding flocks scavenging in the backyards and gardens of small holdings. They were brooded in a shed built in a garden and fed conventional diets ad-libitum from 1 to 28 days of age. The experimental birds were identified and then individually weighed initially, weekly and at periodic slaughters of 70 days for the females and 84 days for the males. All ducks were given the duck plague vaccine and given a dosage of antibiotics on the seventh and twenty-fifth day of age to prevent diseases.

The ducks in experiment 1 were fed in groups with 10 ducks per group from 28 to 63 days old on broken rice based diets offered *ad-libitum* as the major energy source. Based on the protein intake from broken rice, protein from roasted soyabeans was supplemented to balance the protein in the diets. Soyabeans in the experimental diets was replaced by fresh duckweed *ad-libitum* at levels of 30, 45, 60 or 100%, compared to no replacement (control).

The Muscovy ducks in experiment 2 were kept in groups of 5 ducks separated from 28 to 70 days old for females and 28 - 84 days old for males and fed diets based on broken rice *ad-libitum* as the major energy source. Roasted soyabeans were supplemented to balance protein requirements of the control ducks. Based on the control diet, replacement of 60 or 100% soyabeans by duckweed ad-libitum in the experimental diets were fed to the females and males, respectively.

A premix containing trace minerals and vitamins, and common salt was mixed (0.25%) with the control diet (but not in the other diets) for both experiments 1 and 2.

Duckweed used in the experiments was grown on ponds enriched with nutrients from effluent from biodigesters and home wastewater. Duckweed was collected twice daily in the morning and in the afternoon. Before feeding it was cleaned by tap water and put in a big bamboo basket for one hour to drain the excess water from the duckweed to reduce bulk.

Results and discussion

▪ *Experiment 1*

The chemical compositions of the duckweed, broken rice and roasted whole soyabeans as analysed at Cantho University were as follows: DM content 4.7%, on DM basis: CP (N*6.25) 38.6%; EE, 9.8%; NFE, 8.58%; CF, 18.7%; ash, 19% and carotene 1025 mg/kg DM. The broken rice was collected from the rice mill of an identical lot and preserved during the experiment. The sample analysed had the following chemical composition: DM, 86.79%; CP, 8.25%; EE, 1.24%; NFE, 69.77%; CF, 1.75% and ash 1.00%.The whole soyabean seeds were roasted to eliminate the anti-nutritional factors, and ground and had the following chemical composition: DM, 87.01%; CP, 38.3%; EE, 18.3%; NFE, 14%; CF, 8.5% and ash 4.9%.

The effects of dietary treatment on feed and protein intakes are shown in table 1.

Table 1.   *Intake of dietary ingredients, dry matter (DM) and crude protein (CP)*

	D0	D30	D45	D60	D100	SE/Prob
Feed intake, g/d						
Broken rice	82	78	83	82	92	1.31/0.001
Roasted soyabeans	27	19	15	12	0	
Fresh duckweed	0	496	499	505	566	4.88/0.001
Premix and salt	0.25	0	0	0	0	
Total DM	94.7	108	108	105	107	1.11/0.001
Total N*6.25	17.1	22.7	21.6	20.6	17.9	0.16/0.001

The duckweed was consumed readily on all treatments from D30 to D100. The intake of broken rice was depressed slightly on the diet with 70% of the control level of soyabeans (D30) and increased when no soyabeans were given (D100). The intake of duckweed increased as the soyabeans were restricted, reaching an average of 560 g/day. These levels of intake are higher than those (434 to 450 g/day) recorded at Cantho University research station by Becerra et al. (1995) for ducks in a similar trial but fed with reconstituted sugar cane juice instead of broken rice. This may be a reflection of the use in our experiment of duckweed grown on ponds fertilized with slurry from the biodigester-pigsty system which resulted in a higher protein content (38.6% in dry matter compared with 26.3%). Total intake of protein was highest on the 70% soyabean level (D30) and least on the control and D100 diets.

Table 2.   *Liveweights, growth and conversion rates of ducks fed duckweed as replacement for soyabeans in basal diets of broken rice*

	D0	D30	D45	D60	D100	SE/ Prob.
Live weight, g						
Initial	859	851	830	859	842	11.9/ 0.37
Final	1771	1869	1822	1807	1806	22.6/ 0.05
Daily gain	26.1	29.1	28.3	27.1	27.6	0.57/ 0.003
Feed conversion (DM)	3.70	4.21	4.23	4.11	4.17	0.04/ 0.001

The rate of liveweight gain was significantly higher on the D30 and D45 diets than on the control diet. The diet with complete replacement of the soyabeans (D100) supported slightly better (ns) growth than the control diet. Feed conversion was best on the control diet and did not differ between the diets containing duckweed.

Mean values for carcass traits are given in table 3. The weights of chest and thigh muscle tended to be higher on the control diet. There were no differences in weights of the components of the digestive tract nor of the heart and liver.

Table 3. *Mean values for carcass traits of ducks given duckweed as replacement for soyabeans in diets based on broken rice*

	D0	D30	D45	D60	D100	SE/Prob.
Slaughter live weight, g	1870	1865	1851	1819	1821	28.1/ 0.58
Carcass weight, g	1253	1219	1211	1175	1198	25.8/ 0.31
Carcass yield, %	73.5	72.5	72.6	72.2	72.8	0.79/ 0.83
Chest muscle, g	203	166	183	164	175	10.1/ 0.07
Thigh muscle, g	162	156	141	156	153	6.85/ 0.25
Heart, g	14	12	14	12	14	1.01/ 0.49
Liver, g	61	61	56	59	61	2.84/ 0.71
Gizzard, g	55	52	52	57	56	2.26/ 0.51
Small intestine, cm	188	186	186	190	191	4.49/ 0.88
Large intestine, cm	12	13	13	12	13	0.58/ 0.36
Caecum, cm	34	36	34	36	34	1.16/ 0.46

Table 4. *Estimates of feed costs assuming situations of purchase or farm-based production of duckweed (in VND; about 11,000VND=1US$)*

	D0	D30	D45	D60	D100
Feed cost/kg gain					
Duckweed purchased*	11,589	13,364	12,866	12,267	10,923
Duckweed grown by farmer**	11,589	9,492	8,952	8,310	6,498

* Based on prices per kg for roasted soyabeans 5,400, broken rice 1,800, fresh duckweed 200, premix 36,000 and salt 1,000
**Assumes no cost of duckweed as opportunity cost of family household labour (women and children) is usually zero

There would appear to be marked economic benefits to the farmer from using duckweed to replace soyabeans in broken rice diets for fattening ducks in situations where the duckweed is grown on the farm and managed and harvested by household labour (table 4). This emphasises the importance of an integrated farming system as a means of reducing costs and improving the economic competitiveness of the small scale farmer.

▪ *Experiment 2*

Data for feed intake are given in table 5. The duckweed was consumed readily on the Df0, Df60, Dm60 and Dm100 diets. The intakes of duckweed were higher on the Df100 and Dm100 diets than those of Df60 and Dm60, respectively, and the males consumed more than females. Intake was considerably increased as the soyabean was completely replaced by duckweed (297g to 323g for the females and 490 to 540 g for males), but the levels of intakes were lower than those of common ducks (540 vs 560) reported by Men et al. (unpublished) under the same feeding conditions. From the data it can be seen that common ducks, especially the improved local breed when herded, consumed higher levels of duckweed than Muscovies. The intake of broken rice decreased on the Df60 and Dm60 diets, with 40% of the control level of soybeans, and increased when no soybeans were given (Df100 and Dm100). Like for duckweed, the intakes of broken rice were higher for the males than for the females.

Table 5. *Intakes of dietary ingredients, dry matter (DM) and crude protein (CP)*

	Df0	Df45	Df100	Dm0	Dm45	Dm100	SE/Prob
Feed intake, g/d							
28-70 days of age							
Broken rice	77	78	86	106	91	105	1.43/0.001
Roasted SBM	25	11	0	29	16	0	
Duckweed	0	297	323	0	490	540	
Premix + salt	0.25	0	0	0.30	0	0	
28-84 days of age							
Broken rice	-	-	-	115	99	109	1.60/0.001
Roasted SBM	-	-	-	28.5	15.2	0	
Duckweed	-	-	-	0	522	569	
Total DM							
28-70days	88	91	90	117	115	116	1.34/0.001
28-84days	-	-	-	125	124	122	1.52/0.276
Total N*6.25							
28-70days	16	16	13	20	22	18	1.19/0.001
28-84days	-	-	-	20.4	23.5	19.3	0.21/0.001

Table 6. *Live weights, daily gains and feed conversion rates of Muscovy ducks*

	Df0	Df45	Df100	Dm0	Dm45	Dm100	SE/Prob
Live weight, g							
Initial	657	619	650	787	758	775	23.09/0.001
Final							
70 days of age	1638	1563	1563	2303	1946	1935	36.71/0.001
84 days of age	-	-	-	2713	2391	2393	43.54/0.001
Daily gain, g							
28-70days	23.4	22.5	21.7	36.1	28.3	27.6	0.83/0.001
28-84days	-	-	-	34.4	29.2	28.9	0.75/0.001
Feed conversion (DM)							
28-70days	3.76	4.06	4.17	3.24	4.12	4.23	0.16/0.003
28-84days	-	-	-	3.64	4.27	4.21	0.11/0.004

The rate of weight gain was slightly higher on the Df0 diet than on Df60 and Df100 diets, but there was no significant difference for females (23.4 vs 22.5 and 21.7), whereas the gain was significantly higher on the Dm0 diet than on the Dm60 and Dm100 diets for males. The results also show that use of high amounts of duckweed to replace most or all soyabeans in broken rice based diets for fattening male Muscovies, which have a high potential growth rate, decreased weight gains. So whether the high duckweed diets can be fed male Muscovies for breeding needs to be investigated.

Table 7. Mean values for carcass traits of Muscovy ducks given duckweed as replacement for soyabean in broken rice based diets

	Df0	Df60	Df100	Dm0	Dm60	Dm100	SE/Prob
Females at 70 days of age							
Live weight, g	1641	1548	1572				25.12/0.066
Carcass yield, %	73.1	70.1	71.1				0.91/0.109
Chest muscle, g	168	167	168				6.25/0.984
Thigh muscle, g	164	148	156				9.37/0.489
Heart, g	9.5	9.6	9.9				0.33/0.721
Liver, g	38	42	43				3.14/0.505
Males at 84 days of age							
Live weight, g				2758	2451	2495	88.06/0.074
Carcass yield, %				70.0	70.6	68.1	1.17/0.336
Chest muscle, g				277	290	275	13.40/0.701
Thigh muscle, g				272	243	246	9.52/0.102
Heart, g				15.0	13.4	12.6	0.54/0.033
Liver, g				72.0	76.4	78.4	3.82/0.140

Mean values for carcass traits are given in table 7. The carcass yield and thigh muscle tended to be higher on the control diets for both the females and males. There were no significant differences in weights of the components of the digestive tract except the hearts of the males.

Table 8. Estimates of feed costs assuming situations of purchase or farm based production of duckweed (VND; about 11,000VND=1US$)

	Df0	Df60	Df100	Dm0	Dm60	Dm100
Feed cost/ kg gain						
Duckweed purchased	11,574	11,563	10,132	10,499	12,532	10,736
Duckweed grown by farmers	11,574	8,922	7,156	10,499	8,955	6,797

If the farmers use the family household labour and the waste water surfaces to grow duckweed for the substitution of soyabeans in broken rice based diets for feeding Muscovies, they can get clear economic benefits. This is very important for improving living standards and opening up the possibility of an efficient solution for the use of the enormous surplus labour force in the rural areas, especially in the case of the poor farmers.

From the results of experiment 1 it can be seen that fresh duckweed can completely replace roasted soyabeans and a vitamin-mineral premix in broken rice based diets for fattening ducks without any reduction in growth performance or carcass traits. The poorer feed conversion on the diets containing duckweed has no economic significance, as shown in table 4 since duckweed can be grown easily on the farm whereas soyabeans usually have to be purchased. The fact that protein yields of duckweed can be as high as 10 tonnes/ha/year (Preston, 1995) compared with less than one tonne/ha/year for soyabean protein is another advantage for the integrated farming system.

The results of experiment 2 for small farms show that farmers can grow duckweed in an integrated farming system and use it to replace soyabeans and mineral-vitamin premixes completely in broken rice based diets for fattening female Muscovy ducks without any reduction in growth performance and carcass quality. However, when feeding male muscovies with high genetic potential for growth, the complete replacement of duckweed in broken rice based diets reduced weight gain, but the gain was still higher than that of the traditional methods. Replacement with duckweed can be applied in feeding muscovies in the early developer stage because normal growth is possible and inputs can be reduced in the long term.

Trials with duckweed in restricted broken rice based diets

▪ *Experiments 3 & 4*

Methods and materials
Two similar experiments were carried out with 54 common ducks fed duckweed from 28 to 63 days of age and 45 local female Muscovy ducks from 28 to 70 days of age on the experimental farm of Cantho University. The trials included three treatments and three replicates. The diets are shown in table 9.

Table 9.   Experimental diets

Treatment	Feed ingredient, g/day/duck	
	Broken rice	Duckweed
Mixed common ducks		
Ba	*Ad-libitum*	*Ad-libitum*
B80	80	*Ad-libitum*
B60	60	*Ad-libitum*
Female Muscovy ducks		
Bma	*Ad-libitum*	*Ad-libitum*
Bm80	80	*Ad-libitum*
Bm60	60	*Ad-libitum*

The broken rice analysed had DM 86.9% and CP 9.6% of DM and the duckweed had DM 4.9% and CP 39.3% of DM.

Results and discussion
Data for feed intakes are given in table 10.

Table 10. *Intake of dietary ingredients, dry matter (DM) and crude protein (CP)*

Treatment	Feed intake, g/day/duck			
	Broken rice	Duckweed	Total DM	Total CP
Mixed common duck				
Ba	101	765	125	23
B80	78	817	107	22
B60	60	869	94	22
P- value	0.001	0.001	0.001	0.05
Female Muscovy duck				
Bma	104	318	106	15
Bm80	77	324	82	13
Bm60	57	333	66	11
P-value	0.001	0.458	0.001	0.001

The daily intake of duckweed increased as the broken rice was restricted reaching a maximum of 1100 g at 6 weeks of age, which was equivalent to the ducks' own live weights. These levels of intake are higher than those (870 to 560 g) recorded in the report by Men et al. (unpublished) for ducks in a similar feeding trial, but of a different crossbred line. Total intake of protein was slightly decreased on the B60 diet. Total intake of DM was highest on the ad-libitum broken rice diets and least on the B60 diet. The daily intake of duckweed by the female Muscovy ducks increased slightly as the broken rice was restricted, reaching an average of 333g. The level of intake is lower than that (333 vs 870 g) recorded in trial 1 with common ducks. Total intake of DM on the Bm60 diet was lowest (66 vs 106 g for Bma), and was lower than that of common ducks (66 vs 94g). The intake of broken rice was lower than the amount offered because in the first week of the experiment the ducks consumed less feed than was expected.

Table 11. *Live weights, daily gains and conversion rates of common and female Muscovy ducks fed restricted broken rice with duckweed ad-libitum*

Treatment	Live weight, g		Daily gain, g	Conversion, DM
	Initial	Final		
Mixed common duck				
Ba	694	1595	25.8	4.87
B80	686	1537	24.3	4.44
B60	674	1369	19.9	4.77
P value	0.858	0.001	0.001	0.015
Female Muscovy duck				
Bma	639	1681	24.8	4.3
Bm80	646	1483	19.9	4.2
Bm60	644	1388	17.7	3.7
P value	0.969	0.001	0.001	0.008

The daily gain males was gradually decreased as the levels of broken rice were reduced, although there was no significant difference between Ba and B80. The gain was lowest for the B60 diet and significantly different when compared with Ba and B80. Feed conversion was best for the B80 diet, but no difference was found between Ba and B60 diets.

With female Muscovy ducks, the daily gains significantly decreased on the restricted broken rice diets compared with the Bma (control). The gains on the Bma diet is equivalent with that on Ba diet in trial 1. The efficiency of feed conversion of female Muscovies is better than that of common ducks on the same diets.

The results of experiment 3 show that when the price of broken rice is high, providing 80g daily of broken rice for improved ducks given duck-weed *ad-libitum* gives good growth and carcass traits. Growing duckweed for ducks saves capital and can help to solve unemployment in the area.

The results of experiment 4 show that levels of 80g broken rice per duck per day and duckweed *ad-libitum* for fattening female Muscovies decreased their weight gains because the ability of muscovies to consume fresh duckweed is lower than that of common ducks. The Bm80 diet can be used for developers because in this period breeding ducks are not grown so rapidly, but the effects on reproduction need to be investigated.

Effect of energy and protein supplementation of scavenging ducks raised in paddy rice fields

Methods and materials

A total of 800 crossbred Cherry Valley ducklings were brooded and trained to scavenge from 1-28 days of age in the farmer households of a hamlet of Cantho province. After brooding the ducks were divided into 4 flocks with 190 ducks in each and were herded and managed by 4 farmer households. The ducks were let loose in rice fields post harvest from 6.00 h to 17.30 h daily. There were 3 supplemental diets: 50 g/day of a mixture of broken rice and dried fish meal (BFM) for flock 1, 50 g/day broken rice (BR) for flock 2 and 20 g dried fish meal (FM) for flock 3 and compared to no supplementary feed (S0) for flock 4 (control). These diets were given to the ducks from 28 to 70 days of age and given in the evening. At the end of the trial, four ducks (2 males and 2 females) from each flock were slaughtered for carcass evaluation.

Results and discussions

Data from table 12 show that, although having the lowest initial live weight, weight gains of ducks given BFM supplementary feeds were higher that of the unsupplemented flock, and were highest from flock 1 to flock 3. Weight gains were significantly higher for flock 1 and for flock 2 than for flock 3 and flock 4. The feed conversions were higher for flock 1

Table 12. *Mean value for live weights, daily gains and conversion rates of ducks supplemented feeds*

Item	Flock 1 (BFM)	Flock 2 (BR)	Flock 3 (FM)	Flock 4 (S0)	SE/Prob
Live weight, g					
Initial	431	485	480	453	4.94/0.001
Final	1855	1749	1659	1592	13.25/0.001
Daily gain, g	33.9	30.1	28.1	27.1	4.88/0.001
Feed conversion (DM) of supplement feed					
In the day (b. rice)	1.00	1.12	1.20	1.25	
At night	1.28	1.45	0.46	0	
feed cost /kg gain* (VND=1/11000USD)	5139	5027	34457	2446	

* Based on of dried fish, 4,000, broken rice, 1700 VND per kg.

and flock 2, and the costs per kg gain are highest for flock 1, followed by flock 2 and 3. The flock 1 ducks began moulting at 63 days of age, while flock 4 still had the original plumage at 70 days of age. Ducks with full plumage sold at a higher price (12%) than those in flock 4. However, all four flocks were in much better condition than flocks fed and managed by the traditional methods of the farmers in the hamlet. Over the same time period and with the same breed reared, several local farmers finished their ducks at 75 days of age, with average live weights from only 1.2 to 1.3 kg and with poor quality carcases.

Mean values for carcass traits are given in table 13. Live weights at slaughter, carcass weights and yields, and chest muscle tended to be higher for flocks 1 and 2. There were no significant differences in the weights and the lengths of the digestive tract components, except the liver.

Table 13. *Mean values for carcass traits of herded ducks supplemented different feeds*

Item	Flock 1 (BFM)	Flock 2 (BM)	Flock 3 (FM)	Flock 4 (S0)	SE/Prob
Slaughter liveweight, g	1862	1743	1655	1573	62.43/0.024
Carcass weight, g	1267	1218	1118	1058	40.72/0.007
Carcass yield, %	68.0	70.0	67.6	67.3	0.72/0.065
Chest muscle, g	150	140	118	114	8.30/0.018
Thigh muscle, g	168	167	158	140	8.08/0.09
Heart, g	11.8	11.8	10.3	10.3	0.50/0.071
Liver, g	68.3	58.3	51.0	54.8	3.80/0.026
Gizzard, g	63.3	60.3	67.0	60.0	3.48/0.471
Small intestine, cm	191	192	191	181	6.70/0.624
Large intestine, cm	10.8	10.6	10.8	9.3	0.46/0.109
Caeca, cm	33.4	33.3	32.3	32.6	1.49/0.936

From the results of the experiment it is clear that supplements of protein-energy feeds to growing crossbred meat ducks improved the daily liveweight gains and carcass quality, and shortened the raising time. Supplementation is simple to carry out, and an improvement of the traditional scavenging system, and the method meets the increasing demand of the consumers for better quality products.

Discussion and conclusion

From the results of the four experiments it can be seen that fresh duckweed, a locally available unconventional feed source, can completely replace soyabeans in broken rice based diets for fattening crossbred Common ducks and female Muscovy ducks without any problems regarding daily live weight gains, carcass quality and marketing. Replacement decreased the investment by using available wastewater and household labour for managing and collecting duckweed, and increased the economic benefits for the farmers. Carcasses of ducks given duckweed are an attractive yellow colour due to storage in the body of carotenes. However, complete replacement of duckweed in diets for male Muscovy ducks with a genetic potential for high growth rate tended to decrease the weight gain. The high protein content of cultivated duckweed is important because the ducks need to consume large amounts due to the low DM content of duckweed. In fact, growing duckweed is easy for farmers in the Mekong Delta, because the region is an alluvial lowland plain with an inexhaustible water supply and large amounts of decaying organic matter and human wastewater. These conditions make growing duckweed in integrated farming systems very profitable not only for animal feed but also for the treatment of wastewater from animals and humans. Also, this contributes to solving one of the environmental problems that have concerned the government.

An interesting and important observation is the relative capacity of the Common duck breed, compared with the Muscovy duck, to consume duck weed. A strict comparison of the two breeds is not possible as they were not included in the same experiment. The data in table 14 shows the comparison between the breeds on the two main feeding systems: ad libitum or restricted broken rice (confounded with sex in the first case and with location in the second). When the broken rice was restricted, the Common ducks ate almost twice the amount of duck weed as the Muscovy ducks (P=0.001), and even when the rice was given *ad libitum* the intake still tended (P=0.14) to be higher for the Common ducks. The economy of liveweight gain (feed cost/unit LW gain) was best on the restricted rice system, although the rate of gain was slightly lower, and in each case the Common ducks were better than the Muscovy breed. Differences between pig breeds were also observed by Rodriguez (1996) when duckweed was the major protein source; the local Mong Cai breed appeared to be able to consume more of this feed than the exotic (Large White) breed.

As stated earlier, it was not the aim of the experiments to make a strict comparison of the breeds, hence the confounding with location and/or sex. Nevertheless, the observations are interesting and provide a strong justification for future research when the aim is to maximise use of local feeds such as duckweed. The most effective way of maintaining biodiversity is by showing that local (indigenous) breeds may have distinct comparative advantages when local feed resources are used.

Table 14. Comparison of Muscovy and Common duck breeds when fed duckweed and either restricted or ad-libitum broken rice

	Common	Muscovy	SE	Prob
Restricted rice				
Location	Station	Station		
Sex	Mixed	Female		
Live weight, g				
Initial	685	643		
Final	1500	1517		
Daily gain*	20	18		
Intake fresh duck weed				
g/day	817	325		
g/g LW	0.75	0.30	0.039	0.001
Feed cost, VND/kg gain	5427	5796		
Ad libitum rice				
Location	Station	Farm		
Sex	Mixed	Mixed		
Live weight, g				
Initial	845	700		
Final	1826	1751		
Daily gain	28	24		
Intake fresh duck weed				
g/day	517	413		
g/g LW	0.39	0.33	0.023	0.14
Feed cost, VND/kg gain	6500	6900		

Supplementary feeding for scavenging ducks is an improvement of traditional production to improve growth rates and quality of meat ducks. It is a reasonable method to shorten raising time and keep the duck population stable, as with modern rice cultivation opportunities for duck flocks to scavenge are limited. Supplementing 50 g of a broken rice and dried fish meal mixture, or 50 g broken rice per bird per night for herded duck flocks showed that daily weight gains could be improved and also rearing time could be reduced. All supplements resulted in high weight gains and carcass weights when compared with other traditionally reared flocks. The ducks formed their adult plumage 2 weeks earlier than flocks raised by traditional methods. It is clear that if fed a balanced feed of energy and protein (16.8% CP and 11.74 MJ ME/kg DM) ducks grow faster than those fed broken rice or dried fish alone. However, the supplemental cost for the protein-energy diet was highest because of the high price of broken rice and dried fish meal. The farmer can overcome the problem by using duck-weed in an integrated farming system as shown in experiment 1, or by gathering shellfish available in the canals nearby the home for the supplement.

Acknowledgements

This research forms part of the programme of study of the senior author for the Master of Science Degree in "Livestock-Based Integrated Farming Systems for Sustainable Use of Renewable Natural Resources", at the Swedish University of Agricultural Sciences, Uppsala. Development of the basic experimental facilities were facilitated by a grant from the International Foundation for Science (IFS).

I would like to thank Dr Brian Ogle, Dr T. R. Preston and Mr Frands Dolberg for useful guidances and their advisory roles that made this study possible.

References

Anon. 1992. Study on the performance of local ducks reared under scavenging condition (Unpublished data), *Bangladesh Livestock Research Institute*.

Becerra, M., Ogle, B. and Preston, T. R. 1995. Effect of replacing whole boiled soy-beans with *Lemna sp* in the diets of growing ducks. *Livestock Research for Rural Development*. Volume 7, Number 3, 44.8 Kb.

Dean, W. F. 1986. Nutrient requirements of meat-type ducks. *Duck Production Science and World Practice*. (Ed). Farell, D. J. and Stapleton, P. (Ed). University of New England, pp. 31-57.

Leclercq, B. and de Carville, H. 1986. Dietary Energy, Protein and Phosphorus Requirements of Muscovy Ducks. *Duck Production Science and World Practice*. Farell, D. J. and Stapleton, P. (Ed). University of New England, pp. 58-69.

Manda, M. 1992. Paddy Rice Cultivation Using Crossbred Ducks. *Farming Japan* Vol. 26 - 4, pp. 35-42.

Men, B.X., Ogle, B and Preston, T.R. (unpublished). Use of restricted rice in duck-weed based diets for feeding growing Common and Muscovy Ducks. *Cantho University*, Vietnam.

Preston, T. R. 1995. Research, Extension and Training for Sustainable Farming Systems in the Tropics. *Livestock Research for Rural Development*. Volume 7, Number 2, 84 Kb.

Rodriguez J., L. (1996). Appropriate use of local resources in integrated farming as a strategy for sustainable agriculture in Central Vietnam. *Swedish University of Agricultural Sciences, Department of Animal Nutrition and Management*, Uppsala 1996.

Toung, Naren. 1994. Optimum protein Supply and level of inclusion of water spinach *(Ipomoea aquatica)* in sugar cane juice based diets for growing ducks.

Swedish University of Agricultural Sciences, Department of Animal Nutrition and Management, Uppsala 1994, paper 2, 8pp.

Vo Ai Quac. 1990. Paper presented at the *workshop on sustainable agriculture in the lowland.* Sep. 24 - Oct. 7, Bangkok, Thailand.

Zakaria, A. 1992. Advances on Feeding and Management of Ducks in Indonesia. Sustainable Animal Production: *Proceedings of the Sixth AAAP Animal Science Congress,* Vol. II. AHAT, Bangkok, pp. 47-53.

Yeong, S. W. 1986. Utilisation of local feedstuffs in diets of meat and laying ducks in Malaysia. *Duck Production Science and World Practice.* Farell, D. J. and Stapleton, P. (Ed). University of New England, pp. 323-332.

Yeong, S. W. 1992. Advances on Feeding and Management of Ducks in Malaysia. Sustainable Animal Production: *Proceedings of the Sixth AAAP Animal Science Congress,* Vol. II. AHAT, Bangkok, pp. 56-69.

Village Pig Production in Central Vietnam:

Results of a Participatory Rural Appraisal Survey and On-Farm Feeding Trials with Protein Supplementation of Traditional Diets

Nguyen Thi Loc, Brian Ogle** and T.R. Preston****

*Department of Animal Nutrition and Biochemistry,
Hue Agriculture and Forestry University, 24 Phung Hung,
Hue City, Vietnam. E-mail: Loc%Hue%ifs.plants@ox.ac.uk*

**Department of Animal Nutrition and Management,
Swedish University of Agricultural Sciences,
P. O. Box 7024, S-75007, Uppsala, Sweden
E-mail: Brian.Ogle@huv.slu.se*

***Finca Ecologica, University of Agriculture and Forestry,
Thuduc, Ho Chi Minh City, Vietnam
E-mail: thomas%preston%sarec%ifs.plants@ox.ac.uk*

Summary

The study consists of a Participatory Rural Appraisal (PRA) survey focusing on pig production in Binh Dien and Xuan Loc villages in Central Vietnam, followed by an on-farm feeding trial in which traditional diets based on locally available feedstuffs were evaluated and the effects of protein supplementation on pig performance were determined. The results of the PRA survey showed that pig raising is an important source of income for farmers, even the very poor households. Positive factors for pig production are the availability of cheap local feed resources such as cassava, rice bran, molasses and sweet potatoes and low fixed costs (pens and labor). The most under-utilized feed resource is cassava root which is the lowest cost feed in these areas, especially in the harvest season, when the price of fresh cassava roots is only 180-280 VND per kg. The typical diet fed to fattening pigs is based on the following ingredients in order of importance: cooked rice, rice bran, cassava meal, fresh cassava root and sweet potato leaves. Crude protein was very low in the traditional diet (94 - 98 g/pig/day) compared with the standard diet (213 g/pig/day) for Large White x Mong Cai crossbred pigs in Binh Dien and Xuan Loc villages. A feeding trial was carried out in Binh Dien and Xuan Loc villages in which 14 Mong Cai x Large White (MC x LW) weaner pigs (5 farms) were fed a traditional diet, and 12 matching pigs on similar basal diets (6 farms) were given supplements of groundnut and fish meal to provide an additional 100 g/day/pig of protein. The mean daily live weight gain of pigs under the traditional feeding system was low (202 and 230 g/day in each of the two villages), but was significantly increased to 363 and 366 g/day (P=0.001) in Xuan Loc and Binh Dien, respectively, by giving the protein supplement. By end of trial, supplemented pigs weighed 66.5 and 65.1 kg compared with 45.6 and 50.5 on the traditional diet in Xuan Loc and Binh Dien villages, respectively. The net economic benefit after deducting the cost of the protein supplement was VND 800/day equivalent to VND 135,000 for the 150 days fattening cycle.

Key words: Pigs, participatory rural appraisal (PRA), Local feeds, traditional diets, protein supplement.

Introduction

Central Vietnam consists of 13 provinces, covering an area of about 97,000 km² and with a population of 17 million inhabitants. Compared with other regions of the country, Central Vietnam is one of the poorest and least agriculturally developed, mainly due to the severe climate, low level of production, lack of investment funds and unfavorable natural conditions, such as a high frequency of floods, droughts and storms.

In 1994 there were approximately 4.2 million pigs in Central Vietnam and of these around 2,000 were raised in Binh Dien and Xuan Loc villages. The majority (95%) of pigs are owned and raised by peasant farmer families and there are no large-scale private commercial pig farmers, while small scale commercial production units account for less than 5% of the total pigs raised (Nguyen sinh Cuc, 1995). In this context, there is a need to describe and understand the current farmer practice in order to be able to use it as a strategy to identify problems and weak points that require research before new technologies can be promoted (Dolberg 1994).

In Binh Dien and Xuan Loc villages the pig populations have increased rapidly from 586 pigs in 1990 to 2,093 in 1994 (Nguyen Sinh Cuc, 1995). With increasing pig production the feeding system becomes an even more important factor due to the competition between humans and livestock for some feed resources. The live weight of pigs at slaughter is considered to be low due to the use of small local breeds and of feed that is poor in quantity and quality, especially the lack of protein supplements.

The use of restricted amounts of protein to supplement local sources of carbohydrates for growing-fattening pigs has been studied in Vietnam. The amount of protein recommended by Ospina et al (1995) for crossbred Yorkshire x Landrace x Duroc pigs fed cassava root meal throughout the growing-finishing period was 200 g/day. This relatively low level of protein gave acceptable growth performance and carcass quality, and the lowest feed costs/kg gain compared with higher protein levels.

The objectives of the research carried out in villages in the rain-fed, hilly region of Central Vietnam were:

- To identify available feed resources for pigs and prices of local feed-stuffs throughout the year.
- To identify problems, opportunities and potential in pig production in the rural areas.
- To evaluate a simple intervention which, on the basis of the PRA findings, could be expected to lead to improved farmers' incomes.

I. Participatory Rural Appraisal (PRA)

A survey was conducted between February and December 1995 in Binh Dien and Xuan Loc villages as a descriptive study with a predetermined focus on pig production.

Material and methods

Location

The research sites are located in the highland areas of Central Vietnam at an altitude 30-50 meters above sea level. The climate is humid during the hot (March to September) and rainy seasons (October to February). The mean annual temperature is about 28°C with an average rainfall of around 3,000 mm and average humidity between 80 and 85%. A common feature of these provinces is their adjacency to the Truong son mountain range and to the sea. The whole region can be described as having a very narrow and long strip of flat land along the coast and a much larger sloping mountainous area.

Binh dien Village

Binh Dien is located in a larger upland area of Huong tra , Thua thien Hue Province of Central Vietnam. The village is 30 km from Hue city. Binh Dien was established in 1978 and had a total population of 2,957 people in 557 households in 1995. The agricultural area is about 402 ha, consisting of a cultivated area of 127 ha made up of 25 ha of sugar cane, 41 ha of cassava, 40.5 ha of sweet potatoes, 7.5 ha of rice, 8 ha vegetables and beans and 5 ha of groundnuts. The area of natural pasture is about 300 ha.

Xuan Loc village

Xuan Loc is located in an upland area of Phu Loc District, in Thua thien Hue Province of Central Vietnam. The village is 50 km from Hue city. Xuan Loc was established in 1976, and today has a population of 1,990 people divided into two ethnic groups: Lowlanders (kinh) with 1,529 people in 282 families and Van Kieu (Bru-VanKieu) with 461people in 82 families. This

group came to settle in Xuan Loc in 1983 and were originally shifting culti-
vators. The cultivated area in the village is 140 ha, consisting of 75 ha of
cassava, 15 ha of sweet potatoes and 50 ha of rice.

Data collection methodology
The data collection was done by application of Participatory Rural
Appraisal (PRA) methods as described by McCracken et al (1988) and
Chambers (1992a and b) and by reviewing secondary information from an
earlier project. PRA methods included interviews, diagramming exercises,
seasonal calendars, matrix and scoring and matrix and ranking. PRA is a
method used to bring about a good understanding of the rural community
in a short period of time. The essential element of PRA is participation of
the community, assuming members of the community know their own
situation best. The local feed resources, availability, amounts fed, any
limitations, feeding methods, supplements and plant species were identified through farm visits.

The participants in the data collection
The participants were from the local Women's Union and the farming
community of the two villages, while the outsiders included researchers
from Hue University of Agriculture and Forestry and the Canadian International Development Research Center (IDRC), who happened to have a
project in the same area.

Results and discussion
The climate is humid tropical. The characteristics and mean annual climatic
conditions of Binh Dien and Xuan Loc villages are: mean annual temperature about 28°C with an average rainfall of around 3,000 mm and average
humidity between 80 and 85%. A common feature of these provinces is
their adjacency to the Truong son mountain range and to the sea. The
whole region can be described as having a very narrow and long strip of
flat land along the coast and a much larger sloping mountainous area.
    Binh Dien and Xuan Loc villages are fairly typical and representative of
the high land rural areas of Central Vietnam.

Population trends
Human and pig populations have both experienced rapid growth over the
four years from 1990 to 1994, although at different rates (table 1), and with
major differences found between the situation in Central Vietnam and in
the two villages. Since 1990 to 1994 the total human population in Central
Vietnam has increased by 10 percent with an increase for the pig population of 24 percent, while in the two villages human populations have
grown by 9 percent (in Binh Dien village), 25 percent (Xuan Loc village),
but with a much greater increase for the pig population of around two and-

a half times (each village). These increases in Binh Dien and Xuan Loc villages have resulted in expanded competition for the use of cereal grains as animal feed and for human consumption.

Table 1.   *Human and pig populations, 1990 and 1994*

	Humans			Pigs		
	1990	1994	%	1990	1994	%
Central Vietnam	15,689,000	17,284,000	10	3,400,000	4,200,000	24
Binh Dien village	2,718	2,957	9	280	1,013	26.2
Xuan Loc village	1,590	1,990	25	306	1,080	25.3

Pig production systems

In Central Vietnam around 95% of the pig population is kept on family small holdings and pig production plays a very important role in the economy of rural farmers. Typically 1-2 sows and 2-10 fatteners are kept, depending on the family's situation. The major system of keeping pigs is free range, in which inputs are very low and the pigs find most of their feed through scavenging. Pigs are slaughtered when the owner's financial needs and social obligations dictate rather than with regard to their weight and level of production, and also depends on the amount of feed available from village resources. An important initial intervention in free range systems is to house the pigs in small simply constructed units. Farmers have yet to learn the benefit of using a balanced diet for these animals. Feeds used depends on the harvesting season, but as the yields of agricultural products are low the quantities available are limited. The climate affects both animals as well as the vegetation growth and available feeds (Nguyen quang Linh, personnel communication,1995).

In Xuan Loc and Binh Dien villages, the numbers of pigs raised were related to a system of wealth ranking (tables 2 and 3). The households were classified according to three wealth categories: very poor, poor and better-off. The poorest group are classified as those who do not have enough food to eat for 6-7 months of the year, the poor group has insufficient food for 3-4 months, and the better-off group has enough food to eat the whole year around. As shown in the results in tables 2 and 3, pig raising was a common source of income for the farmers, even in very poor households. Positive factors for pig production are that they play an important role as a means of savings and capital investment, and they often provide a substantially higher return than alternative investments. Furthermore, the availability of cheap local feed resources such as cassava, rice bran, molasses and sweet potatoes and low fixed costs (pens and labor) were stated by farmers to give additional advantages and higher net incomes.

Table 2. *Pig ownership in relation to wealth ranking, Binh dien village*

	"Better off"	"Poor"	"Very poor"
Total number of households	55	165	337
Households with pigs, %	64	79	67
Pig population	120	365	528
Number of pigs/family	3.4	2.8	2.3

Table 3. *Pig ownership in relation to wealth ranking, Xuan loc village*

	"Better off"	"Poor"	"Very poor"
Total number of households	33	109	222
Households with pigs, %	100	80	60
Pig population	209	511	360
Number of pigs/family	6.3	5.9	2.7

Pig breeds

In Central Vietnam the pig population was about 4.1 million in 1994 (table 1) with sows comprising about 13.5% of the total. Numbers are increasing by around 10 % per year. The local pig breeds are I, Co, Meo, Tau pha, Trang phu khanh (TPK) and Mong cai (MC). The Mong Cai breed originates in the north east of Vietnam (Dam ha, Ha coi, Tien yen Districts of Quang ninh province), and is commonly crossed with exotic boar breeds such as the Large White from Russia (a kind of Yorkshire), Landrace (Danish), Yorkshire, Cornwall (Hungary), Hampshire or Duroc. The advantages of the local breeds are that they are adapted to the harsh climate and diseases, and are very fertile (Hoang nghia Duyet and Nguyen kim Duong, 1995).

In the two villages, the local pig breeds are predominantly I, Co, Meo and Mong cai (MC). They are characterized by very small body size with a weaning weight of about 4-5 kg and live weight at 1 year is 30-50 kg. Growth rates are typically only around 150 g/day and feed conversion ratios are poor (7-8 kg feed/kg LWG). Mong Cai, have small body size, low growth rates and fat carcasses(Hoang Nghia Duyet and Nguyen kim Duong, 1995). However, they are popular as breeding sows because of their good reproductive performance, adaptation to the climate, and resistance to diseases. According to the work of Rodriguez (1996), Mong Cai at first mating weigh around 50 kg, the number of piglets weaned per litter is 10 and weaned weight is about 8 kg. The common meat animal used in the villages is a crossbred between Mong Cai (MC) and Large White (LW). This cross breed is preferred by the farmers because they attain a slaughter weight of 70 to 80 kg in one year with a feed conversion of 5-5.5 kg feed/kg LWG.

Feeding systems

The effects of the climate in central Vietnam on the livestock are both direct and indirect. One direct effect is that feed intakes are reduced at the higher ambient temperatures. The major indirect effect of the climate is on the quality and quantity of feed available. Throughout the year, the severe climate, poor soils and other unfavorable natural conditions such as floods, droughts, storms, and hot winds have negative effects on the quality and quantity of available feeds for pigs.

Table 4. *Characteristics and usage of some local feed resources for pigs*

Binh dien	Xuan loc						
Local name	Local name	Scientific name	M	P	Y	S	W
Fruits	Fruits						
Mit	Mit	Artocarpus heterophyllus		X	X		
Bau,bi	Bau,bi	Cucurbita spp		X		X	
	Dua hau	Citrullus lanatus		X		X	
	Dudu	Carica papaya		X	X		
Foliage							
Rau muong	Rau muong	Ipomea aquatica Forsk		X	X		
Rau khoai	Rau khoai	Ipomea batatas (L)		X	X		
Than chuoi	Than chuoi	Musa paradisiaca L.		X	X		
Cay cong san	Cay mui		X				X
Oi	Oi	Psidium gujava L.	X				
Co xuoc			X				
Roots and tubers							
San	San	Manihot esculenta crantz		X	X		
Khoai	Khoai, lang	Ipomea batatas (L.)		X		X	
Khoai tia	Khoai tia	Dioscorea alata L		X		X	
Hoang tinh	Hoang tinh	Maranta arudinacea L		X		X	
Khoai so	Khoai so	Colocasia esculenta		X	X		X
Cereals							
Lua ray	Lua ray	Oryza sativa L				X	
Lua nuoc	Lua nuoc	Oryza sp.				X	
Others							
Dau xanh	Dau xanh	Phaseolus aureus (Roxb)		X		X	
Mia	Mia	Saccharum officinarum L.		X		X	
Bong ruou	Bong ruou	Fermented distiler's grain		X	X		
Kho dau lac	Kho dau lac	Arachis hypogaea		X		X	
Ri mat		Saccharum officinarum		X	X		

• M, used as a medicinal plant; • P, used as pig feed; • Y, year round production;
• S, seasonal availability; • W, wild plant

The characteristics and usage of some local feed resources for pigs are shown in table 4. The plants used as livestock feed and as medicine were similar in the two villages. There were only two kinds of plants *Citrullus lanatus* and *Carica papaya* that were used in Xuan loc but not used in Binh dien, while "A" molasses was produced in Binh Dien but not in Xuan Loc.

The main feed resources available are shown in table 5. The price of feeds usually fluctuates considerably as a result of instability in their availability and this affects farmers' incomes. The main feeds in the two villages are rice, cassava and sweet potato. Farmers use their agricultural produce

Table 5 *Main feed resources available for pig production in Binh dien and Xuan loc villages*

Items	Available throughout year	Marketed (%)	Price fluctuations
Rice	yes	0	considerable
Rice bran	yes	0	"
Cassava meal	yes	0	"
Fresh cassava root	no	70	"
Fresh sweet potato root	no	10	"
Vegetables	yes	15	"

for home consumption, livestock feed and for sale in local markets. Annual rice production is not enough to meet demands for home consumption, and additional rice is bought from elsewhere. Most of cassava (70%) is sold to small private traders, the remaining 30% is for feed or food. Sweet potato is mainly for home use as food and feed. Most vegetables (85%) are for family consumption, the remaining 15% is for market.

The cheapest energy source for pigs is fresh cassava root (figure 1), but this is only available in the harvest season (from November to January) when the price may be as low as 180-280 VND/kg. If processing and conservation of fresh cassava roots could be done locally then the feed cost would be reduced and the farmers would have better control of their feed supplies. Most of the fresh cassava grown (68%) is sold to small private traders at a very low price then bought back as cassava meal or by-products at a much higher price (1600 -1800 VND/kg) for feeding pigs. The remaining fresh cassava (32%) is for livestock feed and human consumption.

There are few potential protein supplements available for pigs in the two villages. An unused protein resource is the leaves from cassava. Annually, there is the potential to produce from 28 to 56 tones of crude protein from cassava leaves available at harvest in Binh dien and between 51 and 103 tones in Xuan loc, but these are not normally used as pig feed.

The typical diet fed to fattening pigs is based on the following ingredients in order of importance: cooked rice, rice bran, cassava meal, fresh cassava root and sweet potato leaves. The amounts of each varied between farms and with time. Based on the data collected from 40 farmer house-

Figure 1. Market price of various feeds for pigs in Binh Dien and Xuan Loc villages in 1995

Thousands
VIND/kg

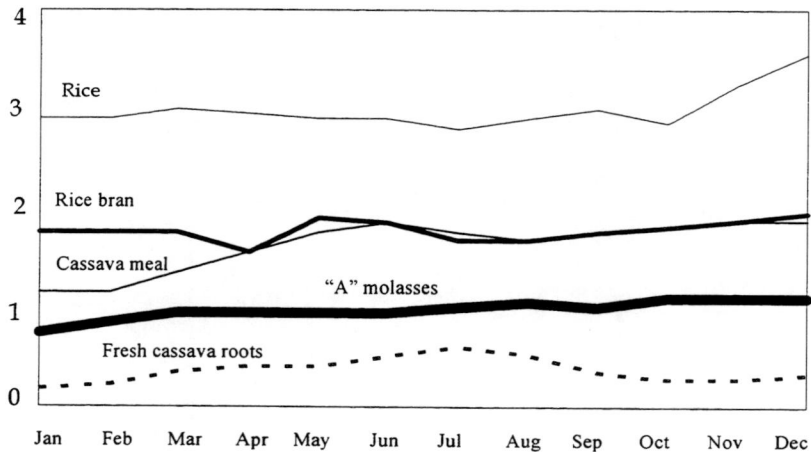

holds raising pigs, on average it was estimated that cassava meal and fresh cassava root accounted for 30-60%, rice bran and rice 35-60% and sweet potato leaves 5-10% of the traditional diets. DM, ME and CP for pigs on traditional diets are presented in table 6. The observations were as follows:

Table 6. Feeding standards and traditional feeding results for Large White x Mong Cai crossbred pigs from 15 to 50 kg in Xuan Loc and Binh Dien villages

Items	Pig live weight 15-50 kg		
	Standard*	Xuan Loc**	Binh Dien**
DM (kg/pig/day)	1.40	1.29	1.35
ME (MJ/pig/day)	18.1	17.2	17.6
C (g/pig/day)	213	94	98

Sources: *Nguyen Van Thuong, 1992; **PRA survey in the two villages.

* ME in the traditional diet (17.5 MJ/pig/day) was lower than that in the standard diet (18.1 MJ/pig/day).
* CP was very low in the traditional diet (94 - 98 g/pig/day) compared with the standard diet (213 g/ pig /day).

These observations indicate a lack of protein in traditional diets for Large White x Mong Cai crossbred pigs in the two villages.

Health care

The following preventive vaccinations are given by veterinarians: Pasteurella, Erysipelas and Hog Cholera. There are some problems with leptospirosis in pigs due to the lack of a Leptospirosis vaccine. The health care system is organized through district extension veterinarians, extension centers and village veterinary technicians.

Marketing

Weaners are bought and sold by farmers at local markets and finished animals are sent to abattoirs. There is a seasonal influence on the price of pigs due to climate and consumer demand e.g. for the annual "Tet" festival (figure 1). The identification of the potential of fresh cassava roots, available at low cost during the peak harvest period, indicated one of the areas where future research should be directed (Nguyen Thi Loc et al., 1996).

Problems in pig production in Binh dien and Xuan loc villages

Figure 2 indicates some of the problems identified as affecting pig production in Binh Dien and Xuan Loc villages such as lack of capital and protein feeds and lack of knowledge of feed processing methods.

Figure 2. *Constraint analysis chart for pig production in Binh Dien and Xuan Loc villages*

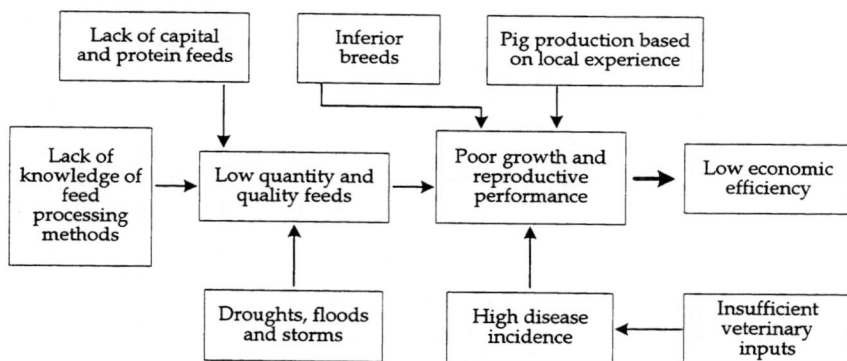

Droughts, floods and storms lead directly to low quantity and quality feeds. All these problems and high disease incidence are the sources of poor growth and reproductive performances of pigs which in turn cause low economic efficiency of pig production in the study areas.

II. Protein supplementation of the traditional diet used by farmers

Hypothesis

Based on the results of the PRA analysis (part 1), it was hypothesized that supplementing the traditional diet with small amounts of protein would lead to biological and economic benefits for pig production in the two villages.

Materials and methods

On-farm feeding trials were carried out in Xuan oc and Binh Dien villages from May to October 1995.

Animals, diets and management

The pigs were Mong Cai x Large White crossbreeds, belonging to the farmers. The farmers also provided their normal management, drugs and vaccines.

A total of 14 pigs owned by 5 farmer households (2 or 4 pigs on each farm) with initial live weights of 10.7 to 20.0 kg in the two villages (3 farmers in Binh Dien; 2 in Xuan Loc) were fed on a traditional (control) diet consisting of cooked rice, rice bran, cassava meal, cassava roots and sweet potatoes (table 5). The amounts of the individual feed ingredients fed varied between farms and also with time, depending on current availability and price. Twelve pigs owned by 6 different farmer households (3 in each village), but of similar genetic background and initial live weight from 9.6 to 23 kg, were fed on the traditional diet supplemented with 190 g/day of groundnut cake (39% CP) and 60 g/day of fish meal (42% CP). It was calculated that the supplement would raise the overall protein supply to 200 g/pig/day. The pigs were fed three times per day. In the case of the supplemented pigs the groundnut cake was soaked overnight and then mixed with the boiled basal feed. The fish meal was fed in the morning after being mixed with the rest of the dietary ingredients.

Data collection

The pigs were weighed in the early morning once each month using a 100 kg capacity portable scale with an accuracy of 0.5 kg. Feed intakes were recorded using a 20 kg capacity portable scale. These records were collected every two weeks, and additional random checks were also made. The major feed resources were identified and representative samples were collected and analyzed for dry matter (DM), crude protein (N*6.25), crude fiber (CF), ether extract (EE) and ash (AOAC 1985).

Results

Feed intake and growth performance

The traditional pig ration consisted of 48% rice and rice bran; 45% cassava root meal and 7% vegetables (% DM basis). For the pigs from 14 kg to 50 kg, the traditional basal diet dry matter intake ranged from 1.2 to 1.4 kg/pig/day. Metabolizable energy (calculated) supplied by the diets ranged   from 16.0 to 18.4 MJ/pig/day. Protein (N*6.25) concentration ranged from 5 to 7% in the dietary DM, with total supply at about 80-110 g per pig per day.

Table 7. *Mean values for initial and final live weight and daily live weight gain for pigs fed on traditional diets with and without a protein (groundnut cake and fish meal) supplement in Binh Dien and Xuan Loc Villages*

| Live weight, kg | Xuan Loc | | Binh Dien | | |
	Traditional	Supplemented	Traditional	Supplemented	SE
Initial	15.3	11.6	15.9	14.7	± 2.0
Final	45.6	66.5	50.5	65.1	± 4.2
ADG	0.202	0.365	0..230	0.363	± 0.020

The mean values for initial and final live weights and daily gains of the pigs on the traditional and supplemented diets, in each of the villages, are shown in table 7. The effects of supplementation and of location (village) are shown in table 8. There was good agreement between villages (292 and 290 g/day live weight gain for the pigs in Xuan Loc and Binh Dien) indicating little difference in the nutritive value of the basal feed resources. Protein supplementation increased live weight gain (adjusted for differences in initial weight) by 83% from 204 to 375 g/day (P=0.001) and final weight from 46 to 68 kg over the 150 days fattening period.

Table 8. *Effect of protein (groundnut cake and fish meal) supplementation of traditional diet and location on growth of pigs*

| Live weight,kg | Protein supplement | | | Villages | | | |
	No	Yes	Prob	Xuan Loc	Binh Dien	Prob	SE
Final	45.9	68.2	0.001	57.9	56.2	0.49	± 1.75
ADG	0.204	0.375	0.001	0.292	0.290	0.84	± 0.0095

Economic evaluation

The economic results are shown in table 9. The value of the additional 171 g/day live weight due to supplementation was estimated at VND 1,700 for an additional feed cost of the supplement of VND 800, giving a net benefit of VND 900 per pig per day (VND 135,000 per pig for the total fattening period).

Table 9. *Effect of protein supplementation of traditional pig diets on net benefit*

	VND/pig/day	VD/cycle (150 days)
Value of extra ADG (171 g)	1700	255000
Cost of protein supplement	800	120000
Net benefit	900	135000

(1USD= 10,000 VND)

Discussion

The importance of protein in the system is shown by the low protein content of the traditional diet and the fact that a supplement of around 100 g/day of protein led to an increase in growth rate of 83%, with subsequent economic benefits.

The variation in growth rates of pigs between households is brought to light by doing on-farm trials. This phenomenon is quite common and has been reported previously from Vietnam by Dolberg (1993). The differences are normally ascribed to variations in management practices among households, which may warrant further studies for their precise identification and explanation. In his analysis of the management factor, Ostergaard (1994) points out that the interactions between farm households are an important aspect to consider, as the decisions about the management of biological or financial subsystems are strongly influenced by the social structure of the farm household and the cultural framework in which it exists. In this case, the trial intervention, which consisted of an equal supplement of protein with variation in energy supply between farmers may therefore be an important factor in explaining the differences.

The growth improvement due to protein supplementation in the present trial was similar to that for pigs fed an extra 200 g/day of protein (obtained from 500 g of a 40% protein soya bean meal) on a traditional feeding system in an on-farm experiment carried out in Colombia (Solarte and Dolberg,1994).

Conclusions

- In Central Vietnam around 95% of the pig population is kept on family small holdings and pig production plays a very important role in the economy of rural farmers.

- The common meat animal used in the villages is a crossbred between Mong Cai (MC) and Large White (LW).

- The typical diet fed to fattening pigs is based on the following ingredients in order of importance: cooked rice, rice bran, cassava meal, fresh cassava root and sweet potato leaves. An approximation of the probable nutrient supply was that protein was the main limiting nutrient with the amount supplied being less than 100 g per pig per day, in most cases.

- Fresh cassava roots, which are available at low cost (180-280VND) during the peak harvest period, offer potential as the basal energy feed for pigs in the Binh Dien and Xuan Loc villages.

- Protein appeared to be the main limiting factor in making better use of local resources for pig production. Limited supplementation of the traditional diet with the equivalent of 100 g protein/pig/day increased live weight gain by 83% and improved economic benefits to the farmers.

Acknowledgments

The authors gratefully acknowledge the help of the People's Committee and the Women's Union of Binh Dien and Xuan Loc villages, and the Women's Union of Thua thien Hue Province in developing the activities with the farmers at the research sites. Thanks to the students and members of staff of the Animal Husbandry Faculty . The help of the PRA team of Hue University is also acknowledged. Thanks are due to the Swedish International Development Agency (SIDA) for providing the financial support that made possible the study.

References

Association of Official Analytical Chemists (AOAC), 1985. Official methods of analysis. 12th ed. Washington, DC.

Chambers R., 1992a. Participatory Rural Appraisals; Past, Present and Future. Forest, Trees and People. Newsletter No.15/16:4-9

Chambers R., 1992b. Rural Appraisal: Rapid, Relaxed and Participatory. Institute of Development Studies. University of Sussex. Brighton, England.

Dolberg F., 1993. Transfer of sustainable technologies in Vietnam. Development of sustainable livestock technologies for ecologically fragile zones in the tropics. Swedish International Development Agency (SIDA) Stockholm Sweden. Internal Report

Dolberg F., 1994. On farm research: A discussion of some practical examples and procedures. Utilization of Tropical Feed Resources. Food and Agriculture Organization of the United Nations. FAO Animal Production and Health paper (In Press). Rome, Italy.

Hoang nghia Duyet and Nguyen kim Duong, 1995 Report in ACIAR-MAFI-MOET animal Science workshop "Exploring approaches to research in the animal sciences in Vietnam" City of Hue, 31 July-3 August 1995.

Le Duc Ngoan et al., 1995. Site Description, Problems, Network Analysis and Proposed R & D Agenda of Xuan Loc village, Phu Loc District, Thua Thien Hue province. A report prepared by PRA team and core-group-IDRC Project-Hue University of Agriculture and Forestry Hue, Vietnam. pp 15.

Rodriguez Lylian, Preston, T.R and Dolberg F., 1996. Participatory Rural Development: "Experiences in Binh Dien and Xuan Loc villages in Central Vietnam". Livestock Research for Rural Development. Volume 8, in press.

McCracken J., Pretty J. Conway G., 1988. An Introduction to Rapid Rural Appraisal for Agricultural Development. Sustainable Agriculture Programme. International Institute for Environment and Development. London, England.

Nguyen sinh Cuc, 1995. Agriculture of Vietnam 1945-1995. Statistical Publishing House, Hanoi, 1995. 225-231 pp.

Nguyen Van Thuong, 1992. The handbook of nutrient composition of animal feed in Vietnam. Agricultural Publishing House, Hanoi, 1992. 237-238 pp

Ospina L, Ogle R B and Preston T R., 1994. Effect of protein supply in cassava root meal based diets on the performance of growing - finishing pigs. M.Sc. thesis. Swedish University of Agricultural Sciences. Uppsala, Sweden. 9 pp

Ostergaard V., 1994. Livestock systems research-appropriate methods of the future. Integrated livestock / crop production systems in the small scale and communal farming sector. Symposia at the Department of Animal Science, University of Zimbabwe, January 26- 28, 1994

Solarte. A; Dolberg. F., 1994. Experiences from two ethnic groups of farmers participating in livestock research in different ecological zones of the Cauca Valley of Colombia Livestock Research for Rural Development. Volume 8, in press.

A Study of the Disease Situation of Rural Scavenging Poultry in the Morogoro Region of Tanzania

Anders Permin

Centre for Experimental Parasitology,
The Royal Veterinary and Agricultural University,
1870 Frederiksberg
Copenhagen, Denmark
E-mail: ape@kvl.dk

Abstract

A cross-sectional prevalence study was conducted in the Morogoro Region, Tanzania during two periods of 3 months from February to April 1995 and from October to December 1995. In total 600 adult chickens were examined for gastrointestinal parasites and antibodies against Newcastle Disease virus and *Salmonella gallinarum/pullorum*. One hundred percent of the animals harboured endoparasites. In total 29 different species were identified. In total 28.5% of the animals had antibodies against *S.gallinarum/pullorum* and 4.2% had antibodies against Newcastle Disease virus. The results indicates a heavy disease pressure on village poultry which might subsequently have an impact on productivity.

Introduction

The majority of the of the 23 million Tanzanian citizens live in rural areas (Kabatange & Katule, 1990). The majority are subsistence farmers with only a small piece of land. Most of these farmers keep poultry but some also have cattle, sheep, goats or pigs. In contrast to cattle, poultry is kept by nearly all social groups including the poorer and landless families. Religious taboos are seldomly related to poultry. In the rapid growing towns the demand for meat is high. Most of the slaughtered cattle, sheep and goats are sold to the markets in the urban areas. Thus the availability of meat is limited in rural areas often leaving poultry as the only animal protein source. Furthermore poultry and eggs are often used as petty cash for small daily needs. Therefore poultry plays an important role in the production systems and for family life in most communities of smallholder farmers (Minga et al., 1989).

In total, it is estimated that smallholders and villagers keep 20 million chickens in Tanzania. Additionally, approximately 3 million commercial chickens are kept in suburban areas (Minga et al., 1989; Kabatange & Katule, 1990). In the villages the poultry is left scavenging around the house during daytime to obtain what feed they may be able to get from the environment often in the form of kitchen offal, insects and seeds (Kabatange & Katule, 1990). The feed is sometimes supplemented by commercial feed and maize bran. At night they are kept in sheds or in the house together with the family for security reasons, and also to secure the egg collection. Diseases are rarely controlled or treated.

By their mode of life on free range, and scavenging habits traditional village poultry are in permanent contact with other flocks, soil and insects. Soil and insects can act as reservoirs or vectors for a range of bacterial and helminth diseases. Matthewman (1977) estimated a loss of animals in scavenging flocks in the range of 80 - 90 % during a year. The daily contact with soil and insects could explain the high frequency and wide distribution of some common diseases thus leading to the high mortality rate. Furthermore, the close contact to other animals favours the transmission of diseases.

For the same reasons, the aim of the present study was to survey the prevalence of viral (Newcastle Disease), bacterial (Salmonella gallinarum/pullorum) and helminth infections in local scavenging poultry.

Materials and methods

Study area and sampling

The poultry population under study consisted of rural scavenging chickens in the Morogoro Region, Tanzania. The region was divided into three areas. The chicken were collected from two different, randomly selected villages from each area. The villages are Langali & Nyandira, Melela & Fulwe and Kiroka & Changa (Table 1). Each set of villages were considered as one population. Three hundred chickens were collected during the rainy season 1995 and other three hundred chickens were collected during the dry season 1995. In order to avoid confounding from age only adult chickens were used for the survey. None of the animals had been vaccinated against any disease.

Table 1. *The selected villages in the Morogoro Region including a chicken population census*

Villages	Human population	Chicken population
	(census February 1995)	
Melela and Fulwe	7990	6227
Changa and Kiroka	6578	7480
Nyandira and Langali	5130	2807

Clinical examination

The chickens were examined for the presence of clinical diseases in the laboratory at the Faculty of Veterinary Medicine, Sokoine University of Agriculture. Blood was taken from each chicken at the time of decapitation and kept in the fridge overnight before the blood was centrifuged and the serum transferred to new tubes.

Parasitological examination

The entire gastrointestinal (GI) tract including the oesophagus was collected from each animal. The GI was divided into four sections and opened in a longitudinal section before being washed carefully through a 100 μm test sieve. The mucosa was scraped in order to collect the helminths embedded in the mucosa. The content was examined under a stereo microscope and all endoparasites were counted and stored in 70% alcohol for further examination under a light microscope. All parasites were identified using the helminthological keys by Soulsby (1982) and Calnek et al. (1991). The parasitological examination is described in detail in a separate paper.

Serological analysis for antibodies against S.gallinarum and S.pullorum

The Rapid Plate Test (RPT) was used to detect antibodies against *S.gallinarum* and *S.pullorum* (Calnek et al., 1991). Serum samples giving any visible agglutination were considered positive.

Serological analysis for antibodies against Newcastle Disease virus

Antibodies against Newcastle Disease virus were determined by using the haemagglutination inhibition test (Allen and Gough, 1974). A serum sample was considered positive when the titre was equal to or greater than $\log_2 1$.

Analytical methods

The prevalence was calculated according to the method described by Thrusfield (1995). The statistical software package Statistix was used for all analyses.

Results

General disease status

No clinical diseases were observed during the visits to the villages, except for ectoparasites and a general poor body condition of the animals. Interviews with farmers and veterinary assistants in the villages indicated a general high mortality, but no exact figures were available.

Table 2. Parasitological and serological findings

	Melela &Fulwe n=200	Langali & Nyandira n=200	Kiroka & Changa n=200	Total n=600
endoparasites:	100%	100%	100%	100%
T. americana	60%	49%	63%	57%
H.gallinarum	59%	88.5%	82.5%	76%
A. galli	26%	32.5%	32.5%	30%
R.echinoboitrida	47%	36.5%	43.5%	44%
NDV	1.5%	1.5%	9.5%	4.2%
S.gallinarum & S.pullorum	28.5%	33.0%	24.0%	28.5%

Parasitological findings

All chickens harboured one or more species of endoparasites the highest being 14 species in one animal. In total 29 different endoparasites were identified. The most common nematodes were *Tetrameres americana* (57%), *Heterakis gallinarum* (76%) and *Ascaridia galli* (30%). The most common cestode was *Raillietina echinoboitrida* (44%). No trematodes were found.

Serological findings

A very low percentage of the animals had antibodies against Newcastle Disease virus (NDV). The overall prevalence was 4.2% with 1.5% in Melela & Fulwe and 1.5% in Langali & Nyandira. There was a significant higher (P<0.05) prevalence of 9.5% in Kiroka & Changa compared to the four other villages.

The overall antibody prevalence of *S.gallinarum/pullorum* was 28.5% . The antibody prevalence was 28.5% in Melela & Fulwe, 33.0% in Langali & Nyandira and 24.0% in Changa & Kiroka. There was no significant (p>0.05) difference between any of the areas.

Discussion

In the present study it was shown that the prevalence of parasitic diseases was high and that there was no significant difference between the three areas, e.g. all the chickens were harbouring parasites, although there was some species variation. Furthermore, a high prevalence of *S.gallinarum/pullorum* was seen in all three areas. Antibodies against Newcastle Disease virus was only seen in 4.2% of the animals with a significant higher prevalence in Changa & Kiroka. Although Newcastle Disease is believed to be the most frequent "killer" of scavenging poultry the prevalence is very low. This might be due to either that the surveyed population consists of survivors from the previous outbreak or adding of new animals to the population since last outbreak.

These results compare with other surveys in Africa although knowledge on the prevalence and significance of diseases in poultry in Africa seems to be rather limited (Pandey et al. 1992). Chryosostome et al. (1995) found that 65% of the chickens in Benin were seropositive for Newcastle Disease virus and 9.6 and 62%, were seropositive for *S.pullorum* and *S.gallinarum*, respectively. Bouzoubaa et al. (1992) tested 23.5% of village chickens in Morocco seropositive for *S.gallinarum* and 6% seropositive for *S.pullorum*. In Mauritania Bell et al. (1990) found that 4.6% of the chickens were seropositive for Newcastle Disease virus and 5.8% were seropositive for *S.gallinarum/pullorum*.

Saad et al. (1989) found that of 123 Sudanese chickens 77.3% harboured endoparasites. Of these 41.0% only hosted one species and that in total only 5 helminth species were found in the chickens. In the Mwanza Region in Tanzania Msanga & Tungaraza (1985) found that 95% of the chickens had endoparasites. In total 100 species were detected. Nine species were recorded by Otaru and Nsengwa (1985) in the Mtwara Region of Tanzania. A study undertaken in Dar es Salaam from 1970 to 1985 showed a single helminth infection in 53.8% of the examined intestinal tracts (Msangi & Mbwambo, 1988).

Parasitic diseases are fairly straightforward to detect, e.g. the direct presence of helminths in the gastrointestinal tract or detection of parasite eggs in the faeces. In the case of viral and bacterial diseases the situation is different, but in the absence of vaccination, detection of antibodies in the serum can be taken as indicative of a history of infection. With the above results there is no doubt that diseases play an important role in village poultry. All three diseases are known to cause mortality, Newcastle Disease being the most severe with mortality rates close to 100%. *S.gallinarum/pullorum* can cause mortality in the range of 50-100%, low hatchability of eggs and weight loss, whereas helminth infections in general cause unthriftiness and reduced productivity.

Whether there is an additive effect of the prevalence of the three diseases is unknown but the fact that all three diseases are present indicates a heavy disease pressure on village chicken. This might be followed by mortality and loss of productivity. The exact impact of each of these diseases on weight gain, egg production and mortality await further studies.

Acknowledgements

The Council for Research in Developing Countries (RUF/Danida) is thanked for the financial support. The Sokoine University of Agriculture (SUA), Morogoro, Tanzania is kindly thanked for making it possible to conduct the study. The assistance from Academic and Technical staff at the Faculty of Veterinary Medicine, SUA is highly appreciated.

References

Allan, W.H. and Gough, R.E. (1974) A standard haemagglutination inhibition test for Newcastle Disease. *Vet.Rec.*, 95: 120-123.

Bell, J.G., Kane, M., Le Jan, C. (1990) An investigation of the disease status of village poultry in Mauritania. *Preventive Veterinary Medicine* 8, 291-294.

Bouzoubaa, K. Lemaiinguer, K., Bell, J.G. (1992) Village chickens as a reservoir of *Salmonella pullorum* and *Salmonella gallinarum* in Morocco. *Preventive Veterinary Medicine* 12, 95-100.

Calnek, B.W., Barnes, H.J., Beard, C.W., Reid, W.M., Yoder, H.W.Jrr. (1991) *Diseases of poultry.* 9th edition. Wolfe publishing.

Chrysostome, C.A.A.M., Bell, J.G., Demey, F., Verholst, A. (1994) Seroprevalences to three diseases in village chickens in Benin. *Preventive Veterinary Medicine* 22, 257-261.

Kabatange, M.A. and Katule, A.M. (1990) Rural poultry production systems in Tanzania. In: *African Network on Rural Poultry Development.*

Matthewman, R.W. (1977). A Survey of Small Livestock Production at the Village level in the Derived Savanna and Lowland forest Zones of South West Nigeria. University of Reading. Department of Agriculture and Horticulture. Study No. 24. ISBN No. 0-70490242-7. pp 40-41.

Minga, U.M., Katule, A.M., Maeda, T., Musasa, J. (1989) Potential and problems of the traditional chicken industry in Tanzania. Proceedings of the 7Th Tanzania Veterinary Association Scientific Conference. Arusha Dec. 1989. Vol. 7, p 207-215.

Msanga, J.F. & Tungaraza, R. (1985) The incidence of external and internal parasites of indigenous poultry in Mwanza municipality, Tanzania. *Tanzania Veterinary Bulletin* 7, 11-14.

Msangi, A.R. & Mbwambo, H.A. (1988) Investigation of poultry helminthiasis in Dar es Salaam from 1970 to 1985. *Tanzania Veterinary Bulletin* 8, 29-32.

Pandey, V.S. (1992) Epidemiology and Economics of village poultry production in Africa: Overview. In: Pandey, V.S. & Demey, F. (Eds) *Village poultry production in Africa.* Rabat, Morocco May 1992.

Saad, M.B., El Sadig, A.A. and Shammat, A.M. (1989) Helminth Parasites of the local Breed of poultry in (Kordofan Region). *Sud. J. Vet. Sc. Anim. Husb.* 28 (2). 54-55.

Soulsby, E.J.L. (1982) *Helminths, Arthropods and Protozoa of Domesticated Animals.* 809 pp. Baillière Tindall. London.

Thrusfield, M. (1995) *Veterinary Epidemiology.* 479pp. Second edition. Blackwell Science.

Recycling and Energy

Wastes as Valuable Resources: Mandatory Recycling for Economic Development

Professor George Chan

Environmental Engineering Consultant,
United Nations University, ZERO-EMISSION PROGRAMME.
E-mail: 100075.3511@compuserve.com

Summary

In the introduction it is emphasised that the objective of recycling wastes as valuable economic resources should be the policy of all countries, rich or poor. The paper goes on to discuss four types of wastes characterised by being rural, urban, agro-industrial and industrial. It is concluded that many economic and environmental problems can be solved effectively and efficiently if the polluters recycle what they call wastes as renewable resources in small individual integrated farms, instead of treating them as useless pollutants.

Key words: Wastes, recycling, integrated farming.

Objectives

The objectives of recycling all wastes as valuable resources for economic development should be the mandatory policy of every country, rich and poor alike, for true as well as viable sustainability, and affordable pollution control. It is the only effective and efficient solution to the existing economic and environmental problems that every government is facing all over the world. Almost every country has enough resources to meet most of the basic and essential needs of its people, but none can afford to waste them unnecessarily.

Some industrialised countries have to import resources from others to satisfy their industrial requirements, and it is much more important for them to recycle all the wastes as much as possible in order not to deprive the exporting countries of their much needed resources. So waste recycling should be everybody's business.

Wastes and poverty

No matter where in the world we have to deal with many kinds of wastes which usually create pollution problems, some of which can be dangerous and kill wildlife or even people. Some wastes are avoidable , and the best answers to these pollutants are NOT to have them, but others are not. They can be ignored in rural areas where nature can deal with them within certain limits, but in any other situation, humans must be responsible for the mess they create. To illustrate how humans have been neglectful, especially in so-called modern societies, most of the world is in a huge mess and it is getting worse all the time. The sick environment is the result of accumulation of wastes, and yet ALL unavoidable wastes can be recycled with available economic technologies to benefit the whole of mankind, and not only alleviate but also eliminate poverty across the world. This paper does NOT deal with global pollution such as aircraft emissions and wars.

Types of wastes

For the right perspective, let us divide the wastes of a peaceful human society into 4 physical categories and find efficient and economic solutions for their ecological recycling to benefit every level of that society:

1. Rural
2. Urban
3. Agro-industrial
4. Industrial

1. Rural wastes

They consist mostly of crop residues, livestock manure, macrophytes, household refuse, sludge and human excreta. All these wastes can be recycled on site, as space is not a problem in the countryside. The best situation is where the family lives and works in an Integrated Farming System (IFS) of livestock, aquaculture, agriculture and produce processing, with the residues of one process used as input for subsequent ones. The most favourable and profitable environmental conditions for such a system are in the wet tropics, which can bring a highly rewarding income to the family with the minimum of investment and even work, as it can provide most of the means of production such as fertilizer, feed and fuel on site. The ecological balance is almost perfect, with only carbon dioxide as an undesirable emission to the atmosphere, but this is well compensated by the carbon sink provided by the abundance of crops -- figure 1.

Such systems have been in continuous operation for centuries in certain Asian countries, and are responsible for their continuous rural and agro-industrial development success to start with, and their spectacular industrial expansion in recent years. There is also enough recycling capacity in the integrated farms, or space in other decentralized composting or sewage

Figure 1. Wastes as valuable resources

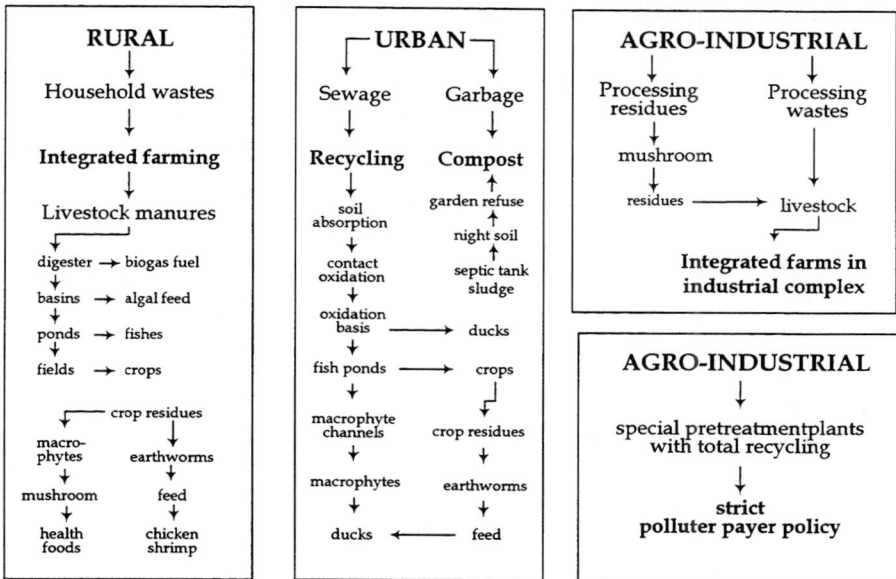

stations as shown below, to deal with wastes coming from the urban, agro-industrial and even industrial sectors provided that they do not contain any non-biodegradable or toxic chemicals.

2. *Urban wastes*

The absence of space in urban areas makes it impossible to deal with both solid and liquid wastes generated by the dense population, which must pay for the transport of their wastes to the rural areas. In small towns with scarcity of water we should use dry toilets, nightsoil cans, improved pits and septic tanks for human excreta and possibly sludge, provided that the appropriate trucks are available to empty them when necessary and take them to proper composting stations strategically located in the nearby rural areas. These stations also deal with other urban wastes such as municipal garbage, and commercial and garden refuse, which are all recycled as compost.

The quality of the compost depends on the system used. The conventional one of turning the mixture over manually or mechanically for aeration produces poor compost, because over half the nitrogen content is lost to the atmosphere as gases, and is not popular with farmers who still have to add chemical nitrate. Composting which uses special bacteria to accelerate the transformation is of better quality but is much more expensive. The best system could be one presently using a controlled environment, preferably located next to an integrated farm to use the "free" energy, that retains the nitrogen while taking only 3-4 weeks for the total transformation process, with sorting out of plastics, metal cans and bottles done afterwards. These inert materials, coming out of the composting process unchanged, have also created new recycling industries in the countryside. Any leachate is collected and treated separately in an adjacent integrated farm.

In towns with adequate water supply and most cities, the human excreta and sludge are transported in a sewerage system with gravity flow from every catchment area to a decentralized sewage treatment plant in the surrounding countryside. It uses innovative biosystems with biomass nutrient recycling at integrated farms nearby to utilise the final effluent without any discharge into the environment.

3.   Agro-industrial wastes

Every country should aim at adopting agro-industries to meet the immediate needs of its population first, before thinking of producing goods for other countries. Every factory should only be located in the countryside and given enough space to deal with all its processing residues and wastes on surrounding or nearby individual integrated farms, with the required farm infrastructure provided by the factory itself and every farm preferably run by the spouse of a factory worker. The integrated farmers receive their daily share of both solid and liquid wastes from the factory. The solid wastes are used as substrate for mushroom or earthworm culture, breaking down the lignocellulose to improve the digestibility, before feeding the residues to livestock. The liquid wastes are used to wash the pens or, after treatment in oxidation ponds, used to fertilize fish and macrophyte ponds. No artificial feed is given to fish which eat natural plankton only.

The gaseous wastes should be recycled and reused on the premises or stored in proper cylinders for sale. As a typical example, the fermentation gases from a brewery can be recovered and used in their bottling plants or mixed with their draught beer, or bottled for sale as dry ice or for the soft drink or soda industry.

The agro-industries include commercial farms with or without processing plants for preservation or added value, abattoirs, canneries, and various factories producing flour, juices, liquors, sauces or other foods and goods from the primary farm produce.

4.   Industrial wastes

The siting of industries in urban areas should be totally prohibited because of scarcity of space and their potential risks to densely populated communities. They should only be allowed in industrial parks, with the workers' flats built on the outskirts. All industrial wastes are pre-treated in special plants approved by the local authority, with the gases emitted into the atmosphere and effluents discharged into the sewerage system strictly monitored daily or even more often to prevent contamination of air, soil, water and food. All the inhabitants also undergo constant health checks. The mandatory policy of "Polluter Pays" should apply totally and mercilessly to all the industrialists concerned.

More importantly, the toxic industries should be isolated in specific areas far from human habitation, and every factory should have enough space to contain any spillage if one of its containers should split open and to treat fully all its wastes on site, without releasing any of its pollutants outside its premises. Moreover, no such toxic industry should be allowed to transfer its dangerous operations to other countries.

Conclusion

World-wide, big and medium farms use many costly inputs such as fossil fuels, artificial feeds, chemical fertilizers and toxic pesticides to boost up production, but they are not viable economically and require substantial government or other subsidies to survive, while the environment is continuously being degraded because of increasing waste pollution. Such economic and environmental problems can be solved effectively and efficiently if the polluters recycle what they call wastes as renewable resources in small individual integrated farms, instead of treating them as useless pollutants.

The Introduction of Low-Cost Polyethylene Tube Biodigesters on Small-Scale Farms in Vietnam

Bui Xuan An[*1], Thomas R. Preston** and Frands Dolberg***

* *University of Agriculture & Forestry,*
Thu Duc, Ho Chi Minh City, Viet Nam
E-mail: an%sarec%ifs.plants@ox.ac.uk

** *University of Agriculture & Forestry,*
Thu Duc, Ho Chi Minh City, Viet Nam
E-mail: thomas.preston%sarec%ifs.plants@ox.ac.uk

*** *Department of Political Science,*
University of Arhus, 8000c, Denmark
E-mail: frands@po.ia.dk

Summary

In order to evaluate polyethylene tubular digester development in Vietnam, interviews were carried out in Thuan An district and at two extension centers. Data of design parameters, gas production, economic aspects, farmers' participation, technical problems and methodologies of biogas development were collected. The technology was appealing to the rural people because of its low costs, fast payback, simplicity and positive effect on pollution. The finding pointed to the importance of farmers' participation for technical feedback, plant maintenance, plant repair and teaching of other farmers. The dissemination of the technology needs the selection of real farmers with high fuel demands as demonstrators, the support of a credit system for the poor farmers and strengthening farmer-extension-scientist relations. Follow-up research should be focused on on-farm studies, particularly the use of the slurry.

Key words: Plastic tube biodigester, Low cost, farmer impact, women's lives, sustainable development.

Introduction

In recent years, the conversion of biomass to methane for use as an energy source has excited interest throughout the world. Biogas digestion was

---

[1] Corresponding author

introduced into developing countries as a low-cost alternative source of energy to partially alleviate the problem of acute energy shortage for households. However, few farmers used biogas in practice. The poor acceptability of the digesters is believed to be due mainly to the high cost of the digesters, difficulty in installing them and difficulty in getting spare parts. The biogas programs developed quickly in some developing countries only under substantial support from governments and aid agencies (Gunnerson,, 1986; Kristoferson and Bokalders, 1991; Marchaim,, 1992; Karki,, 1996). Besides, the replacement of worn-out parts posed another technical problem, in addition to the fact that such spare parts were not always locally available.

Many developing countries, such as Colombia, Ethiopia, Tanzania, Vietnam, Cambodia and Bangladesh promoted the low-cost biodigester technology aiming at reducing the production cost by using local materials and simplifying its installation and operation (Solarte, 1995; Chater, 1986; Hieu et al, 1994; Sarwatt, 1995; Soeurn, 1994; Khan, 1996). To this end it was decided to use a continuous-flow flexible tube biodigester based on the bag digester model as described by Pound et al (1981) and later simplified by Preston and co-workers first in Ethiopia (Preston, unpubl.), Colombia (Botero and Preston, 1987) and later in Vietnam (An et al, 1994). Within three years, more than 800 polyethylene digesters were installed in Vietnam, mainly paid for by farmers (An and Preston, 1995).

The objectives of this study were to assess the effects of low-cost biodigesters in small farms in Vietnam and to identify experiences, effects, constraints and problems associated with this technology.

Materials and methods

Parameters of digesters
Data of design parameters and cost were collected from 194 biodigesters installed from April 1993 to December 1995 around Ho Chi Minh City.

Influence of biodigesters on the farmers' lives
Open-ended interviews (Casley and Kumar, 1988) were carried out at 35 small farms with biodigesters in the Thuan An district, 40 km north of Ho Chi Minh City. In the selected area there were both upland and lowland ecosystems with 1800 mm rainfall, an average temperature of 27°C with small difference between seasons, sugarcane and cassava as the main crops and pigs as the most common animal (Thuan An people's committee report, 1994; unpubl.). The questionnaire for farmers and their wives contained questions such as:

- Use of the biodigesters: cooking parameters, economics of biogas uses, uses of effluent for fertilizer and other uses.

- Farmers' participation: where did they get the information, payment of the digester, conditions before and after biogas use, opinions and suggestions.

- Digester life: technical problems, when did problems occur, who fixed them, how was it done, what materials were needed.

Input and output from digesters

- Manure inputs were weighed directly on most of the farms. In cases when weighing was impossible, the amounts were estimated by comparing with others farms with similar numbers and ages of animals. The amounts of water were weighed directly or estimated from multiplying water speed and washing time.

- Gas production: by industrial gas-meter model k-875-1, Yazaki Keiki Co., Japan connecting to the gas outlet of digesters.

- pH: by digital pH-meter.

- Dry matter (DM) of the manure was measured by drying at 105°C until constant weight in a forced drought oven.

- COD: Chemical Oxygen Demand (the amount of oxygen consumed for the oxidation of the reductive substances contained in a liter of liquid waste sample), using standard methods (HMSO, 1986).

The data and samples were taken for two consecutive days in the rainy season (from May to October) from 31 biodigesters randomly selected around Ho Chi Minh City. The average temperature in the area was 27.5°C and temperature difference between day and night was 10°C.

Interviews on extension and demonstration farms

The topic-focused interviews (Casley and Kumar, 1988) were carried out with two extensionist groups, one at BaVi, HaTay province and the other at ThuDuc, Ho Chi Minh City. The topics concerned selection and status of demonstration farms, list and rank of problems in the biogas development. The data were then verified by farmer's interviews and field observations.

Results and discussion

Design parameters and cost of digesters around Ho Chi Minh City

The data are presented in table 1. The average length of the digesters was 10.2 m with an estimated digesta volume of approximately 5.1 m³ (length x 0.5 m³). The material cost was slightly more than US$25 for a family digester.

Table 1. *Mean values for some design parameters and cost of 194 digesters installed around Ho Chi Minh City*

	Mean	Range
Length (m)	10.2	4 - 30
Digester liquid volume (m3)	5.1	2 - 15
Distance to kitchen (m)	23	8 - 71
Material cost (US$)	25.4	14 - 82
Time to first gas production (days)	17	1 - 60
Digesters in rural areas(%)	91	
Floating digesters (%)	5	

In most developing countries, when the subsidies from governments are reduced the number of plants built each year falls dramatically (Ellis and Hanson, 1989; Qiu et al 1990; Desai, 1992; Karki et al, 1994). The most important problem in biogas programs in developing countries has been the prices of digester plants. For exempla: The price of a concrete digester plant installed for an average family in Vietnam varied from 180 to 340 US$ (Thong, 1989). This size of investment is considered unaffordable by average farm families (An et al, 1994). Chinese designers tried to reduce the cost of red-mud digesters to 25-30 US$/$m^3$ (Gunnerson and Stuckey 1986), but it was still high in comparison with the polyethylene, digesters (5 US$/$m^3$). This is obviously one important feature which makes the polyethylene digesters attractive and no farmer in the present study complained about the price.

Among the polyethylene digesters installed, 5% of them were floated in ponds adding an innovative feature to the development. According to Khoi et al (1989), in the Mekong Delta where most land is low, the application of concrete digesters was very difficult especially when the water level went up. The floating digesters solved this problem and as they also required little space they were very well suited for low-lying areas. More than 90% of the plants were installed in rural areas indicating the good fit of the technology under these conditions.

Influence of biogas on small farms in the ThuanAn district

The effects of the introduction of digesters on small farms are presented in tables 2-4.

Table 2. *Economic aspect of biogas introduction in 31 small farms in ThuanAn district, Vietnam*

	Mean	Range
Cooking time (hour)	4.4	1 - 9
Fuel saved in cooking (US$/month)	6.5	1.8-13.6
Biogas plant cost (US$/unit)	34.8	18 - 53
Number of pigs/farm	10.7	0 - 40
Payback time (month)	5.4	2 - 19

Table 3. *Farmers' participation and opinions on plastic biodigesters in ThuanAn district, Vietnam*

	Alternatives	No.[a]
Getting first information from	Neighbours or relatives	32
	Mass media	3
Payment of the digester plants	Farmers paid totally	33
	Partially (demonstration)	2
Using slurry for	Plants	3
	Ponds	3
	Nothing	31
Status of gas production	Enough gas	26
	Little gas	5
	No gas	4
Advantages of biogas	Saves money	34
	Less pollution	35
	Easy cooking	35

a) Number of farmers

Table 4. *Input and output of 31 digesters working at small farms around Ho Chi Minh City, Vietnam*

	Mean	Range
Size of family	5.9	3 - 12
Manure loading (kg/d)	16	2 - 27
Ratio Water/manure	5.1	2.9 - 8.1
Loading rates (kgDM/m^3)	0.7	0.1 -1.2
Temperature of loading(°C)	26.4	25.7 - 28.5
Temperature of effluent (°C)	27.0	26.0 - 29.1
pH of loading	6.7	6.4 - 7.1
pH of effluent	7.2	6.8 - 7.5
Gas production (l/unit/day)	1235	689 - 2237
Vol. Gas/capita (l/person/day)	223	68 - 377
Methane ratio (%)[a]	56	45 - 62
COD[b] of loading (g/litre)	35.6	22.4- 46.0
COD of effluent (mg/litre)	13.5	8.8 - 23.9
COD removal rate (%)	62	42 - 79

[a] From 9 digesters

[b] COD = Chemical Oxygen Demand (the amount of oxygen consumed for the oxidation of the reductive substances contained in a litre of liquid waste sample by a strong oxidizer).

According to the annual report of ThuanAn people's committee (1994, unpubl.), most of the farms with biodigesters belonged to the medium income group (sufficient food all year round). In this group animal production is a very important component the farming systems and a sufficient number of animals is important for the dissemination of biodigesters. The expense for the digester plant was paid back within slightly more than 5 months, so most of the farmers found a great benefit from installing them.

Among 35 farmers interviewed, four of them were poor (not enough food in certain months) The most important thing for them is food and they could not afford a sufficient number of animals for feeding manure to the digester. They wanted to borrow money to be able to raise animals. Four farmers had no gas when the interview was carried out. Three of them did not have animals because they found raising animals unprofitable if they had to borrow money from local lenders at 5-10% monthly interest. This was an important aspect as especially resource-poor farmers cannot support the digester installation and keep animals, although they know the advantages of biogas. The average manure DM percentage was 25% and the loading rates ranged from 0.1 to 1.2 kg DM/m^{3}digester liquid volume. Previously, animal manure was an environmental problem in villages in the district, mainly in crowded and lowland areas where it caused pollution in the air, water and soil. After installation of the digesters, all 35 families recognized better environmental conditions, less smell, fewer flies, cleaner waste water, etc. Summarizing details of the experiments conducted with pig slurries, Pain et al (1990) concluded that the digestion reduced smell emission considerably although not completely. According to the women who were responsible for food preparation, use of biogas meant that they could attend to other work, while cooking. This is in contrast to the situation when they use solid fuels such as fire wood, which require much closer supervision. The women stressed that they could now cook in a clean environment free of smoke. Their pots and pans were clean and they did not have to spend time on tedious cleaning. They stated that they could cook all food items on gas.

Input and output of small farm digesters

In our study, biodigestion decreased COD from 35610 mg/lit in the inlet to 13470 mg/l in the effluent indicating a process efficiency of 62% (COD removal rate). The volume of gas per capita per day was enough for cooking three meals and was about 200 litres. The loading rates were low and gas production could be improved by increasing the amount of manure fed to the digesters. Beside cooking meals, using gas, five farmers cooked animal feeds, three made wines, one made cakes and two prepared tea and coffee in their cafeterias. This demonstrates that there are several reasons for uptake, as discussed by Dolberg (1993).

Technical problems with the digesters

Main causes of damage to the digesters were the sun, falling objects, people and animals (table 5).

Table 5. *Technical problems with polyethylene tube digesters in ThuanAn district, Vietnam*

Damaged by	Location of damages			Total
	Digester	*Reservoir*	*Others*	
Sun	4			4
Falling objects	2	1		3
People	2	1		3
Animal	1	1		2
Material quality	1		1	2
Wind		2		2
Overloading	1			1
Total	11	5	1	17
Self help[c]	6	5	1	12

[c] Farmers fixed digesters by themselves

In cases when the digesters had been totally exposed to the sun, the plastic film was broken after 2 years. Seven digesters had films older than 2 years and four of them had been changed by technicians or farmers. The material cost for changing was about 15 US$ and one work-day was needed. Most digesters installed during 1995 were protected by roofs made from local materials, mainly palm leaves. Also, simple fences were made around the digesters to prevent damage from animals or people.

In figures 1 and 2, the analyses of technical problems are presented. Slightly more than 40% of the biodigester plants had problems especially with the plastic tubes. An interesting observation was that in 70% of the cases (12/17) the farmers could correct the problems by themselves and only in 30% of the cases did they need help from technicians. Repairs were mainly simple and farmers could teach each other. The first farmers who had digesters installed more than 2 years ago needed help from technicians, while farmers who had installed their digesters within the last year could resolve their problems by themselves. They had received information, experience and guidance from their neighbors. With increasing age of the plants more problems would be expected. Nevertheless, as more plants are installed in a village there would be more experienced farmers to do repairs and the help required of technicians would therefore be less. Also if there are good written instructions summarizing experiences from users, demand for the technical personnel will be less. This result shows that technical problems with the polyethylene digesters were resolved more easily than with other materials, such as concrete, steel and red mud. In

Figure 1. *Technical problems and their solutions according to damaged places in plastic tube digesters in ThuanAn district, Vietnam*

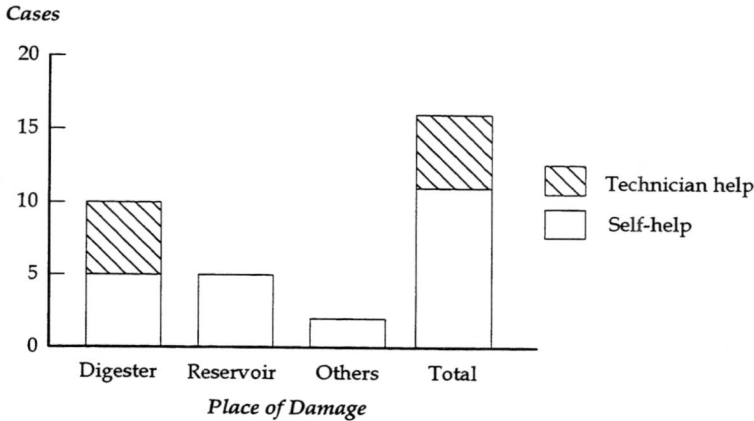

Figure 2. *Technical problems and their solutions according to age of plastic tube digesters in ThuanAn district, Vietnam*

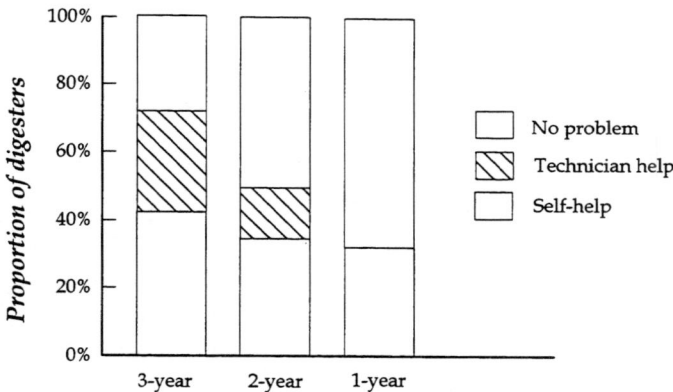

many developing countries, the biogas programs have failed because of inefficient maintenance due to lack of technical personnel (Kristoferson and Bokhalders, 1991). When the farmers do not take care of the digesters, only a small problem can cause gas production to cease, making the farmers disappointed. The participation of the farmers has played an essential role in the dissemination of the technology. Some digesters which were not studied were installed by farmers themselves in the district.

Problems in the extension of biogas technology

There are some constraints and problems in the dissemination of biogas technology in developing countries. The question is how to solve them and what priorities to make. Some of the biggest problems at Bavi and Thuduc are listed in table 6 in order of priority. In Bavi, the most important problem was inappropriate selection of demonstration farms (where the main income was not from farming activities) which resulted in low attention by the farmers to the operation of the plants (table 7).

Table 6. *Comparison of demonstration farms and digesters installed at two extensionist groups in Vietnam*

Item		BaVi	ThuDuc
Demonstration farms:	Total participants	7	8
	Main income from agriculture	1	6
	Government employees	6	0
	Enough fire wood	3	0
	Enough wood & lack of manure	1	0
Demonstration digesters:	Still working after 2 years	2	6
	Enough gas produced	0	6

Table 7. *Problems in plastic tube biodigester development in order of priority according to the extensionists in two extension centres in Vietnam*

Priority	BaVi	ThuDuc
1	Extension methodology	Investment of poor farmers
2	Installation technology	Plastic quality
3	Unstable animal production	Unstable animal production
4	Investment of farmers	Technical maintenance
5	Plastic availability	Efficiency of gas production & use
6	Plastic quality	

The selection of demonstration farms is expected to promote the degree of farmer participation in digester introduction and provide technical feedback. In the first year, the Thuduc group installed 60 digesters with the motto of "farmers pay" in order to strengthen their motivation. Full-time farmers (most activities are on-farm) with high demands for fuel were selected as demonstrators. They paid more time to their farming activities and were more motivated to look after the digesters carefully, and treated the digesters as carefully as animals. Several meetings between farmers and extensionists were held. Many small but important innovations were learned from farmers when extensionists spent time working and discussing with them. After 3 years more than 200 units have been installed by the Thuduc group and the technology has been improving.

Although the biogas technology has been developing steadily around Ho Chi Minh City, there are still many questions from farmers, such as amounts of loading of on-land and floating digesters, how to prolong plastic life under farm conditions, how to use slurry for crops if the fields are far from the digester, incorporation of fish ponds and other uses of the gas, etc. The other problems, such as investment problems of poor farmers, variable animal production and plastic quality were also mentioned. Many aspects involved in the technology should be studied carefully under real farm conditions. Sustainable use of natural renewable resources will be facilitated when the feed is grown, the animals are fed and the excreta are recycled on the farm in ways that reduce the use of imported inputs including energy (Preston, 1995a). This idea has been displayed in integrated farming systems in many developing countries in South East Asia. In this respect, Dolberg (1994) pointed out the need to develop the ability of researchers to be sensitive to the farmer's perspective and convert to feedback from farmers into hypotheses for research and new possible solutions, which would then have to go through the same iterative process of trial and error. On-farm work will accelerate the research process and make it move faster than if the scientists confine themselves to the research station and laboratory. In order to realize this process, the professional agriculturists in developing countries should be re-trained for sustainable tropical agriculture in their home countries (Preston, 1995b).

Allowing some time for the farmers to "digest" the biodigester technology is essential. It took about 3 months from the time the first digester was installed as a demonstration to the moment when the first digester was purchased by a farmer. It took an additional 6 months for the first digester to be installed by a farmer by himself (An and Preston, 1995). It is essential to strengthen the relationship between farmers and scientists in order to receive the feedback. According to Dolberg (1995a), an important condition for success of that approach is that the leading scientists take it seriously and are prepared to spend time in the field with farmers, showing how to deal with feedback from farmers and to convert it into researchable problems.

It should be noted that the technology of the polyethylene tubular digesters is not fully developed and the technology depends very much on natural, as well as socioeconomic conditions. Therefore, it is necessary to study on-farm conditions in different areas to improve the technology. An exchange of experiences between institutions should take place which should improve results. Communication between the institutions and between technical personnel is not sufficient. A network of all institutions and people involved in the biogas technology should be built over the country and overseas. Some recommendations for future developments and research of biogas programs in Vietnam based on the foregoing criteria were made by Dolberg (1995b).

Conclusions and recommendations

The polyethylene tubular film biodigester technology is a cheap and simple way to produce gas for small-scale farms in Vietnam. It is appealing to rural people because of the low investment, fast payback, simple technology, positive effects on the environment and women's lives in rural areas. The farmers' participation is essential in technology feedback, maintenance, repair and education of others farmers. The extension of the technology requires the farmers' motivation which can be ensured by selecting full-time farmers with high fuel demands for demonstrations, supporting credit systems to poor farmers and strengthening farmer-extension-scientist relations. In future, research should start by involving farmers, creating feedback from the farmers and letting this feedback serve as a foundation for the formulation of research problems. One immediate problem to attend to is the use of the slurry.

Acknowledgments

The authors gratefully acknowledge the help of farmers in ThuanAn district, Binh, Suc and staff of the Goat and Rabbit experimental station in BaVi for their valuable participation. We also would like to thank the Swedish International Development Authority for funding this study. Appreciation is extended to the staff of the Animal Nutrition Department of the University of Agriculture and Forestry, HoChiMinh, Vietnam and the Department of Animal Nutrition and Management of the Swedish University of Agricultural Sciences in Uppsala, Sweden who helped in any way. Special thanks to Duc Anh, Khang, Trung, Minh, and Diep for their cooperation.

References

An Bui Xuan, Man Ngo Van, Khang Duong Nguyen, Anh Nguyen Duc & Preston, T. R., 1994. Installation and Performance of low-cost polyethylene tube biodigesters on small scale farms in Vietnam. Proc. National Seminar-workshop in sustainable Livestock Prod. on local feed resources. Agric. Pub. House Ho Chi Minh, pp.95-103.

An Bui Xuan and Preston, T. R., 1995. Low-cost polyethylene tube biodigesters on small scale farms in Vietnam. Electronic Proc. 2nd Intl. Conference on Increasing Animal Production with Local Resources, Zhanjiang, China, p. 11.

Botero, R. and Preston, T. R., 1986.Low-cost biodigester for production of fuel and fertilizer from manure (Spanish). Manuscrito ineditado CIPAV, Cali, Colombia, pp1-20.

Casley, D.J. and Kumar, K., 1988. The collection, analysis, and use of monitoring and evaluation data. Johns Hopkins Univ. Press, Baltimore and London, 174 pp.

Chater, S., 1986. New biogas digester for African small holders. ILCA Newsletter 1986, 5:4.

Desai, A.V., 1992. Alternative energy in the Third World- a reappraisal of subsidies. World Development Oxford, 20: 959-965.

Dolberg, F., 1993. Transfer of sustainable technologies in Vietnam. Development of Sustainable Livestock Technologies for Ecologically Fragile Zones in The Tropics.

SIDA MSc course in sustainable livestock production systems Report.

Dolberg, F., 1994. The farmer-extension-scientist interface: a discussion of some key issues. Proc. National Seminar-workshop in sustainable Livestock Prod. on local feed resources. Agric. Pub. House Ho Chi Minh, pp.118-122.

Dolberg, F., 1995a. On-farm Research: A discussion of some practical examples and procedures. In: T. R. Preston, Tropical Animal Feeding: A manual for research workers. FAO Anim. Prod. And Health, Rome, 126: 253-264.

Dolberg, F., 1995b. Development of sustainable livestock technologies for ecologically fragile zones in the tropics. SIDA MSc course in sustainable livestock production systems' Report, pp. 5-6.

Ellis, G. and Hanson, B. 1989. Evaluating appropriate technology in practice. J. Contemporary Asia, 19: 33-47.

Gunnerson, C. G. and Stuckey D. C., 1986. Anaerobic Digestion- Principles and Practices for Biogas Systems. The World Bank Technical Paper # 49 , Washington, D.C., pp 93-100.

Hieu Luu Trong, Ly Le Viet, Ogle, B. and Preston T. R., 1994. Intensifying livestock and fuel production in Vietnam by making better use of local resources. Proc. National Seminar-workshop in sustainable Livestock Prod. on local feed resources. Agric. Pub. House Ho Chi Minh, pp 9-16.

HMSO, 1977. Chemical Oxygen Demand (Dicromate Value) of Polluted and Wastewater. In: Methods of the Examination of Waters and Associated Materials. HMSO, London, UK.

Karki, A.B., Gautam, K.M. and Joshi, S.R., 1994. Present structure of biogas sector in Nepal, In: Foo E.L. (Eng-leong.foo@mtc.ki.se), "Ecotech 94" Electronic conference, Jun. 1994.

Khan S. R., 1996. Low cost biodigesters. Program for Research on Poverty Alleviation, Grameen Trust Report, Feb-1996.

Khoi Nguyen Van, Vinh Huynh Thi and Luu Huynh Thi Ngoc, 1989. Evaluation of biogas digesters in Cantho City. Proc. First Nat. workshop on biogas application in Vietnam. Polytechnic Univ. of Hochiminh City, pp 28-35.

Kristoferson, L. A. and Bokhalders, V., 1991. Renewable energy technologies- their applications in Developing Countries. Intermediate Technology Publications, London, pp 112-117.

Marchaim, U., 1992. Biogas Processes for Sustainable Development. Bull. FAO Agric. Services, Rome, 95: 165-193.

Pain, B.F., Misselbrook, T.H. and Crarkson, C.R., 1990. Odor and ammonia emissions following the spreading of anaerobically-digested pig slurry on grassland. Biological Wastes, 34:259-276.

Pound B., Bordas, F. and Preston, T. R., 1981. Characteristics of production and function of a 15 cubic meter Red-Mud PVC biogas digester. Tropical Anim. Prod. 6: 146-153.

Preston, T.R., 1995a. Tropical Animal Feeding: A practical manual for research workers. FAO Anim. Prod. And Health, Rome, 126: 155-166.

Preston, T.R., 1995b. Research, extension and training for sustainable farming systems in the tropics. Electronic Proc. 2nd Intl.. Conference on Increasing Animal Production with Local Resources, Zhanjiang, China, p.3.

Qiu, D.X., Gu, S.H., Liange, B.F. and Wang, G.H. 1990. Diffusion and innovation in the Chinese biogas program. World Development Oxford, 18: 555-563.

Sarwatt, S. V., Lekule, F. P. and Preston, T. R., 1995. Biodigesters as means for introducing appropriate technologies to poor farmers in Tanzania. Electronic Proc. 2nd Intl. Conference. on Increasing Animal Production with Local Resources, Zhanjiang, China, p.6.

Soeurn Than, 1994. Low cost biodigesters in Cambodia. Proc. National Seminar-workshop in sustainable Livestock Prod. on local feed resources. Agric. Pub. House Ho Chi Minh, pp.109-112.

Solarte, A., 1995. Sustainable livestock systems based on local resources: CIPAVs experiences. Electronic Proc. 2nd Intl. Conference on Increasing Animal Production with Local Resources, Zhanjiang, China, p.2

Thong Hoang Van, 1989. Some experiences on the development and the application on biogas digesters in Dongnai province. Proc. First Nat. Workshop on Biogas Application in Vietnam: Polytechnic Univ. Press. Hochiminh City, pp 60-69.

Participatory Rural Development: Experiences in Binh Dien and Xuan Loc Villages in Central Vietnam

Lylian Rodríguez J*, Thomas R Preston** and Frands Dolberg***

*Fundación Centro Para la Investigación en Sistemas Sostenibles de Producción Agropecuaria CIPAV, AA: 20591 Cali, Colombia E-mail: cipav@mafalda.univalle.edu.co

**Finca Ecologica, University of Agriculture and Forestry, Ho Chi Minh City, Vietnam E-mail: thomas%preston%sarec%ifs.plants@ox.ac.uk

***Department of Political Science, University of Aarhus, 8000 Aarhus C, Denmark E-mail: frands@po.ia.dk

Summary

In the past, especially in developing countries, research has been conducted mostly in experiment stations or universities and the results have rarely been of benefit to the small scale farmer. They were often detrimental in terms of impact on the environment. A more appropriate approach is to focus the research on the reality facing the small scale farmer and complement this with supporting problem-based experimentation in the research station and university.

This study was carried out in two villages in a rain-fed hilly region in Central Vietnam within a broad conceptual framework of sustainable development. The original idea of evaluating an intervention based on restricted milking of the local cattle was abandoned in the light of the insistence of the farmers that the expected benefits were too long term and they had other more immediate priorities.

In contrast, discussions about the potential benefits from introducing low-cost biodigester technology were enthusiastically received, especially by the women. Over fifty biodigesters were installed in Xuan Loc and Binh Dien villages and around Hue city. Officials both of the People's Committee and the Women's Union took an active part in the programme and became "trainers" of other farmers, as well as "maintenance technicians", offering help to neighbours who had problems with their biodigesters. An evaluation of the impact of the intervention was made by the women farmers who had participated together with officials of local aid agencies. The result was a frank discussion of the problems and of the mistakes but with a final endorsement of the advantages of the technology. During the project activities there was concern on both sides (outsiders and villagers) about the follow up of the technology after the project ended. Therefore, the Women's Union in Binh Dien village developed a proposal project with the objective to secure development funds to ensure an extension of project activities, so that monitoring would continue of the biodigesters already installed and to facilitate the introduction of the technology in neighbouring villages. Visits by staff members of the Canadian Embassy were made to the village and the proposal was approved by the Canadian Embassy at the beginning of December 1995. The traditional diet for pigs in the region was found to be lacking in protein (Nguyen Thi Loc et al 1996). Conventional protein supplements are only available in the market at a distance in Hue and are expensive. Biodigesters produce nitrogen-rich effluent and are a logical source of the required nutrients for growing duck weed as a local source of protein. Thus there was a potential connection between the biodigesters (being installed primarily as a source of fuel) and the need to improve the diet of the pigs. As a result of the project activities and based on farmers' wishes research to document the local pig breeds became a priority. A survey was done to get some baseline data and a project was started at the end of August with 10 families, who were to receive one Mong Cai gilt per family. In retrospect as the process of research unfolded, it had much in common with the participatory learning approach for sustainable agriculture described by Pretty (1995).

Key words: Vietnam, women, on farm research, low cost plastic biodigester, extension, effluent, duck weed, lemna, integration, ponds, nitrogen, protein, indigenous breeds, Mong Cai, energy.

Introduction

In the past, especially in developing countries, research has been done mostly in experimental stations or universities but the results have rarely been of benefit to the small scale farmer and often were harmful in terms of impact on the environment. A more appropriate approach is to focus the research on the reality facing the small scale farmer and complement this with supporting problem-led experimentation in the experiment station and university.

Another problem is that excessive centralization and inflexible management tend to suffocate new initiatives. The reward systems for researchers are usually based on scientific publications which often discourages them from working in the field, where research is less controllable and the topics may be seen as less scientific. The analysis by Gupta and colleagues (1989) shows how this translates into a repeated focus on "modern" methods of farming and widespread ignorance of alternative resource-conserving technologies and practices.

On-farm research has many advantages. Farmers have always experimented to produce locally-adapted technologies, practices, crops and livestock (Chambers et al 1989; Brouwers 1993; Scoones and Thompson 1994). They are continuous adaptors of technology and their systems are rarely static from year to year. Richards (1989, 1992) has linked this process of adaptation to a performance, in which the actors change the nature of the performance according to the specific conditions. The problem is that many researchers commonly do not understand or even accept that farmers can be "researchers". They assume that farmers are conservative and bound by tradition. Static and unchanging practices can therefore, upon investigation at a particular time, be characterized, analysed and so developed. But such an analysis can give nothing better than a snapshot of a complex and changing reality. It is important therefore, to begin to see technologies in a different light, not as fixed prescriptions but as indicators of what can be achieved. What agriculture needs is a willingness among professionals to learn from farmers, Pretty (1995).

The "Farmers first and last model (FFL)" is an alternative to the transfer-of- technology model (TOT), and is based on farmers' perceptions and priorities rather than on the scientist's professional preferences, criteria and priorities. The starting point is the scientists learning from and understanding of the resources, needs and problems of the resource-poor farmers and the research stations and laboratories play referral and consultancy roles. This model is characterized by the use of informal survey methods, research and development within the farms and with the farmers and the final evaluation criteria is that a technology is adopted (Chambers and Ghildyal 1985).

Compared to a conventional experiment station approach to research, starting in villages with farmers is more comprehensive. It makes demands on skills beyond the capacity of the individual scientist and a team of scientists has to be formed, representing more than one discipline (Dolberg 1990), for instance in this project the authors come from different backgrounds and have tried to learn and understand from social sciences by doing explorative research in the villages.

Materials and methods

The research site

The research was conducted from February to December 1995 in the Socialist Republic of Vietnam. The country has an area of 329,560 km² and its population is over 70 million people. The criteria to select the villages were:

Binh Dien village has been involved since 1990 with a SIDA-SAREC research project so there was a mutual interest to carry out the work there. In addition, the MSc batch of the first author was involved in an evaluation visit in 1994 and had installed two biodigesters which created interest among the local people in this new technology.

Xuan Loc village was selected despite it being a relatively new work area but the Agriculture University of Hue was conducting a research project funded by IDRC, Canada.

Bien Dien village

Binh Dien village is in the uplands in Huong Tra District in Thua Thien, Hue Province in the central area of Vietnam. It is located 35 km to the south-west of Hue city.

Xuan Loc Village

Xuan Loc village is located in the uplands in Phu Loc District in Thua Thien, Hue Province, to the south-west of Hue City (45 km distant).

Organizational setup

The Socialist Republic of Vietnam is divided into 53 provinces. These are divided into Districts, the Districts into Villages or Communes and these are divided into Hamlets or Groups.

The villages have a strong organizational system. The main person in the village is the chairman of the People's Committee. A large number of organizations are involved in the People's Committee, such as the Communist party, the Police, a village military commander, Veteran's Union, Farmer's Union, Women's Union, Youth Union, Finance group, Land office, Tax office and members from the Communication and Cultural office.

General characteristics

Population

Table 1 shows the general population characteristics in Xuan Loc and Binh Dien villages.

Table 1. *General characteristics of Binh Dien and Xuan Loc villages*

	Binh Dien	Xuan Loc
Population	2,957	1,990
Households	557	364
Members/family	5.3	5.5

Source: Binh Dien and Xuan Loc villages, 1995

Binh Dien and Xuan Loc villages were established after the liberation and reunification of Vietnam (1975) and the people come from different areas and different backgrounds. In Xuan Loc there are people from different ethnic groups. More than 77 % of the population belong to the lowland

"Kinh group" that come from the coastal plain areas of Phu Loc District; 23 % belong to the "Van Kieu group", whose origin is Quang Tri Province where the main means of livelihood is shifting cultivation (Le Duc Ngoan et al 1995).

Land Distribution

Land use is shown in table 2 and the allocation to crops of cultivated land is shown in table 3. In Binh Dien and Xuan Loc the forest is divided into natural forest and planted forest. In the case of Xuan Loc village about 45% of the land is occupied by natural forest (tropical rain forest), but day by day it is shrinking as a result of human activities such as extraction of firewood, timber wood, rattan, grass, etc. Further, there is a tobacco planting practice whereby forest is cleared every year for tobacco planting.

Table 2. *Land use in Binh Dien and Xuan Loc villages*

	Binh Dien	Xuan Loc
Total area, ha	2,498	4,236
Forest, ha	1,796	2,739
Grassland, ha	300	1,231
Agricultural land, ha	402	266
Cultivated area, ha	127	140

Source: Binh Dien and Xuan Loc villages, 1995

Table 3. *Allocation to crops of cultivated land in Binh Dien and Xuan Loc villages*

	Binh Dien	Xuan Loc
Cultivated area, ha	127	140
Sugar cane	25	0
Rice	7.5	50
Cassava	41	75
Sweet potato	40.5	15
Vegetables, groundnuts	13	nd

Source: Binh Dien and Xuan Loc villages, 1995.

Animal population

The animal population in 1995 is shown in Table 4.

Table 4. *Animal population 1995 in Binh Dien and Xuan Loc villages*

	Binh Dien	Xuan Loc
Pigs	1,200	1,056
Cattle	800	1,281
Buffalo	49	81
Goats	10	0
Ducks	450	nd

Source: Binh Dien and Xuan Loc villages, 1995.

Climatic conditions

Both villages have the typical climatic conditions of the zone between the coastal plain area of central Vietnam and the Truong Son high mountain range. The data shown in table 5 are from the nearest meteorological stations.

It is possible to distinguish two seasons: the dry and the rainy season. The rain is concentrated mainly from September to December. The temperature falls in those months. Storms and floods are frequent in this period, which make the climatic conditions even more complex. Total annual rainfall ranges from 2,000 to 2,7000 mm in 1995.

Table 5. Climatic conditions in Binh Dien and Xuan Loc villages

	Binh Dien	Xuan Loc
Altitude, masl	30	50
Temperature, °C	13-29	20-28.7
Relative humidity, %	80-85	78-92
Rainfall, mm/year	2,021	2,681

Source: Forecast Station Service - Thua Thien Hue Province, 1995

Water resources

Drinking water is mainly from wells but also from the rivers, streams and springs. In Xuan Loc village about 10% of the households have deep wells.

Electricity

In Binh Dien village there are a few farmers (< 30) that have access to electric power from a hydroelectric scheme but it is only during some time of the day and in the rainy season. In the rest of the year there is no power. In Xuan Loc village there is no electrical power, but there is one farmer using a small generator. In both villages there is widespread use of motor car batteries for lighting and to run a radio and in a few cases a TV.

Education

In Binh Dien village there are 3 schools: a secondary school with 275 pupils, a primary school with 400 and a kindergarten with 175 pupils. In Xuan Loc there is only one elementary school (grades 1-5) and the nearest secondary school is 15 km from the centre of the village.

Health service

In Binh Dien there is a health centre (2 doctors, 2 nurses and 2 assistants) and a small hospital. In Xuan Loc there is only one small health centre.

As can be seen Binh Dien must be characterised as the better endowed village with regard to modern institutions and facilities.

Objectives

The overall objective of the research was to devise ways of making better use of the local resources. As it has been argued (Pretty 1995b) that the dominant positivist research paradigm has strong limitations when research is conducted in an open system like the on-farm situation, it was an important objective in terms of research strategy to test the farmer first and last model- FFL - (Chambers and Ghildyal 1985) and the participatory learning model (Pretty 1995) as applied to livestock research. The immediate objectives were:

To understand the system of production in the selected hamlets and the role of the cattle in this system.

To study the possibility to establish a milk production programme to develop the "multi-purpose" (Preston and Leng 1987) use of the local cattle.

Methodology

The methods used were:

- Secondary information.
- Participatory work with the community.
- Feedback workshops with the farmers and the leaders of the community.
- As a result of the participatory work with the community on-farm research was carried out through appropriate interventions.
- To select a group of farmers to conduct the research in accordance with the decision of the community.

Results and discussion

Action programme

The field work started by the end of February 1995, but exploratory visits were made to both villages in 1994. Binh Dien was visited in August 1994 and Xuan Loc in November 1994 to introduce the researcher (the first author) and to discuss possible ideas to develop in the future work. The initial idea was to understand the role of the local cattle in the system of production and the possibilities for initiation of a milk production programme. The first discussions with the members of the People's Committee and the Women's Union indicated there was interest in the development of new ideas.

Participatory work with the community such as visits, field work, workshops, informal meetings, and interviews were used during this phase in order to obtain the confidence of the authorities and farmers and to exchange ideas and to develop the project. An interpreter was used most of the time because of the differences in language.

There was good support from the People's Committee and the Women's Union organizations in both villages but it was markedly stronger in Binh Dien village. Some farmer leaders participated actively in the process and worked with the project to monitor and follow up the introduction of the new technologies.

Visits were made to the farmers who raise cattle and also farmers involved in a project in the use of ensiled cassava for fattening pigs (Nguyen Thi Loc et al 1996). After discussions with the village leaders it was agreed, as the first step, to set up biodigesters in the households of those farmers participating in the project as a token of their participation in the project and as pilot demonstration plants in each village.

The results are shown villagewise because the procedure and development of the project was quite different in both villages.

Xuan Loc village

Biogas digesters

During the first visit the possibility was discussed to build a small house for the project-researcher (the first author) and the People's Committee agreed. The University was given the responsibility for developing the idea with the community and to look for the cheapest and simplest way to build it. There were many aspects to take into account in order to make the decision such as:

- there was an amount of money available for lodging of the student
- the advantage of living and working in the village
- the most important -- the possibility to develop a small-scale integrated farm and start with the biodigesters

Despite the fact that the community wanted to choose a different place, the house was built taking into account other aspects less important for the researcher and the project, but these were reasons that were outside the project control. At least one of the objectives was met, namely the demonstration of the potential for recycling manure. The biodigester was built the same day that the researcher took up residence and was functioning before the house was finally completed.

It was the biodigester that served as the real initiation of the technical/biological part of the project, since as soon as it began to function it became the centre of interest for the women in the village. Many people visited the house to see the biodigester to learn about it and for curiosity as well.

The SAREC-MSC programme had another project in the village on the use of ensiled cassava roots to feed pigs (Nguyen Thi Loc et al 1996). This project involved 12 families and credit was supplied from the project to buy 4 pigs and protein supplements for each family. In view of the interest in the biodigester technology, it was decided to install demonstration units in each of these households as a starting point to encourage participation by the community in project activities. The advantage of the low cost bio-digester technology (Bui Xuan An et al 1996) is that the simplicity of the system facilitates maximum participation in the installation process - an ideal way of "learning by doing".

Thus the next step in project activities was the installation of biodiges-ters in ten households in the village. In the process of installation of the biodigesters, and the compilation of general data about the families (table 6), there were opportunities to discuss with them their ideas about the es-tablishment of the milking programme with the local cattle.

Table 6. *General information of families involved in Xuan Loc biodigester project*

	Average	SE
Family size	5.6	±0.6
No. children	3.1	±0.6
Biodigester		
Length, m	10.0	±0.0
Total volume, m3	5.4	±0.0
Distance digester-kitchen, m	9.0	±1.2
No. burners	2	±0
No. animals		
- Sows	0.7	±0.2
- Piglets	1	±0.7
- Fattening Pigs	5.1	±0.7
- Cattle	10.4	±2.5
- Buffaloes	0.7	±0.5
Firewood for people/day, kg	5.1	±0.5
Firewood for pigs/day, kg	7.6	±0.8

The data summarized in Table 6 were collected before the start of the project. It can be seen that firewood is used not only for people. The feeds for pigs are traditionally cooked daily and it takes even more firewood than for satisfying the needs of the family. The data can be used to estimate that at least 1000 tonnes of firewood are used annually to cook the feed for the pigs and that 678 tonnes of firewood are used to cook for the 364 households in Xuan Loc village. One kg of firewood on average is worth from VND 200 to VND 300 (opportunity cost) which means that without the forest the villagers would have to spend between VND 340 to 500 mil-lions (US$34 000 to 50 000) to buy fuel. In other words this is what the for-est provides them free of cost today.

Establishment of a milk program with the local cattle

In addition to the general reaction to the idea of milk production, gained during the installation of the biodigesters, visits were made to several farmers who kept cattle as their major activity. The aim was to understand the system of cattle production in the village.

The local breed is almost 100% "Chinese yellow" characterised by small body size (adult body weight of 180-200 kg), and light bone structure. As part of a genetic improvement programme, the Vietnamese government is promoting crossbreeding by supplying Red Sindhi semen to be used on local animals. Several of the richer families had purchased animals of the same cross (Yellow cattle x Red Sindhi).

Table 7. Cattle population according to wealth categories in Xuan Loc village

	Very poor	Poor	Better-off
No. of households	222	109	33
Households having cattle or buffaloes (%)	63	95	93
No. of cattle/buffalo/household	2-4	5-6	10-12

Source: Le Duc Ngoan et al, 1995

Almost all the families in the village own some cattle (see table 7) and as expected the richest people raised the largest numbers. The main purpose of raising cattle is for meat, manure for the crops and as a way to save money. Traditionally the management system is based on free grazing in the forest. It is mainly the children who look after the cattle (when they go to school normally one goes in the morning and another in the afternoon) or sometimes a "cowboy" is hired to do this job. Every day the herd is taken to the forest for grazing and during the summer many farmers may keep the herd there for 2 weeks or a month according to the weather. In the rainy season, cattle are grazed near to the house because of the cool weather.

Cows usually have less than one calf/year weighing 7-10 kg at birth. The calves are suckled for 7 to 8 months by which time they may weigh 60 to 80 kg. After two years they can reach 180 kg. Natural mating is wide-spread in the village, as normally the cattle are kept together and there is no control. Farmers who want to improve their cattle use artificial insemi-nation (AI) obtaining semen from the station in Hue. One of the farmers is trained as an inseminator.

Most farmers have a simple pen where the cattle are kept at night time and during the rainy season. Supplementation is not common. Some farm-ers feed salt and some offer rice straw during the rainy season, and occa-sionally some cassava leaves when the harvest time comes, but generally there is no supplementation.

The idea of adding a new "purpose", i.e. milk production to the local cattle was discussed. It was explained that to do this meant some requirements such as: supplementation of the cow and the calf, some changes in the traditional management system "bringing the cow near to the house" and "separating the calf". Fresh milk is not traditionally consumed in the family. When it is consumed it is in the condensed and sweetened form, usually taken with coffee. Even then, it is mainly reserved for old people, children and those who are ill. Nevertheless, there is "supply and demand" at local level, which means that it would be possible to start a local production programme at some time.

It is important to mention that in Hue there is a project promoting exotic cows such as Holstein with milk yields up to 15 litres daily, but with very high inputs. Farmers in that project were visited in April 1995 and concern was expressed such as: "we can't recover the money that we invested in feeds". In fact the same 2 farmers belonging to that project were visited by the MSc students of the first author's batch in August 1994 and after 8 months (in April 1995) there were no more farmers involved. The "fresh milk shop" was visited as well where the price of milk was 4,000 VND per litre and that manifested that there is a market for fresh milk in Hue. These observations seem to support the assertion that "specialized livestock may produce less than local animals where the available feed resources are of low nutritive value and high energy and high protein concentrates are expensive" (Preston and Murgueitio 1994)..... however, in that situation.... multipurpose livestock offer advantages over specialized animals where there are great risks due to dependence on a single product. This is one of the important reasons why dual purpose or multipurpose cattle are so common throughout Latin America (Sere and Vaccaro 1985; Vaccaro 1986). It seems valid to suggest these aspects must be taking into account for future developments in other countries as well.

Other aspects, which will need attention are such as marketing and preservation of milk according to local conditions.

Finally, after carefully analysing the situation, the decision was taken not to go ahead with the milk production programme as pilot results showed very low yields in the local cattle and marketing of the milk was a problem.

Another important development arose in the process of installing the first biodigesters as half of these were established in Binh Dien village where the pig-cassava project was based (Nguyen Thi Loc 1996). It became apparent that, in this village, there was more enthusiasm and participation. A second important decision (from the point of view of impact) was to work in both villages - in Binh Dien as well as in Xuan Loc. In fact, because of strong support from village leaders it was decided to concentrate most of the work in Binh Dien village.

The use of effluent from the biogas digesters to produce duck weed as a protein supplement for the traditional diet

The traditional diet for pigs in the region was found to be lacking in protein (Nguyen Thi Loc et al 1996). Conventional protein supplements are only available in the market in Hue and are expensive. Duck weed can contain up to 40% protein in the dry matter when raised on fertilized ponds (Leng et al 1995) and can be grown almost anywhere in the tropics where there is water. Biodigesters produce nitrogen-rich effluent and are a logical source of the required nutrients for growing duck weed as a local source of protein. Thus there was a potential connection (see figure 1) between the biodigesters (being installed primarily as a source of fuel) and the need to improve the diet of the pigs.

This idea was enthusiastically received by the farmers especially the women. And so, in Xuan Loc, 10 families were selected through the Women's Union to join in this "pig" project which in turn was linked with the biodigester project. Funds were given to the Women's Union to develop a credit system for those ten farmers in order to buy the material to make the biodigester and to buy the pigs and to improve (or prepare) the pond to grow the duck weed. Later the money was to be repaid to the Women's Union to establish a revolving fund to give an opportunity for more women farmers to participate in these activities.

There was no experience in the village in the growing of aquatic plants. But several farmers learned quickly how to grow them, and to keep them in good condition (fertilized with biodigester effluent), and that the duck weed plants could be used as a high quality protein supplement not only for pigs, but also for ducks and chickens. The farmers appreciated that to introduce such a new technology implied a "learning process" which would take time and would not necessarily be suitable for everyone. Experiences in Colombia (Espinel 1994) showed that in such a process some farmer "leaders" will continue with the idea even after the project finishes and that others will realize the importance of the idea and eventually follow the example of the "leaders".

It became apparent during the development of the "duck weed" project that there were many factors which influenced duck weed production - some controllable and others determined by climatic conditions (flooding!!). Management was found to be the most important factor --e.g.: the levels of effluent to be used, water exchange and the need to renew the seed. Very little was known on the use of the biodigester effluent to produce duck weed, so this aspect was a logical subject for "on-station" research (Rodriguez and Preston, 1996).

The role of leaders and local organizations in "on-farm research" and technology transfer

Close collaboration, confidence and "clarity" between researchers, advisors and local leaders, institutions and organizations is fundamental for successful "on-farm research and technology adaptation". There are many factors affecting the process specially when an "outsider" comes to the village with some ideas but little understanding of the real situation or even more important when there are cultural and language differences between the outsider and the target group. In Vietnam, where "the local" organization is well developed, leaders and organizations such as the Women's Union's can move masses and thus activate the process. Equally they can "stop it" and then there is no impact. There are many internal (in the village) factors affecting the role of these institutions such as for instance lack of communication among organizations, competition for power, self interest and created interests - all of which can make the situation even more complex and will affect the work environment.

Binh Dien village

Table 8. General information of families involved in Binh Dien biodigesters project

	Average	SE
Family size	4.9	±0.33
No. Children	2.7	±0.34
Biogasdigester length	9.7	±0.28
m3 liquid volume	5.63	±0.16
Distance Digester-Kitchen	9.8	±1.06
No. Burners	2	±0
No. Animals		
- Sows	0.1	±0.094
- Piglets	0	±0
- Fattening Pigs	3.5	±0.62
- Cattle	3	±2.638
- Buffaloes	0.1	±0.094
Firewood for people, kg	7.1	±1.226
Firewood for pigs, kg	8.3	±1.239

Biogas digesters

The biodigester project in Binh Dien was the same as in Xuan Loc. It was a component of the SIDA-SAREC M.Sc. project on the use of ensiled cassava to feed pigs (Nguyen Thi Loc et al 1996). There were 12 families and credit was supplied from the project through the Women's Union to buy 4 pigs and protein supplements. The materials for the biodigesters were donated as it was considered they would be used as a demonstration of this new technology. It was also a way of compensating the farmers for taking on the role of "researchers", doing the extra work in keeping records, controlling the feeds and generally carrying out many of the functions which in the research station would be done by skilled technicians. Throughout the process, emphasis was on the type of "participation" that can be classified according to Pretty (1995, p.173) as "interactive participation"-- in installation of biodigesters, evaluating the results, providing feedback, discussing adjustments to the technologies and, always, "learning by doing" was a basic philosophy. Data on the families in the Binh Dien component of the project are in table 8.

In Binh Dien village there was a very strong participation by the Women's Union, the leaders of the People's Committee and some farmers who, from the beginning, made the process more dynamic and by their enthusiasm facilitated the feedback between the target group and the project, thus enriching the work in this village.

Establishment of a milk programme with local cattle

Visits were made to several farms to survey the system of cattle production in the village. The cattle population is shown in table 9. Distribution was less equitable than in Xuan Loc. Cattle numbers increased much more rapidly among the wealthier families.

Table 9. *Cattle population according to wealth categories in Binh Dien village*

	Poor		Medium		Better-off	
	1990	1994	1990	1994	1990	1994
No. of households	225	237	60	165	15	55
% households having cattle	0	0	40	30	60	55
No. of cattle	0	40	30	360	70	300

Source: Binh Dien Village, 1995

The management was the same as in Xuan Loc but in this case there were some farmers interested to try to start the project. Workshops were held to discuss with the farmers and leaders possible ways to begin the project. The proposal was to use a supplement of rice straw treated with urea (that has been shown as successful technology in China and Bangladesh, Dolberg et al 1995) with some tree leaves (e.g.: from the Jack fruit tree) for cow and calf and some additional rice bran for the cow. A group of farmers that had 1 or 2 cows recently calved agreed to build a simple place to milk the cow and a pen to keep the calf away from the cow. Several farmers milked the cows but the yield was too low -- only 50 to 100 ml in the first days. Discussions were held with the Women's Union and farmers and many factors were analysed. Finally there was a meeting to decide the following steps and the farmers gave their reasons for stopping the project:

- It was a very new technology for them
- There is no tradition to milk the cows
- The traditional management is so easy and when we milk the cows makes the management complicated especially working with only one or two cows from the herd
- Marketing is still a big constraint.
- The production of local cows appears to be very low
- There are priorities that can generate better results and have more effect on the family
- In the future it could be a good idea to be developed but maybe it is not the right time

Discussions were also held about "How did they want to work and invest their efforts?" and the following points were made:

- Pig production is a very important source of income in the family
- There are many problems to buy weaner piglets for fattening and even if they are available they can bring many diseases to the village, so breeding sows could be a good project because there were only around 20 sows in the village.
- The local "Mong Cai" is the right breed to use because it has a very good reproduction, produces many piglets and is well accustomed to use local feeds such as green plants.
- The Mong Cai sows can be crossed with white (improved) boars to produce crossbred pigs for fattening.

Consequently, research to document production traits of the local breeds became a priority. A survey was done to get some baseline data and a project was started by the end of August with 10 families, who were to receive one Mong Cai gilt per family. It took the Women's Union a long time to buy the right breed because it is becoming more difficult to find pure or high grade gilts of the Mong Cai breed. The last family to get a gilt received it in November 1995. The Women's Union agreed to be responsible for supervising the follow up of the project and to collect the data on the reproductive performance of the Mong Cai sows. The farmers will repay the credit and the Women's Union will manage it as a revolving fund to get more farmers involved in this project.

The use of effluent from the biogas digesters to produce duck weed as protein supplement for the traditional diet

The comments made under this heading in the Xuan Loc project are equally applicable to Binh Dien, and will not be repeated.

The role of leaders and organizations in "on farm research" and technology transfer

The chairman of the village, the leaders of the People's Committee and the Women's Union especially, were actively involved in the development of the project, in providing new ideas, adopting and adapting the technologies, questioning the researcher's ideas and were concerned for the continuity of the project and not only for their own development but also for the development of surrounding villages. This experience confirms the importance of close collaboration, confidence and clarity between researchers, advisors and local leaders and institutions.

Introduction of new ideas to farmers outside the project and other villages

In Binh Dien village where the process was more active the Women's Union took the initiative to introduce the biodigesters, duck weed and cassava silage to other villages, so there was a meeting with more than 60 women from the 15 villages belonging to Huong Tra District where women leaders from Binh Dien Village, the researchers (Nguyen Thi Loc and Lylian Rodriguez), one of the advisors (T R Preston) and the principal researcher in the biodigester project in Vietnam (Bui Xuan An) were together to share the experiences during the process of establishing the project in Binh Dien village and other places in Vietnam.

Low cost plastic biogasdigester in Binh Dien and Xuan Loc villages

Background on the fuel and forest situation

Energy is a fundamental factor for economic development, but normally the energetic models are based on non-renewable resources. There are many kinds of energy such as:

- Hydraulic energy
- Wind energy
- Solar radiation
- Biomass (through pyrolisis and gasification)

During the course of this century the world energy consumption per inhabitant has grown 16 times. Today the industrial countries with 32% of the world population consume 82% of the planet's energy. On average, a person from an industrialized country consumes 20 times more energy than a person in Africa. It is clear that the economic development model is what drives energy consumption.

In many developing countries there is a serious shortage of fuel and the energy crisis is a daily reality for most families. Cooking is one of the most energy-consuming activities, yet is often inefficient. The open fire is still very common. Today the decline of the forest in developing countries is frequently mentioned in the mass media. The forests in Africa, Asia and Latin America are disappearing and infertile land is advancing in its place with serious consequences. Deforestation has many causes. Poor people are migrating and inhabiting, cultivating and using new forest areas. In some areas they use slash and burn methods and this is another factor rapidly depleting the forest areas. War has been an important cause of deforestation. However, the daily consumption of fuel (tables 6 and 8) must not be underestimated when considering causes of deforestation. It is not unusual for a family to have to spend the greater part of their day gathering fuel for their home. At times dozens of kilometres need to be covered to find fuel (Nystrom 1988).

Facing this situation the use of renewable energy sources like solar energy for lighting and low cost plastic biodigesters to give an efficient use of the manure in the farming system to produce gas for cooking and effluent to fertilize ponds for fish, aquatic plants and crops will bring advantages to the family and to the environment.

Low cost biodigesters in Binh Dien and Xuan Loc villages

The objectives of the biodigester programmes in developing countries should be to establish minimum cost systems, using only local materials and with a simple technology so that farmers themselves can readily learn to install and manage the biodigesters. It was therefore decided to use a continuous-flow flexible tube biodigester based on the "Taiwan" model as described by Pound et al (1981) and later simplified by Preston and co-workers first in Ethiopia (Preston 1985, unpublished data), in Colombia (Botero and Preston 1986) and later in Viet Nam (Bui Xuan An et al 1993).

In Binh Dien and Xuan Loc villages, more than 50 biogas digesters were installed as part of the project activities, with an average cost (for materials) of US$29.00, including two burners (see Table 10).

Gas production was measured using a Japanese gas flow metre (Model number: K875-/YAZAKI KEIKI Company) which was installed on a rotational basis in the households of 16 families. The inputs (manure and water) and outputs (effluent) were recorded (see Table 11). The data showed a significant difference (p=0.000) in the proportion of water/ manure used to load the biodigester between the two villages. In fact the difference is in the amount of manure used to load the biodigester, but not in the amount of water and it is because the availability of cow manure is higher in Xuan Loc than in Binh Dien village. The difference in gas production per kg of DM can be attributed to the fact that, in Binh Dien, farmers use mainly pig manure. There is also the fact that latrines were linked with the biodigester when the gas production was measured in Binh Dien village. However, there are not enough data to draw firm conclusions on this aspect. Further research is needed, although, according to the literature the data are reliable because the gas production per kg of DM of manure could be around 150 litres (Bui Xuan An et al 1996).

Table 10. Cost of a Plastic Biogas Digester in Hue-Vietnam

	VND/Unit	Units/digester	Total
Plastic+transport HCMC-Hue	208,000	1	208,000
Transp Hue-Xuan Loc/Binh Dien	13,500	1	13,500
Ceramic Pipes (100 mm id)	11,000	2	22,000
PVC Elbows (21 mm id)	800	3	2,400
PVC pipe (21 mm id)	2,000	3	6,000
PVC Union (Male/female) (21mm)	1,600	1	1,600
PVC "T" (21 mm id)	900	3	2,700
PVC Union (21 mm id)	500	1	500
Hose pipe (21 mm id)	1,640	10	16,400
Gl pipe (21 mm id)	1,800	8	14,400
Ball tap (21 mm)	12,000	1	12,000
Gl elbow (21 mm id)	1,500	1	1,500
PVC glue	6,000	0.1	600
Car inner tubes (worn)	15,000	0.5	7,500
			VND321,100*

*US$=29.2

Table 11. Input and output data for plastic tube biodigesters in Binh Dien and Xuan Loc villages

	Binh Dien	Xuan Loc	SE/P
Gas, litres/m3 liquid volume	91.9	138	±25/0.238
Manure, kg/day	8.8	22.5	±2.15/0.002
Water, kg/day	76.8	65.2	±6/0.197
Water/Manure	9.3	3.08	±0.45/0.000
Gas, litres/kg DM	248	153	±32.5/0.085

Analyses of N content, Chemical Oxygen Demand (COD) and colonies of *E. coli* were made for the material loaded and effluent in 20 biodigesters. The results are shown in table 12 and the effect on the reduction of COD and *E.coli* is clear. McGarry and Stainforth (1978) showed that by recycling human and animal wastes in biodigesters at internal temperature of 30-35°C, it is possible to destroy up to 95% of the eggs of parasites and almost all the bacteria and protozoa that cause gastrointestinal diseases.

Table 12. Material loaded and effluent composition

	Material Loaded	Effluent	SE/P
N, % in DM	2.85	2.49	±0.45/0.6
COD, mg/kg	29,291	7,906	±2,108/0.000
Coli, Colonies/g	261 x 105	1.5 x 105	±38x105/0.000

For N content the data are not reliable, because it would be expected that the content of N in the DM of the effluent is higher than in the input manure. It is important to mention this point because it highlights the difficulties of obtaining reliable data when the samples are just sent to a laboratory to be analysed and it shows the need for the researcher to do her/his own analysis.

Stoves
The biodigester plant includes a simple stove made with galvanized pipe of 21 mm internal diameter. It has 2 burners made with the same sort of pipe and 2 ball tabs with the same diameter. It seems to work quite well but it can be improved. Measurements were taken fitting the gas metre between the storage bag and the stove and the results were on average that in 1 hour it is possible to boil 6 litres of water and the amount of gas used is 26 litres/litre of water, in other words 156 litres/hour which agrees with the literature (150 litres of gas per hour (CVC et al 1987; Botero and Preston 1986). The farmers developed many ideas to avoid the effects of wind and to have a more convenient place to put the pans. Each farmer had different ideas and those that improved the stoves said that they were much better. In fact a lot of research has been done about improved stoves for firewood, charcoal and others sorts of fuel, but only a little in stoves for low cost plastic biodigesters. In India a study was carried out among user populations and it was found that fuel saving range is from 32 to 42% using the improved stoves (Ramachandra 1994).

Latrines linked with the biodigester
The leader of the Women's Union from Binh Dien village visited the project in Xuan Loc where the project house was built and immediately appreciated the significance of having a toilet connected with the biodigester system. She returned to Binh Dien convinced of the advantages of developing the idea in her own village. A credit system was therefore organized through the Women's Union to establish toilets as components of the biodigesters in families already participating in the biodigester project.

Future follow up!
Over 50 biogas digesters were installed during the project of which 20 were donated for the project (10 per village) to farmers involved in Nguyen Thi Loc's (1996) project and the 30 left were financed through a credit system from the Women's Union. All the money used for the project was given to the Women's Union and they developed a credit system to work with the farmers with the idea to use the money in a revolving research fund (Solarte et al 1994) thereby giving the opportunity to more farmers to participate in such projects in the future.

During the project activities there was concern on both sides (outsiders and villagers) about the follow up of the technology after the project ended. Therefore, the Women's Union developed a proposal and presented it to the representative of CIDA (Canadian International Development Agency). The objective of the proposal was to secure development funds to ensure an extension of project activities, so that monitoring would continue of the biodigesters already installed and to facilitate the introduction of the technology in neighbouring villages. Visits by staff members of the Canadian Embassy were made to the village and the proposal was approved by the Canadian Embassy by the beginning of December 1995.

In fact there are many technical and non-technical aspects that must be included in the follow up of the project such as:

- Technical aspects: proportion of water and manure to be loaded
- Proportion of pig to cattle manure
- Gas production in the dry and the wet season
- Different uses of effluent to irrigate and fertilize trees, crops, fish and to produce water plants as a source of protein
- Installation procedures (to link the pig pen, latrine, biodigester), the width of the trench for specific kinds of soils
- Effect of linking latrines with the biodigesters
- Management and maintenance: fence, shade or not?
- Longevity of the plastic
- Stoves' design

Concerning the introduction of the technology:

- How to introduce the technology to a new area?
- Which aspects to take into account when introducing the technology to a new area?
- Who are the right people to start?
- Technicians' attitudes
- Credit systems for the poor
- Is the biodigester suitable for the poorest?

Evaluation of the project in Binh Dien and Xuan Loc Villages

By the end of the project a meeting was held with the village leaders and farmers to evaluate the results of the year's activities. The aim was to identify and discuss the problems and the benefits, and to plan future activities. The outcome of the evaluation is summarized below:

Binh Dien Village

The farmers' thinking shows the reality in the village. In Binh Dien many families have to buy firewood because the forest is far from their houses or because there is not enough labour in their families, so the benefit that they can get from the biodigester is very important, but there are also many farmers whose only source of income is from the forest, cutting firewood for their families and selling at village level or in the city. There is a clear understanding of their situation and the comments reflect their observations and they show many aspects that need to be improved from a practical point of view and some which can be subjects of further research.

A very important aspect in table 13 is the lack of manure in poor farms. This is where the biodigester can be important, but not as a first step to start to improve the standard of living because very poor farmers have other priorities and the strategy much be different to improve their conditions. The value of effluent to irrigate the trees or vegetables in the home garden or to produce aquatic plants such as duck weed is appreciated.

Table 13. Evaluation of the biodigesters in Binh Dien village

Benefits	Problems	Future Activities
◊ Saving time	◊ Flood damages digesters	◊ Perfect installation proceedings
◊ Cleaner kitchen		◊ The walls of trench will be more inclined
◊ Less labour spent for cooking	◊ Animals can damage the plastic	
◊ Good health for the cook	◊ Lack of manure in poor farms	◊ Technical follow up!
		◊ The fence always needed
◊ Saving money for buying and labour for collecting fire wood	◊ Cattle dung products less gas than pig manure	◊ Do not let the waste into the digester when the animals are sick (applying antibiotics)
◊ Using effluent for watering duck weed, crops and trees	◊ Antibiotics used to treat pigs reduce the gas production	◊ Combination between the farmers' contribution and outsiders' help
	◊ Cattle dung can form a hard layer on the top of the liquid in the digester	◊ The fence always needed
◊ Preventing bad smell		
◊ Preventing flies and mosquitoes	◊ Biogas plan must be "well" connected with animal house and the kitchen	

Source: Binh Dien village, 1995

At the end of the project 25 biodigesters were visited and 68 % were in good conditions and 32% had problems or were not working. The reasons were: 12% had problems because of accidents (especially animals that fall into the trench), 8% due to problems at installation, 8% because of management or lack of manure and 4% (i.e. one) was destroyed due to flooding.

In table 14 the farmers' reactions are presented to the introduction of a new plant, i.e. duck weed as source of protein in the village. In fact several farmers having biodigesters, but outside the project started to grow duck weed because they realized the nutritive value for their animals not only for pigs, but ducks and chickens as well. It is important to mention that, traditionally, duck weed is sold at the market in Hue.

Table 14. *Evaluation of the introduction of duck weed in Binh Dien*

Benefits	Problems	Future Activities
◊ Pigs fed with duck weed grow very fast	◊ Heavy rain lets duck weed "escape", dilutes nutrients in pond ◊ New for us: "lack of experience"	◊ Good drainage system ◊ The pond must be close to the biodigester. ◊ Continue the project

Source: Binh Dien village, 1995

In Binh Dien village following the decision not to go ahead with the milk program a project was initiated with the indigenous breed Mong Cai pig breed. The farmers, therefore, also discussed some benefits, problems and future activities in that regard (table 15).

Table 15. *Evaluation of Mong Cai breeding sows*

Benefits	Problems	Future activities
◊ Adapted and grow very well ◊ Can use crop residues	◊ Difficult to find	◊ Training on rearing of breeding sows ◊ Continue the project

Source: Binh Dien village, 1995

Xuan Loc Village

The results of the evaluation are shown in table 16.

In Xuan Loc the situation was quite different and there are many factors to be analysed:

There is plenty of forest and the forest is one of the main sources of people's sustenance.

Living conditions are not as good as in Binh Dien which means that maybe biodigesters are a suitable technology for some part of the population, but for the very poor people another strategy must be developed in order to improve their standard of living before biodigesters have a role to play.

There was less participation in this village during the project and less enthusiasm from the leaders. Although the leader of the Women's Union was very active there was no strong support from the People's Committee.

Climatic conditions are a constraint, because when floods come the bio-digester can be taken away, in reality only one biodigester was destroyed during the last wet season, but the possibility must be considered when people chose the site to setup the pig pen and biodigester. However, this is a risk not only for the biodigester.

Table 16. Evaluation of biodigesters in Xuan Loc village

Benefits	Problems	Future activities
◊ Enough manure, enough gas.	◊ Rain decreases gas production.	◊ Improve gas production
◊ Better than fire wood for cooking	◊ When there is not enough gas for cooking we are less efficient!.	◊ Protection (fences)
◊ Clean kitchen.	◊ Feeding cattle dung can form hard layer in the top of the liquid in the biodigester	◊ Installation strategies for small families.
◊ Good for small households.		◊ Regular maintenance.
◊ Saving labour for fire wood collection.	◊ Flood can break the trench	
	◊ The technology has less impact in up land region where fire wood is available	

Source: Xuan Loc village, 1995

At the end of the project 20 biodigesters were visited and 60 % were in good condition and 40% had problems or were not working, the reasons were: 10 % had problems because of accidents (specially animals that fall in), 10% because they had problems at installation, 15 % because of poor management or lack of manure and 5 % (which means one) were destroyed by the flood.

After the evaluation a workshop was held (funded by FAO) in both villages to give some practical recommendations about biodigesters, introduce another protein source for livestock (a tree - *Trichanthera gigantea*) that has given good results in Colombia and in the south and north of Vietnam and the duck weed was evaluated (table 17).

Table 17. Evaluation of the introduction of duck weed in Xuan Loc

Benefits	Problems	Future activities
◊ Animals fed duck weed grow very fast.	◊ Flood drives out the duck weed	◊ Plant again after December
◊ High yield.	◊ less yield in rainy season.	
◊ Easy to manage		

Source: Xuan Loc village, 1995

Survey about Mong Cai sow performance in Binh Dien village

Background

The population of local breeds of livestock is day by day decreasing due to lack of infrastructure and of economic incentives to encourage access to, and use of, this genetic material for breeding. To date, almost all efforts have been directed towards the introduction of "improved" and "new" breeds with very high genetic quality, but with an equal dependence on high inputs and with poor adaptability to the less than ideal conditions prevailing in rural areas of most developing countries.

In Binh Dien village, pig production plays an important role in the farming system but there is a shortage of piglets which brings many problems for the village because farmers have to buy the piglets from different places. The problems include:

- Disease transmission
- Poor quality (bad health and poor development)
- High price (many middlemen!!)

In the village there are around 20 families that raise sows for breeding, but annually farmers in the village require to fatten upwards of 1000 pigs. It is evident that there is a place to develop a larger population of breeding sows. For this reason a survey was done in order to understand the traditional management system, the breed used and its performance. Sixteen families were interviewed and the following information was obtained:

Breeds

In Binh Dien there is a long history of using local breeds for reproduction purposes, principally the Mong Cai or Cornwall, which is not a local breed, but well adapted because it has been in Vietnam for many years (Molenat et al 1991). Despite this, the Government in 1994 established a policy, which prohibited farmers to use the Cornwall breed.

The Mong Cai breed is important for breeding but the males have a very low price which makes it difficult for farmers to keep the pure breed as the demand is only for females. Thirteen families in our sample had a history of raising Mong Cai sows. Performance parameters are summarized in table 18.

According to our interviews Mong Cai sows can have 8 litters. Data are summarised in table 18. The data show a very good picture of the reproductive performance of Mong Cai sows which compares well with typical performance traits that are obtained on commercial farms with breeds of so-called high genetic merit receiving feeds of high nutrient density. For instance the farrowing interval, which is a good measure of the reproductive performance, is 181 days; while in high quality breeds it is of the order

Table 18. *Performance of Mong Cai sows in Binh Dien Village*

Items	Average	n
1st mating weight, kg	51.7	16
Live weight gain to 1st mating, kg/day	0.223	16
No. of piglets/litter	11	49
No. of piglets weaned/litter	10	49
Mortality, %	8	48
Weight of piglet weaned, kg	8	47
Lactating days	51	47
Open days	16	46
Farrowing interval, days	181	47
Income/litter, VND	800,000.00	7

Source: Binh Dien, 1995

of 175 days (Sarria et al 1994). But in the latter case, the conditions are very different. Mortality to weaning in high technology systems can be very high upwards of 10% . Yet in the baseline data for the Mong Cai the mortality to weaning is less than 10%.

Housing
The management is very simple. Often the pig pen is an open one with enough space to keep the sow and the piglets in a comfortable position. Most farmers use a packed clay-floor. A few have a partially concrete and packed-clay floor. Some farmers have a fence around the pen, often a live" fence of cassava or bamboo, so as to restrict the area in which the piglets can scavenge. The emphasis is usually on some form of housing which is cheap and yet comfortable for the sow and her litter.

Feeding system
Farmers use the traditional locally available resources to feed the sows. The results of a survey are presented in table 19. They treat the sows almost like people. In Vietnam sticky rice is used for the celebrations or special occasions, but from table 19 it appears farmers feed some sticky rice, green beans or eggs when the sows have farrowed and some days before. However, it is important to point our that the amounts fed are very low. For example: 1 egg per 3 days or 100 gm of green beans per day the first week after farrowing.

In general, farmers try to provide some protein-rich supplement in order to keep the lactating sow in good condition. However, the amount is very limited and in general the feeding system is according to the crop season and the family condition. The interesting finding is the adaptation and good performance of the Mong Cai sows under relatively poor conditions of nutrition and management.

Table 19. *Feed resources used for breeding pigs in Binh Dien village*

Family	Growing Sow	Pregnant Sow	Lactating Sow
Mr Nguyen Van Ve	Rice Rice bran Cassava meal Wine by-product Vegetables (Sweet potato an ipomea leaves)	Rice Rice bran Cassava meal Wine by-product Vegetables (Sweet potato an ipomea leaves)	Rice Rice bran Cassava meal Wine by-product Vegetables (Sweet potato an ipomea leaves)
Mr Tom	Rice, Rice Bran, Sweet potato leaves, Bon e flour Cassava (not often) After the 4th Month: supplement some fish flour and mineral premex.	Rice, Rice Bran, Sweet potato leaves, Bone flour Cassava (no often) After the 4th Month: supplement some fish flour and mineral premex.	supplement: green beans and fish meal After 30 days the piglets are fed with rice soup and bone flour
Mr Vo Dat	Rice, Rice bran, Wine by-product, Vegetables	Rice, Rice bran, Wine by-product, Vegetables	Supplement: Green been Sticked rice, 20 days after the farrow. After they reduce and increase wine by-product
Mr Gan	Rice, Rice bran, vegetables According with the family conditions	Rice, Rice bran, Wine by-product, Vegetables	Increase the amount of rice and add sticky rice
Mrs Bui thi Chi	Wine residue, Rice, Vegetables, Salted fish and shrimp heads	Wine residue, Rice Vegetables, Salted fish and shrimp heads	They increase the amount and add some sticky rice.
Mrs Luong Tri	Rice, Rice bran, Vegetables Small fish	Rice, Rice bran, Vegetables Small fish	For 10 days green been Sticky rice
Mrs Chung	Rice, Rice bran, Vegetables Jack fruit seed, Cassava	Rice, Rice bran, Vegetables Jack fruit seed, Cassava	Supplement: Salted fish Sticky rice with sugar
Mrs Dieu	Rice, Cassava, Vegetables Some times rice bran	Rice, Cassava, Vegetables Some times rice bran	Supplement: Green been Sticky rice
Mr s Nguyen thi Muc	Rice, Rice Bran, Vegetables	1 week before farrowing Sticky rice, green beans Duck eggs	for two weeks Green bean Duck egg
Mrs Le Thi Tam	Rice, Rice bran, Vegetables	Rice, Rice bran, Vegetables	Supplement: Sticky rice Green been, Groundnut cake

Source: Binh Dien, 1995

Some farmers use specific strategies like for instance after the weaning they wait for two heats before mating which gives some time for the sow to recover its body condition.

Breeding System

Artificial insemination is a common practice in this village. Farmers can obtain semen from an Insemination Centre in Hue city and they have learned artificial insemination. It is a very effective and cheap system. In the village, boars are not available.

In the Department of Agriculture of Thua Thien-Hue, Mr Nguyen Viet Dan (personal communication) explained that the Vietnamese government has a plan to destroy the breeds that have very low production like some of the local breeds. In 1975 Mong Cai pigs were brought from Tam Dao, Quang Ninh, Hai Phong Provinces in the north, when 300 Mong Cai sows were taken to a state farm, but in 1980 this farm collapsed and some of the sows were given to the farmers. There are no statistics about the number of animals in Hue Province, but it has still a very low population. Actually, the government plan is to keep the Mong Cai for breeding and to cross it with an exotic breed such as Large White for fattening, but the problem is availability of semen because in the province there are only 3 "insemination centres" (farms!) that will be describes below:

Namgia (Hue city): State farm, 7 Large White boars and a young Mong Cai (not yet mature to produce semen). 40 doses Large white (VND 6000/dose) per day are distributed to 4 workers in 4 districts (Phu Loc, Thuan An, An Lo and Bao Vinh). The veterinarian or farmer learn how to do the insemination and they do it at village level.

Nguyen Hau (Quang Phuoc village, Tam Gian Lagoon): It is a farm owned by the Agricultural Department, but rented and administered by a farmer from this village. 25 doses are available per day, but there is no Mong Cai Semen. However the farmer raises 3 Mong Cai sows and is planning to have a Mong Cai boar.

Van Huu Nguyen: It is a private farm with a Mong Cai and some Yorkshire boars. There is semen available, but according to the farmer almost all Thua Tien Hue province depend on this Mong Cai boar and some other provinces also. The availability of Mong Cai semen is very low: only 2 doses each 3 days and the prices are VND 8,000 per Mong Cai and VND 4,000 per dose of Yorkshire. There is a big demand for Mong Cai in his village and in general in the province and other provinces.

It is evident that there is need to develop simple centres at village level to increase the population of the Mong Cai breed. This local breed has very good reproductive performance and is well adapted to the local conditions and local resources.

Observations at village level about the efficiency of the Mong Cai breed in the use of local resources were the motivation for carrying out an on-station experiment to explore the digestion parameters and N metabolism of Mong Cai and exotic breeds and their crosses (Rodriguez and Preston, 1996b).

Factors that interact in on-farm and on-station research

Research on behalf of developing countries has been done ostensibly with the objective to improve the standard of living of people. Unfortunately, much of the research in the industrial countries is done on how humans can destroy humans, either directly in development of weapons of war or indirectly through the destruction of local cultures and self reliance through exports of inappropriate technologies.

Research and development activities must be viewed in the perspective of interactions, the most important of which is that between imported technology and the local situation. Leng (1995) has shown very clearly how many components interact to influence overall performance of ruminants on a specific diet. In the laboratory, with controlled conditions, one of the characteristics of feeds, such as digestibility, often has the greatest impact on animal productivity, but this is not the case in most field situations.

According to our experience in this project this interaction can be demonstrated in figure 2 where on farm research or much better the participatory learning process (Pretty, 1995b) is the starting point to get a clear understanding of the situation and to define research priorities, but there are many aspects interacting and the road ahead is not as straight as that:

- Human factors of the outsiders and the target group: culture, communication and attitude are essential factors that can stop or make the process work.

- Environmental factors such as rainfall distribution, temperature, storms, floods will affect the results and lead some scientists to decline undertaking "on-farm" research, but they will then miss the reality.

- Technical aspects such as: communication with other scientists working in the same way, facilities to get relevant information, references and more specific aspects such as laboratory facilities that are a constraint in developing countries. Knowledge of their existence may lead the researchers to allocate them a higher priority than they deserve in order to obtain reliable data. But the other side of the coin is, what is the meaning of reliable data produced out of context?

- Economic resources such as access to adequate credit which is also one of the main constraints because it is not available and the knowledge (Hashemi, 1996) about it is very weak in both outsiders and in the target groups, contributing to the complexity of a solution.

- Feed resources (availability, quality, prices)

- Animals (health conditions, breeds, age)

- Water (quality, availability), for instance for biodigesters water is an important resource.
- Soil (type, fertility)

Those aspects are interacting positively or negatively in the process. However, the experiences of the present work lead us to conclude that these issues are best addressed within a framework of FFL and learning approaches, where the research starts in the villages and the station and the laboratories play referral and consultancy roles.

What is the way ahead?

At the beginning of this paper data were shown about the animal population in Binh Dien and Xuan Loc villages, but based on the insights we have now gained it is important to think deeply about some aspects such as the change of the animal population in both villages through time. It is evident that animal production plays a important role in these two villages. It seems that the cattle numbers are increasing very fast in Xuan Loc village (table 20), which through grazing in the forest is at the expense of the environment.

There is not much difference in the development in pig numbers over time, but there are clearly more poultry (not shown in the tables) in Binh Dien, probably reflecting the easier access to the market, making it easier to sell poultry products

In Table 20 and 21 data are shown about the distribution of cattle and pigs according to wealth ranking of the households. There are no date about poultry distribution in the village, but according to our observations poultry is especially important for the poorest people, because poultry need much less capital than pigs and cattle.

The data confirm that cattle are mainly in the hands of the wealthiest people, while the pigs are more equally distributed. Thus in Xuan Loc village in 1994 of a total cattle population of 1453 animals 1210 or 83% were owned by 39% of the households categorized as poor and better off. The 61% very poor households had 17% of the cattle. With regard to pigs, the situation is less biased. In 1994 19% of the pigs belonged to the 9% better off households, but 47% to the 30% poor households and 33% of the pigs to the 61% very poor households. The same broad picture with regard to ownership of animals by wealth ranking of households can be seen in Bien Dinh village. However it is interesting that in that village the majority of pigs are owned by the very poor households.

At this point it is pertinent - in a context of self criticism - to point out that most animal production research in developing countries has been on large ruminants and within large ruminants on crossbred milking cows. We considered ourselves to be advanced in relation to that situation by initially wanting to do research on the local cattle, exploring the potential of adding another (milking) purpose to their uses in support of the general idea of multipurpose cattle being more appropriate (Preston and Leng, 1987). However, the data in tables 20 and 21 clearly show that had we stuck to this original idea, we would only have done research for the wealthier sections of the villages and hardly touched on problems of relevance for the poor and - the environment, as the poor live off the environment to a great degree.

Table 20. *Distribution of Cattle and pig populations according to wealth rank in Xuan Loc village in 1990 and 1994*

	Total Fam.	%	% Fam.Popul. Cattle	Cattle	% Fam.Popul. Pigs	Pigs
Better off						
1990	22	10	100	110	100	66
1994	33	9	100	760	100	209
Poor						
1990	50	22	55	85	70	140
1994	109	30	70	450	80	511
Very poor						
1990	155	68	10	30	50	100
1994	222	61	40	243	60	360

Source: Nguyen Thi Loc et al 1996

Table 21. *Distribution of Cattle and pig population according to wealth rank in Binh Dien village in 1990 and 1994*

	Total Fam.	%	% Fam.Popul. Cattle	Cattle	% Fam.Popul. Pigs	Pigs
Better off						
1990	15	3	90	120	90	60
1994	55	10	55	300	64	120
Poor						
1990	60	11	60	84	60	120
1994	165	30	30	360	79	365
Very poor						
1990	447	86	10	0	30	100
1994	337	60	3	39	67	528

Source: Nguyen Thi Loc et al 1996

We overlooked poultry. There is recent evidence that species such as poultry needs further research. Ongoing work in Bangladesh shows that properly designed poultry projects (Saleque A and Mustafa S 1996, Askov Jensen H 1996, Nielsen H 1996) have a potential to reach the poorest in a village that even Grameen Bank type loans do not have (Hashemi, personal communication).

Conclusion

We have found it fruitful to apply a participatory approach in this research. If there is true participation the real necessities can be highlighted and new ideas can be developed and allowed to influence the research. Participation is a mutual learning process where "outsiders", local authorities and farmers can increase their awareness of what to do to achieve change. But what is true participation? There are many kinds of participation from passive participation, where people are involved merely by being told what is to happen, to self-mobilization, where people take initiatives independent of external institutions (Pretty 1995). Through the project activities it has been shown that participation is also a learning process, based principally on confidence among outsiders and the target group.

On-farm research is a dynamic process. Farmers have always experimented to produce locally-adapted technologies. Farmers are often excellent "researchers" and "extensionists". In this way, research and extension go together which is the best. When the research is done on-farm the process can be faster and there is a "natural selection" of technologies and priorities. There can, therefore, be less waste of time and money. Applied in this way the "Farmers first and last model (FFL)" is an alternative to the transfer-of- technology model (TOT). It is based on farmers' perceptions and priorities rather than on the scientist's professional preferences, criteria and priorities.

In this project there was a clear example of how we "outsiders" (Chambers, 1983) think about "appropriate technologies" to be applied at village level and the result was a "learning" from farmers and the project changed from, milk production as an additional purpose for the local cows to biodigesters to duck weed as a source of protein to local breeds on pigs and, finally, to get and overall view of the socio-economic situation of the village. Definitely it is a way to really, but not completely, understand the village situation. There must be an active process where outsiders try to understand the situation, offer alternatives which may have some impact in the village, using an iterative process of trial and error (Dolberg, 1994) in which villagers participate actively making criticisms and suggestions to the outsiders, giving ideas which may change the researcher's objectives. The starting point must be around this approach, it can not be achieved only with participation in information giving (Pretty 1995) where people participate by answering questions posed by extractive researchers using questionnaire surveys or similar approaches in which people do not have the opportunity to influence proceedings. What agriculture needs is a willingness among professionals to learn from farmers.

Close collaboration and confidence between researchers, advisors and local leaders, institutions and organizations are important preconditions for successful on-farm research and technology adaptation. There are many factors affecting the process especially when an outsider comes to the village with some ideas, but little understanding of the real situation or even more important, when there are cultural and language differences between the outsider and the target group. In Vietnam, where the local organization is well developed, leaders and organizations such as the Women's Union's can move masses and thus activate the process. Equally they can stop it and then there is no impact. There are many internal village factors affecting the role of these institutions such as for instance lack of communication among organizations, competition for power, self interest and created interests - all of which can make the situation even more complex and it will affect the work environment.

In some aspects, especially when new technologies are being developed or adapted, it is advantageous if on-farm research is complemented with "on station" research. But there must be a clear understanding of, and link with, the reality of the farmer situation.

Milk production as another "purpose" for the local cattle may have a potential. However, it may not be a first priority. Very few very poor people keep cattle, but they live off the environment by collecting and selling firewood. More work with pigs and poultry and appropriate credit arrangements are more likely to benefit them and thereby the environment.

Biodigesters can play a pivotal role in integrated farming systems by facilitating control of pollution and at the same time adding value to livestock excreta through production of biogas and improved nutrient status of the effluent as fertilizer for ponds and crop land.

Biodigester impact is variable, adoption and successful results depend on aspects such as location, availability of fuel, the way the technology is introduced, supported, adapted and the technicians' attitudes.

Traditional diets for pigs lack protein and conventional protein supplements are expensive or not available. Therefore when the biodigester is part of the system producing gas and effluent rich in nitrogen duck weed can be grown to improve the pig and poultry diet.

The present work suggests local breeds are better adapted to local environments and local resources and may out-perform the "improved" breeds under these circumstances.

Further research is needed in areas such as:

- The role of poultry in the farming system for the poorest people
- Comparison of local and exotic pig breeds in the context of integrated farming systems.
- The way to introduce the biodigester technology in different social and ecological conditions.

- Technical aspects of low cost plastic biodigesters: use of effluent as fertilizer, stoves for low cost plastic biodigesters, effect of use of latrines.

Those aspects are based on farmers' ideas and experiences. They are not the researcher's idea, or they have become so after a learning process at village level. However, sources of protein for livestock like aquatic plants and tree leaves need to be added to the list of future research topics.

The role played by livestock in farming systems for poor farmers is multi-faceted and synergistic and must be seen not as a primary form of production but rather in terms of its overall contribution to the total farming system and to the immediate needs of the family.

Following this exposure, we argue that integrated farming systems offer unique opportunities for maintaining and extending biodiversity. This is an area deserving much more research. However, "context influences content" (Gupta, personal communication) and it seems very important to conduct the comparisons in typical contexts (on-farm) with typical feeds (those farmers use).

The well-being of poor farmers can be improved by bringing together the experiences and efforts of farmers, scientists, researchers, and students in different countries with similar eco-sociological circumstances.

Learning by all scientists and farmers is a fundamental issue: Are we new professionals prepared to work according to this approach?

Acknowledgements

The first author would like to express her gratitude to the Swedish Agency for Research Cooperation with Developing Countries (SAREC) for financing my studies for a Master's degree.

Sincere thanks to all the farmers in Binh Dien and Xuan Loc villages in Vietnam for sharing their enormous experience, for their friendship, hospitality, patience and continuous participation in this project. We were able to become good friends despite the language and cultural differences which was a great and unforgettable experience.

Special thanks go to the Women's Union in Thua Thien Hue Province: to Madam Loc, Madam Thu and all the staff members for their collaboration; and the Women's Union at village level -- Mrs Hoang Thi Hien in Xuan Loc and Mrs Phan Thi Duong Chi in Binh Dien Village for their collaboration, friendship and interest in improving the standard of living of their communities. It is an experience to adopt and adapt in my (first author's) country, Colombia.

The first author is especially indebted to Mrs Phan Thi Duong Chi the leader of the Women's Union in Binh Dien village, who became a good friend and without whom it would not have been possible to achieve what was achieved. Thanks to her enthusiasm and to her capability to develop ideas and to share them with her people, thanks to her patience, we became excellent friends and she was able to understand me even in Vietnamese!. I think if both of us could speak the same language it would be great, but it is an example of sharing "convictions".

Special thanks also to Mrs Nguyen Thi Loc, classmate and friend, for her strong support, patience, for sharing her experience and for making me feel at home while I was in "Hue-Vietnam" , thanks to her family.

Thanks also to Mr Bui Xuan An for his friendship, for sharing with me his skills and wide experience.

Special thanks to Mr Le Quyet and his family in Binh Dien -- one of "my families" in Vietnam for their hospitality and friendship and constant support to the project.

Thanks to Nguyen Van Bang for his support, patient, enthusiasm and hard working to start the project.

Thanks to the "farmers' biodigester team": Mrs Chi, Mr Quyet and Mr Son in Binh Dien and Mr Thang and Mr Thuong (a student) in Xuan Loc, and Mr Bang, interpreter - it was great to work hard, exchange ideas and learning, sometimes without speaking but "doing".

Thanks to Veronika Brzeski for her friendship, for sharing many things with me and for her important contribution to the project. Thanks to Dianne Hentschel for her interest and contribution to the project.

Thanks to the staff members of the Agriculture University of Hue that in some way were involved in the project specially to Mrs Hang, Mrs Lu, Mr Hoang, Mr Ngoan, Mr Duong and Mr An.

Gratitude is expressed to Chi Ninh and Em Sa for their friendship and help.

Finally, thanks to everyone who was involved in the project, which hopefully will continue wherever we are.

References

Botero R and Preston T R 1986 Low-cost biodigester for production of fuel and fertilizer from animal excreta: A manual for its installation, operation and utilization CIPAV: Cali, Colombia

Brouwers J H A 1993 Rural People Response to Soil Fertility Decline. The Adja case (Benin). Wageningen Agriculture University Papers 93.4. Wageningen Agriculture University, The Netherlands

Bui Xuan An et al 1993 Quy trinh lap dat ham u khi dot bang nylon. Circular in Vietnamese: UAF, Ho Chi Minh City

Bui Xuan An, Preston T R and Dolberg F 1996 The introduction of low-cost plastic polyethylene tube biodigesters on small scale farms in Vietnam. Livestock Research for Rural Development, Volume 8, Number 2, in press

Chambers R 1983 Rural Development: Putting the Last First. Longman

Chambers R and Ghildyal 1985 Agriculture research for resource-poor farmers: the Farmers-First-and-Last model. Discussion paper no 203. Institute of Development Studies. University of Sussex. Brighton, England

Chambers R, Pacey A and Thrupp L A (Editors) 1989 Farmer First: Farmer innovation and Agriculture Research. Intermediate Technology Publications Ltd, London

CVC, GTZ and OEKOTOP, 1987 Difusion de la tecnologia de biogas en Colombia. Cali, Colombia, pp43-10

Dolberg F 1990 Farmer-scientist interaction. Can we identify procedures, which work?. In: Metodologias de investigacion con pequeños agricultures y Tecnicas de investigacion en Fincas. Facultad de Ciencias Agropecuarias. Octuber 26, 1990 Palmira, Colombia

Dolberg F 1993 The farmer extension-scientist interface a discussion of some key issues. In: Preston, T.R., Ogle, B., Ly, L.V. and Hieu, L.T., 1994. Sustainable livestock production on local feed resources. Proceedings of the Regional Workshop held November 22-27, 1993 in Hanoi/Ho Chi Minh City, Vietnam. University of Agriculture and Forestry of Ho Chi Minh City, pp. 105 - 109.

Dolberg F and Finlayson P 1995 Treated Straw for beef production in China. World Animal Review, 82:14-24

Espinel R 1994 Sociedad y economia de campesinos cafeteros de la cordillera occidental en el Norte de el Valle de el Cauca. Factores que inciden en la construccion de sistemas agrarios.Fundacion CIPAV. Maestria en Desarrollo Sostenible de Sistemas Agrarios

Gupta A K, Patel N T and Sanh R N 1989 Review of post-graduate research in agriculture (1973-1984): Are we building appropriate skills for tomorrow? Centre for Management in Agriculture, IIM, Ahmedabad, India

Hashemi S 1996 Grameen's programme for research on poverty alleviation and biodigester experiments in Grameen. Development workers course: Integrated Farming in Human Development. Landboskole, Tune March 25-29, Denmark

Jensen A H 1996 Semi-Scavenging Poultry Flock. These proceedings.

Le Duc Ngoan et al 1995 Site Description, Problems Network Analysis and Proposed R & D Agenda of Xuan Loc village, Phu Loc District, Thua Thien Hue Province. A report prepared by PRA team and core group - IDRC Project - Hue University of Agriculture and Forestry Hue, Vietnam. pp 48

Leng R A, Stambolie J H and Bell R 1995 Duckweed - a potential high-protein feed resource for domestic animals and fish. Livestock Research for Rural Development. Volume 7, Number 1:36 kb

Leng R A 1995 Appropriate Technologies for field investigations in Ruminant Livestock Nutrition in Developing Countries. Proceedings of a Workshop: Agricultural Science for Biodiversity and Sustainability in Developing countries, April 2-7, Tune Landboskole, Denmark pp:57-78

McGarry M and Stainforth J 1978 Compost, fertilizer and biogas production from human and farm wastes in the People's Republic of China. IDRC-TS8e, Ottawa, Canada

Molenat M and Tran The Thong 1991 La Production porcine au Viet Nam et son amelioration. Etudes et Syntheses de l'IEMVT. No. 38 115pp

Nguyen Thi Loc 1996 Evaluation of protein supplementation of traditional diets and cassava root silage for local crossbred pigs under village conditions in central Vietnam. Department of Animal Nutrition and Management, Swedish University of Agricultural Sciences. M.Sc. thesis.

Nguyen Thi Loc, Preston T R and Ogle B 1996 Evaluation of protein supplementation of traditional diets and cassava silage for local cross-breed pigs under village conditions in Central Vietnam. Livestock Research for Rural Development, Volume 8, Number 2, in press

Nielsen Hanne 1996 Socio-Economic Impact of Smallholder Livestock Development Project Bangladesh. These proceedings.

Nystrom Maria 1988 Kitchen and Stove: The Selection of Technology and Design. Lund University, Sweden pp:136

Pound B, Bordas F and Preston T R 1981 Characteristics of production and function of a 15 cubic metre Red-Mud PVC biogas digester. Tropical Animal Production 6: 146-153

Preston T R and Leng R A 1987 Matching Ruminant Production Systems with Available Resources in the Tropics and Subtropics. PENAMBUL Books Ltd: Armidale NSW, Australia

Preston T R and Murgueitio E 1994 Strategy for Sustainable Livestock Production in the Tropics.2nd Edition. CONDRIT Ltd: Cali Colombia pp89

Pretty J N 1995 Regenerating Agriculture: Policies and Practice for Sustainability and self-reliance. Earthscan Publications, London, England pp320

Pretty J N 1995b Participatory Learning For Sustainable Agriculture. World Development, Vol 23, No. 8, pp 1247-1263, London

Ramachandra T V 1994 Getting into Hot, Hot Water. Centre for Ecological Sciences, Indian Institute of Science. GLOW Volume 14-December. pp:7-11

Richards P 1992 Rural Development and Local Knowledge: The case of rice in central Sierra Leone. Discussion paper presented at the IIED/IDS Beyond Farmer First: Rural people's Knowledge, Agricultural Research and Extension Practice Workshop, 27-29 October

Richards P 1989 Agriculture as performance. In: Farmer First. Farmer innovation and Agricultural Research.(Editors: Chambers R, Pacey A and Thrupp L A). IT Publications, London

Rodríguez L and Cuellar P 1994 Evaluación de la Hacienda Arizona como Sistema Integrado de Producción. CIPAV: Cali, Colombia

Saleque A and Mustafa S 1996 Landless Women and Poultry: The BRAC Model in Bangladesh. These proceedings.

Sarria P, Gomez M E, Rodríguez L, Molina J P, Mlina C H, Murgueitio E 1994 Pruebas de campo en el tropico con el uso de biomasa para sistemas integrados y sostenibles de produccion animal. CIPAV, Cali, Colombia

Sere C and Vaccaro Lucia 1985 Milk production from dual purpose systems in tropical Latin American. In: International Conference on milk production in developing countries (Editor: A J Smith). University of Edinburgh, Edinburgh pp459-475.

Solarte A, Dolberg F and Corrales Elcy 1994 Descriptive study of the local resources used as animal feed and animal production systems in a rural community on the Pacific coast of Colombia. Dissertation, MSC Thesis. Swedish University of Agricultural Sciences. pp12

Scoones I and Thompson J 1994 Beyond Farmer First. IT Publications, London

Vaccaro L 1986 Sistemas de Produccion Bovina Dominantes en el Tropico Latinoamericano.In: International Seminar: Sistemas de Producion Doble Proposito en el Tropico. Bogota, Colombia. Mimeo 17 p

Manure Management in Two Zones in the Cauca Valley of Colombia

Tania Beteta H., T. R. Preston** and A. Solarte***

*Universidad Nacional Agraria,
P.O.Box 453 Managua, Nicaragua
E-mail: vblandon@ibw.com.ni,

**University of Agriculture and Forestry,
Ho Chi Minh City, Vietnam
E-mail: thomas%preston%sarec%ifs.plants@ox.ac.uk

***CIPAV, Cali, AA 20591, Colombia
E-mail: fhv@cali.cetcol.net

Abstract

Case studies were carried out in two areas of the Cauca Valley in Colombia: on six small scale family farms in El Palmar village, in Dagua municipality and on one commercial farm in Jamundi municipality.

The objectives were: (i) to document livestock manure management and (ii) to determine the input-output ratios and the social benefits associated with the different ways of utilization.

The methods used included structured and informal discussions with the farm families during regular visits; collection of data in joint activities by the researcher and the farmers; verification of the data during extended stays (one week) in the village and on the commercial farm.

There was close integration of livestock and crops with most species of farm animals represented. Cattle, pigs and chickens were kept on all farms. Buffaloes were also present on the commercial farm and used for draught as well as milk and meat production.

The manure was recycled or transformed according to three main processes: biodigesters, earthworm culture and directly as fertilizer. Most social and environmental benefits were associated with the use of biodigesters. Savings of firewood on the six farms in El Palmar village were estimated as almost 3 tonnes per month as a result of recycling manure through biodigesters.

A comparison of the fixed dome of Chinese design and plastic tube biodigesters showed few differences in operating parameters, but the investment costs were much higher and the repayment period much longer for the fixed dome system.

The rate of transformation of cattle manure by earthworms was relatively uniform with 100 kg manure yielding between 0.9 and 2.6 kg of earthworms and 49 to 57 kg of humus.

On all the farms there was a strong appreciation of the need to conserve the environment and promote biodiversity. The advantages from integrating the various elements of the farming system were recognised in terms of social (closer involvement of family members in the production process) as well as economic benefits.

Key words: Farming system, manure, biodigesters, earthworms, humus, environment and socio-economic benefits.

Introduction

One of the main advantages of integrated farming is the opportunity to convert by-products and wastes from one activity into inputs for another. This form of horizontal integration has environmental as well as economic benefits.

Livestock excreta is frequently a source of environmental pollution in intensive, specialised animal agriculture. By contrast, the efficient recycling of manure in integrated farming systems can lead to increased profits and reduced environmental damage.

Many countries have developed a long and varied experience in recycling manure through biodigester systems. Examples are China, Nepal and India, where it has been promoted by many governmental and non-governmental organizations, through direct subsidies, cheap credit and supporting research. Marchaim (1992) reported that there are about 5 million family size biodigesters operating in China. According to Bhakta (1991), Krishna (1994) and Van Nes (1994) there are over 6000 units in Nepal. Production of biogas not only saves firewood, but also has the potential to improve soil fertility. After digestion, all nutrients and more than half the organic matter are still available in the effluent.

Another alternative is to recycle manure through earthworm production. The most common specie used to transform the manure into humus is Eisenia foetida. This process yields protein for animals (the worms) as well as fertilizer and soil conditioner in the from of humus (Soto, 1993).

The objectives of this study were:

- To study and document the ways of managing the manure from livestock on integrated farms in the Cauca Valley in Colombia.

- To determine the input-output ratios and the economic and social benefits associated with the different ways of utilization of manure.

Materials and methods

Location
The study was done on 6 small farms in a village in the Andean foothills and on one medium scale commercial farm in the Cauca valley.

The village
This part of the study was done in the El Palmar village, situated in the municipality of Dagua, Department of Cauca Valley. El Palmar is located 1,450 metres above sea level, in the foothills of the western chain of the Andean hills, 36 kilometres from Cali city. The average rainfall in this area is 1,550 mm per year, the average temperature is 18° C and the relative humidity 86% (CVC, 1995). The major part of the population depends on agricultural activities, the basis of the economy being agriculture and livestock production. The families in the municipality have the following public services: electricity, water, collection of garbage and a telephone line (URPA 1995). The municipality has its own water sources and the community has been working to preserve it through various activities. Many governmental and non-governmental organizations have been working in the zone of the study, in order to find appropriate ways of reducing environmental problems.

The research was initiated on seven farms. The criteria for farmer selection was: the experience the farmer had in recycling manure, the possibility of getting information and the availability of the farmer for collaborating with the researcher. Eventually, one farm was discontinued because of lack of interest on the part of the farmer.

The commercial farm
The Arizona farm in Jamundi was chosen because of its extensive experience of manure recycling over the past 8 years. The Jamundi Municipality is located around 25 kilometres south of Cali, in the Department of Cauca Valley, Colombia, at 923 metres above sea level, with an average temperature of 24°C, relative humidity of 80% and annual rainfall of 1,800 mm (CVC, 1995). The high rainfall months are from March to May and September to November. The soils are different from those commonly found in the Cauca Valley, having a high clay content with a high water level.

Data collection
Each farm was taken as a separate case study, but the format of the evaluation was the same for each.

In El Palmar the data collection was undertaken in two periods (February to June 1995; and June to November 1995); the Arizona farm study was done from September to November 1995. The methodologies were:

El Palmar (period 1)

In this part of the study general information was collected about each farm, previous information from Solarte et al (1994) was checked and discussions were held with each farmer about the research proposal and possible modifications to be made. Modifications were to consider each farm as a case study, not to include experiments with duck weed as it was not a common crop in the farms studied and because the farmers had already started to work with other water plants, especially azolla and water hyacinth.

For this phase of the study the methods used were based on participatory rural appraisal techniques (Aluja and McDowell 1984; Chambers et al 1989; Chambers 1993a,b and Berdegué et al 1994) according to the following schedule:

- Visits to the village usually three days every week, although sometimes according to requirements the visits were daily. One week per month was spent in the village working with the farmers and collecting general and specific information and trying to monitor previously collected data and verify the information obtained through the normal weekly visits.

- Visits were also made to organizations such as UMATA (Unidad Municipal de Asistencia Tecnica Agropecuaria), EMCALI (Empresas Municipales de Cali), URPA (Unidad Regional de Planificacion Agropecuaria) and the mayor's office, to get more secondary information on the village.

- A questionnaire was used with the objective of collecting base line information that would be needed, although it was not used directly in the field. This questionnaire was used during both periods.

- Formal and informal meetings with the farmers and with other researchers from CIPAV working in the village, and sharing of experiences with farmers from other regions where they were also working to develop sustainable systems of production.

El Palmar (period 2)

The following information was collected from June to November.

Manure production

The production of manure during a 24 hour period was determined every month on each farm. In the case of cow manure, account was taken of the number of animals and the time they were confined in the cowshed, and adjustments made to convert the data into manure production per animal per day. For pig manure the total weight was the production per day because they were confined permanently. In the case of chicken manure this was determined every two months after the birds were sold. The total weight of litter and excrement was recorded. Samples were then separated into small and large particles using a sieve with mesh size 5*5 mm. The fine material was recorded as manure. Manure production from rabbits was determined at the same time, because it is a common practice in a farm which has rabbits, to mix rabbit manure with chicken manure.

Biodigesters

Three of the farms had the "fixed dome" dual compartment type of bio-digester (CVC et al 1987); the others had the plastic tube continuous flow model (Botero and Preston 1986; Bui Xuan An et al 1994; Solarte 1995). The inlets to the biodigesters were connected directly to the outlet drain of the pig and cattle pens. All the pig excreta (faeces and urine) and the washing from the floor of the cattle pen passed into the biodigester. On some of the farms a portion of the cattle manure also was put into the digesters. Esti-mates of the amount of manure entering the biodigesters were derived from the monthly weighing of the manure and, in the case of cow manure, on information from the farmer about the proportions of the total added to the biodigester. The amount of water added to the digester was calculated on a daily basis from the time spent washing the pen floors and the rate of discharge of water from the hose. Some farmers used buckets of water to wash the floors. In this case the volume and number of buckets used were recorded. The amount of slurry produced was determined during a 24 hour period once per week. A graduated container of 120 litres was situ-ated at the exit of the outlet box of the digester and the amount produced in 24 hours was recorded. The gas flow was measured with a meter con-nected permanently in the pipe line between the digester and the reservoir (for the plastic digesters) and between the digester and the burner for the fixed dome digesters.

Earthworms

The total amounts of manure and the weights of worms used to inoculate each bed were measured at the beginning of each cycle. At the end of the cycle the total harvested material was weighed and fed to the laying birds. The residue, after the birds had eaten the worms, gave an estimate of the relative proportions of worms and organic fertilizer or humus (the worm casts), the data were adjusted to the total available area for earthworms.

Other uses of the manure

The amounts of manure used directly as fertilizer, or not used at all, were ascertained by direct conversation with the farmers and direct observation.

Arizona farm

The measurements were the same as in the village farms but were taken weekly during the three months September to November. The methods were similar except for the earthworms which were harvested by applying a thin layer of manure on the top of the bed and collecting the worms that became concentrated in this layer.

Analysis of data

Statistical analysis

Comparisons were made of the performance of the two types of biodigester in El Palmar village, using analysis of variance and the general linear model of the Minitab statistical software (version 9.2) (Minitab Inc., 3081 Enterprise Drive, State College PA, 16801-3008, USA).

Economical evaluation

Costs were mainly the investment in the biodigesters and the labour required for each activity. The fertilizer value of the manure and biodigester slurry was estimated from the content of N, P and K applying prices for these elements on the basis of those for equivalent amounts of chemical fertilizer. The value of the biogas was estimated on the basis that 1 cubic metre is equivalent to 0.5 kg of liquid propane (CVC et al 1990). The available nitrogen was estimated considering the nitrogen content in manure and slurry and the daily production.

Environmental evaluation

The environmental evaluation was made on the basis of the amount of firewood that was saved.

Social evaluation

The social evaluation was based on the degree of appreciation felt by the farmer and the family (eg: facility to make a hot drink in the early morning; cleanliness of the kitchen)

Results and discussion

Manure use in the farming system in each case study

El Palmar village

El Cambio

A small proportion of the manure from horses and cows is used as the substrate for earthworms but the greater part of the production is applied directly to crops destined for use as animal feed especially the "nacedero"

(Trichantera gigantea) tree and sugar cane. The washings from the cow-shed are the feedstock for a plastic tube biodigester (now 6 years old) and to this is added 50% of the pig manure; the other 50% of the manure is wasted. Manure from laying hens and ducks is not recycled as they are managed in a semi- scavenging system. The effluent from the biodigester is applied by gravity to vegetables (Cucurbita maxima, Sechium edule, Daucus carota, Lactuca sativa, Raphanus sativas, Capsicum annum, and Allium cepo). The vegetables are associated with sweet smelling and col-ourful flower species to help to avoid diseases transmitted by insects. Humus is applied to the forage trees, which are used for animal feed, and to crops grown for sale and family consumption such as coffee, banana, plantain and Ananas sativus. The earthworms are used to feed laying hens. A new activity is the growing of maize in association with beans. Field di-visions are made with "live" fences chiefly the multi-purpose trees, Trichantera gigantea and Gliricidia sepium. The family derives direct bene-fit from the animals in the farming system by consumption of meat and eggs, and indirectly through the sale of products in the market. The biogas is used for cooking for 10 people on average.

La Esfera

Manure from cows is the substrate for earthworms (about 50% is used for this purpose) and the remainder is processed in a fixed dome biodigester (now 3 years old). All the pig manure also goes to the biodigester. Manure from laying hens and chickens (in confinement) is applied directly to the pasture area. The effluent from the biodigester is applied by gravity to sugar cane. Humus from the earthworms is applied to forage trees, used as animal feed, and to crops for sale and family consumption (eg: coffee and plantains). The earthworms are fed to laying hens. The family benefits di-rectly from the animal component as food (meat and eggs) and through the sale of products in the market.

Biogas is used for cooking for 5 people, for heating chickens during the brooding period, for heating water when the birds are slaughtered and for drying coffee during the rainy season.

El Porvenir

Manure from the horses is not used. Manure from the cows (the smaller part) is a substrate for earthworms; the remainder is produced and left in the field during the grazing. Some 50% of the pig manure is used as feed-stock for a fixed dome biodigester (installed 4 months ago); the remainder is not used. Manure from the laying hens and broilers (in confinement) is not used. The effluent from the biodigester goes by gravity to the wooded area which is planted with bamboo (Bambusa guadua) although this is not done in a rational way. The humus from the earthworms is applied to the forage trees, used for animal feed, and to crops grown for sale and for family consumption (eg: plantains). The earthworms are used to feed laying hens and sometimes the earthworms and humus are applied in the field without harvesting the former. Chemical fertilizer is still used for the coffee and the pasture. The family benefits directly from the animal component (meat and eggs) and indirectly by selling in the market. Biogas is used exclusively for cooking for 15 people including labour, family and visitors.

Villavictoria

Manure from the cows is applied directly to the field (about 90%), and to crops such as: "nacedero" (Trichantera gigantea), grown for animal feed, sugar cane, "Boton de oro" (Tithonia diversifolia), "Matarratón" (Gliricidia sepium), "Písamo" (Erythrina fusca), Leucaena (Leucaena leucocephala), Bore (Alocasia macrorhiza). The remaining 10% is processed by earthworms to provide humus for nurseries. Sometimes the earthworms are harvested but not always. The washings from the cowshed and all the pig manure are used as feedstock for a plastic tube biodigester (installed 3.5 months ago). Manure, in the form of poultry litter, from the laying hens and the broilers is used as a supplement for the milking cows. The effluent from the biodigester is applied by gravity to the pasture area. From the animal component the family benefits directly (meat and eggs) and indirectly by selling in the market. Biogas is used for cooking and for boiling milk to made cheese for sale and home consumption.

El Recuerdo

All the manure from cows and pigs is used as feedstock for a fixed dome biodigester (installed 8 months ago). Manure from the laying hens (in confinement) is mixed with rabbit manure and applied to the pasture and to crops grown for sale and family consumption. The effluent from the biodigester is applied by gravity to the vegetables in the home garden and to fruit crops (pineapple, banana, plantain and Ananas sativus). The family benefits directly from the animal component (meat and eggs) and indirectly by selling in the market. The biogas is used for cooking for 10 people including labourers as well as the family members and for heating water required when chickens are slaughtered.

Sindamanoy

Most of the cow manure production (98%) is used as substrate for earthworms; the remainder (washings from the floor) is collected and applied directly to crops used for animal feed, especially "nacedero" (Trichantera gigantea), "Boton de oro" (Tithonia diversifolia) and sugar cane. All the pig manure is used as feedstock for a plastic tube biodigester (installed 8 months ago). Half of the chicken manure is applied directly to crops for animal feed and for sale and home consumption; the other 50% is used as a supplement for milking cows. The effluent from the biodigester is applied by gravity to the vegetables (Cucurbita maxima, Sechium edule, Daucus carota, Lactuca sativa, Raphanus sativas, Capsicum annum and Allium cepo); and also as fertilizer for aquatic plants. The humus from the earthworms is applied to crops for animal feed and to crops for sale and family consumption. The earthworms are used to feed laying hens. The family benefits from the animal component directly, by consumption of meat and eggs, and indirectly by selling in the market. This farm produces every week the food for the members of three families (value estimated per family is $ 20 US) living in the city as well as those taking care of the farm. The biogas is used for cooking, and for heating water during the slaughter of the poultry.

Arizona (the commercial farm)

Most of the manure from cows and buffaloes is used as substrate for earthworms. One part of the washings from the cowshed is used as feed-stock for two plastic tube biodigesters (installed 8 and 4 months pre-viously) together with the solids and most of the washings from the pig pens. The rest of the washings is mixed with irrigation water, and with the effluent from the biodigesters, and applied as fertilizer to the pastures. A proportion of this nutrient-rich water is applied to the azolla ponds. The chicken manure is also used to fertilize the pastures. One part of the humus is applied to forage trees, grown for animal feed, but most of it is sold as fertilizer. Animal products are sold (the major part), but a part is consumed by the family. Biogas is exclusively used for heating the "creep" areas of the pigs (20 burners operating for 12 hours daily).

Livestock component

In El Palmar village, it was noted that the proportions of the different species of animals, were quite similar, ranging from the smallest farm (Villavictoria) to the biggest (Porvenir).

On Arizona farm the livestock component was made up of crossbred (Holstein x Zebu) cows for dual purpose milk and beef production, pigs for breeding and fattening and industrial chicken raising. A small herd of buffalo (Riverine type) is kept for triple purpose draught, milk and meat production.

Land Use

El Palmar Village

There was a wide range of uses of the land. At one extreme, 90% was de-voted to crops for sale or family consumption and only 11% for producing feed for animals (Recuerdo farm); at the other extreme, 95% was in forage crops/pastures for direct use by animals and only 5% in crops for sale and consumption (Villavictoria). There was no relation (r = 0.09) between the size of the farm and the proportion of the area allocated to forage crops for animals. This tendency depended on the criteria of the farmer and also, historically, this zone had been planted to monoculture coffee. Due to low prices, the coffee trees had been gradually eliminated over the years and the subsequent introduction of alternative crops had followed no particular pattern.

The ornamental area includes the house, and flower and fruit garden. Although small in area, it is important socially. Valencia (1995) has called this area the "engine house" because it is the principal point of activity of the women, where in addition to taking care of the family, they can also do productive work and derive income. Such activities also facilitate the transferring of knowledge to the new generation.

Four of the six farms use only organic fertilizer; on the remaining two farms, chemical fertilizer is still used on the coffee crop.

Arizona

In Hacienda Arizona the whole land area is dedicated to animal production. In addition to cultivated pastures (49%), important crops for livestock are sugar cane (38%) and forage trees (9%). The latter (Erythrina fusca) are grown in association with African Star Grass for grazing. There is 1 ha (4%) devoted to ponds for aquatic plant production (mainly azolla).

Fertilisation is only with organic products derived from within the farming system.

Pattern of usage of the components in the recycling processes

Manure management

Mean values for manure production for cattle on the six farms in El Palmar village were 20 ± 1.55 kg/day. The three principal uses of manure were:

- Cultivation of earthworms
- As substrate for biodigesters
- Direct fertilisation of the crops

On two farms in El Palmar village more than half the manure was used for earthworm production while on two farms over half the manure went to the biodigesters, and the remaining two farms more than half was used directly as fertilizer. On only one farm was a significant proportion (31%) not recycled. In Arizona farm, approximately one third of the manure went to each activity with a species allocation whereby the biodigesters were mostly fed with pig manure and the earthworms with cattle manure.

Biodigesters

Mean values for operating characteristics of the fixed dome and plastic tube biodigesters in El Palmar village are shown in Tables 1 and 2.

Table 1. *Biodigester production parameters in El Palmar village*

Parameter	Plastic biodigester	Fixed dome biodigester	SE ; Prob.
Liquid volume (LV) , m3:			
Range	8-9	9.4-15.7	
Mean	8.6	11.5	± 0.7 ; 0.01
Biogas, litres/day	1720	2200	± 308 ; 0.5
Production / cubic metre of LV:			
Biogs, litres / day	200	191	± 33 ; 0.4
Manure , kg /day	2.1	4.6	± 0.6 ; 0.01
Water, day	14	16	± 1.7 ; 0.5
Water: manure, litres/kg	7.1	3.1	± 1.7 ; 0.5
Effluent, litres / day	14.8	16	
Retention time, days	65	50	± 6.1 ; 0.08

Table 2. *Economic summaries for plastic and fixed dome biodigester*

Parameter	Plastic	Fixed dome	SE ; Prob.
Liquid volume, m³	8.66	11.5	± 0.8; 0.01
Initial investment, $	246	2133	± 172 ; .002
Investment / m³ LV, $	28.8	189	± 10.5 ; 0.001
$ gas propane / day, US	1.03	1.2	± 0.02 ; 0.6
$ NPK / day, US	0.11	0.07	± 0.009 ; 0.01
Total value / day, US	1.14	1.3	± 0.01 ; 0.5
Repayment, months	8.5	74	± 1.7 ; 0.03

LV= Liquid volume

Data for the large plastic tube biodigesters on Arizona farm are in Table 3.

Compared with the plastic tube, the fixed dome biodigesters had a greater liquid volume, and for each cubic metre of liquid volume slightly less gas was produced, more manure and less water was used and the hydraulic retention time was shorter. The period required to repay the investment was 9 times longer for the fixed dome biodigesters (74 months) than for the plastic tube model (8.5 months). CVC (1987) estimates a repayment time of 72 months, and Van Nes (1994) reported in Nepal a repayment time of between 72 and 144 months for fixed dome biodigesters erected without subsidy. The large plastic tube biodigesters in Arizona produced, per cubic metre liquid volume, less gas, used more manure, used a higher ratio of water to manure and had a much shorter hydraulic retention time and a shorter payback period (2.25 months) than either of the two types of biodigesters in El Palmar (Tables 1 and 2).

Table 3. Summary of data of Arizona farm's biodigester

Parameters	Value	SE
Liquid volume, m³	150	
Biogas, litres / day	15300	487
Production / cubic metre of LV		
Biogas, litres / day	102	6
Manure, kg / day	10	0.4
Water, litres / day	160	35
Water / manure, litres/ kg	16.2	2.2
Slurry, litres / day	173	23
Hydraulic retention times, days	6	1.1
Economics		
Initial investment	840	
Investment / m³ of liquid volume, US	5.6	
$ gas propane / day, US	9.2	
$ NPK / day, US	9.9	
Total value / m³ of liquid volume, US	19	
Repayment time, months	2.25	

LV= Liquid volume

The most common use for the gas on the farms in El Palmar village was for cooking. However, there were reports of its use to boil milk in making cheese (Villavictoria), to boil water for slaughter of chickens (on the 4 farms with broiler production), and to provide heat for raising chickens and drying coffee (La Esfera farm). On Arizona farm the main use of the biogas was for heating in the pig farrowing and rearing pens (20 burners in use for an average of twelve hours daily).

In the farms in El Palmar village the slurry from the biodigesters was directly applied to the crops by gravity, a procedure favoured by the sloping ground. On Arizona farm, it was mixed with irrigation water and pumped on to cultivated pastures for the dairy herd. In the village farms and on Arizona farm a small amount is applied to the ponds as a source of nutrients for water plants: for Azolla on Arizona and water hyacinth in the village farms.

Appreciating the value of the biodigester effluent

One experience, that gives an idea of how the farmers learned to appreciate the value of the effluent, occurred in El Cambio farm where the owner had an agreement with a neighbouring farmer who, all his life, had grown horticultural crops but using chemical fertilizer. The two farmers began to work together (the owner of El Cambio provided the land and the biodigester effluent; the neighbour provided the seeds and the labour). After some days, when the vegetables started to grow, the wife of the owner of El Cambio complained of too low gas production. On checking the bio-digester, it was found that the outlet had been closed so the effluent would not escape and be wasted, but would be stored for use on the vegetables.

Earthworms

Five farms in El Palmar village, and Arizona farm, recycled manure through earthworms (Table 4).

Table 4. Earthworm production per cycle

Farm	Substrate	Production	Humus		Worms		Worms/humus
	(kg)	(kg)	(kg)	%	(kg)	%	(g/kg)
Porvenir	15700	9420	9043	57	376	2.6	42
El Cambio	18941	8940	8769	50	171	0.9	19
Sindam.	13500	7965	7726	57	239	1.8	31
Villavict	2300	1150	1127	49	23	1	20
La Esfera	15040	8272	8112	54	160	1.06	20
Arizona	861795	482605	468128	54	14478	1.68	31

On the small farms the amount of manure processed annually ranged from 2.3 to 19 tonnes. The products were earthworms (0.9 to 2.6% by weight of the original substrate) and humus (49 to 57% of the substrate). The losses, presumably as moisture and gases, ranged from 41 to 50%. On Arizona farm the corresponding figures were 861 tonnes processed giving rise to 1.68 % of earthworms and 54% of humus of the original substrate. The production of earthworms relative to humus (g earthworms/kg fresh humus) ranged from 19 to 42 g/kg in the village and was 31 g/kg in Arizona. According to Fuentes (1987) the optimum temperature required for earthworm production is 18-20° C, which is close to the average temperature in El Palmar village. Experiences in Medellin of Corporación Penca de Sabila (1995, personal communication) reported that the yield of earthworms (earthworms plus humus) is from 50 to 60% of the substrate, similar to the range obtained in the village. Ortiz (1994) reported a range of 40-55%. The relation of grams of earthworms per kg of humus obtained in this study is similar to that reported by Ortiz (1994) of 30 g of earthworms per kg of humus.

Savings of electricity and fuel prices

In Table 5 is shown the savings arising from use of biogas for cooking or heating. The comparison is with the equivalent amounts of electricity. The savings range from 0.93 to 3.62 US$ per month per household in the village and on Arizona farm it was US$10.62/month.

Table 5. *Money saved per month on the study farms by replacing electricity with biogas*

Farm	Biogas use (hours/day)	US$/month
Sindamanoy	7	1.64
Villavictoria	4	1.2
Cambio	4	0.93
Porvenir	12	3.62
Esfera	8.5	2.56
Recuerdo	8	2.18
Arizona	12.5	10.62

Other consumption figures and prices are presented in tables 6 and 7.

Table 6. *Consumption (Kwh) and price of electricity for different stoves and candles*

Equipment	Kwh/month	Kwh/d	US dollars/Kwh
Stove / 2 burners	225	7.5	0.025
Stove / 3 burners	245	8.16	"
Stove / 4 burner & oven	290	9.66	"
Stove / 4 burner without oven	262	8.73	"
Candle 100 W/ 6pm - 6 am	24	0.8	"

Source: EMCALI (Empresas Municipales de Cali), energy service, 1995

Table 7. *Market price of fuel and organic products*

Products	Unit	Colombia $ U.S	Nicaragua $ U.S.
Firewood	kg	0.015	0.099
Propane gas	25 pounds	7.5	4.97
Coal	kg	0.1	0.20
Humus	kg	1.5	2 - 5
Cows manure	kg		0.70
Chicken manure	kg	0.175	0.3
Pig manure	kg		0.60
Petrol	Gallon	1.5	2.25
Sawdust	45 kg	0.5	0.4
Urea	45 kg	10.5	10

Source: Market survey

Available nitrogen on the farms

The estimates of the amounts of organic nitrogen fertilizer available on each farm per year are in the range from 200 to 1000 kg/ha/year in El Palmar village. On Arizona farm the figure is 2,000 kg/ha/year. Even the lowest quantity available (200 kg N/ha/year) exceeds average rates of use in developing countries. For Colombia, FAO (1995) reported 101 Kg/ha in 1990. This is a further advantage of the close integration between livestock and crop production.

Environmental situation

All the farms are trying to preserve the environment, avoiding the use of fire wood for cooking, or reducing its use by replacing it with biogas. Even when only dead wood is used, time is required to collect it from the fields and if left in situ it is recycled as organic matter to the soil. The women and children in the Villavictoria and El Cambio farms had a programme to re-forest the area near the water source in the village.

According to MARENA (Ministerio de Recursos Naturales y del Ambiente) (1995), in Nicaragua, the per capita consumption of firewood in rural areas is 1.8 kg per day. A similar figure was reported by Ramachandra (1994) in a study in Karnataka, India where per capita fuel wood, consumption was 1.92 kg per day. Only two of the six study farms in El Palmar continue to use firewood, estimated at a consumption of 15 and 50 kg/month. If the 51 people in the 6 farms had to depend entirely on firewood the consumption would have been almost 3 tonnes/month, according to the projections of MARENA (1995). The benefit to the environment from using biogas on the six farms is thus considerable.

The only negative features are the continued use of chemical fertilizer on Porvenir and El Recuerdo farms, mainly because of labour constraints. Also some pig manure is not yet recycled in El Cambio farm although the farmer expected to start to do it in December, like on El Porvenir farm.

Social situation

According to the farmers and their families there are benefits in three main areas:

- For the women the conditions for cooking are much better since there is less or even no smoke in the kitchen;
- Less work is required as use of firewood means labour to collect it and to tend the fire;
- There are no bad smells around the farm when wastes are recycled through biodigesters.

Similar observations were made in Karnataka state in India (Ramachandra, 1994) and in Nepal (Van 1994).

A general comment was that integrating the production system helped to integrate the family, and that living conditions were now more accept-able, thus reducing the attractions of the city.

Conclusions

- The most common ways of manure recycling were as substrate for bio-digesters and earthworms and as organic fertilizer, either directly or after transformation in the biodigesters or by earthworms.

- On all the farms studied, each activity appeared to be profitable in its own right. However, most social benefits were associated with the use of biodigesters, especially for the women.

- There appeared to be few differences in operating parameters between fixed dome (Chinese design) and plastic tube biodigesters but the pay-back period was nine times as long (74 compared with 8.5 months) for the former.

- There was a wide variety of knowledge applied in the recycling processes according to the needs and creativity of the farmers.

- There was a strong relationship between the integration of the farm and the integration of the family and this facilitated the transfer of knowledge to the new generation, enabling them to develop and improve it.

Recommendation

The monitoring of farm activities is a time consuming process. Yet it is an essential step in assessing the productivity of the overall farming system and the relative benefits associated with the component sub-systems.

The present study related to one part of data collected over a period of 5 months between June and November 1995. A more extended study period over the full year would provide more comprehensive data so as to study a wider range of factors, such as for example the contrasts between wet and dry seasons for recycling processes.

Acknowledgements

I would like to thank the Swedish Agency for Research Cooperation with Developing Countries (SAREC) and the people of Sweden who directly and indirectly help to support this kind of project in developing countries.

I am indebted to the six families in: "Sindamanoy", "Villavictoria", "El Porvenir", "El Cambio", "La Esfera" and "El recuerdo" farms in El Palmar Villages and Alfonso Madriñan in Hacienda Arizona in the Cauca Valley in Colombia, for their hospitality and valuable support.

My sincere thanks to Dr Brian Ogle for his continuous assistance and guidance during the course of this thesis.

I gratefully extend my sincere thanks to Dr Thomas R. Preston for his valuable help in this study and his dedication to sustainable agriculture in developing countries.

My sincere thanks to my adviser M.Sc Antonio Solarte for his continuous help, guidance and friendship. I also thank CIPAV's team for their collaboration, hospitality and friendship.

References

Aluja, A. and McDowell, R.E., 1984. Decision marking by livestock/crop small holders in the state of Veracruz, Mexico. Cornell International Agriculture. Ithaca, New York, pp 3-5.

Berdegué, J. and Larraín, B., 1994. Como trabajan los campesinos. Producción Agropecuaria Campesina. Santiago de Chile, pp 5-38.

Bhakta, H. K., 1991. Appropriate technology for rural development: a case study. South Asia Partnership, Kamalady Nepal.

Botero, R. and Preston, T. R., 1986. Biodigestor de bajo costo para producción de combustible y fertilizante a partir de excretas. Cali, Colombia, pp 5-10.

Bui, Xuan An, Khang, Duong Nguyen, Anh, Nguyen Duc, and Preston, T. R., 1994. Installation and performance of low-cost polyethylene tubular digesters in small farms in Vietnam. In: Proceedings of National Seminar-workshop "Sustainable Livestock Production on Local Feed Resources", Ho Chi Minh City, 22-27, November.

Chambers, R., 1993a. Rural Development. Putting the last first. New York, pp 75-91.

Chambers, R., Pacey, A. and Ann, L. T., 1989. Farmer first. Farmer innovation and agricultural research. London, pp 9-11.

Chambers, R., 1993b. Challenging the Professions. Frontier for rural development. London, pp 15-25.

CVC., GTZ. and OEKOTOP., 1987. Difusión de la tecnología the biogás en Colombia. Cali, Colombia, pp 43-110.

CVC. and GTZ., 1990. El Biodigestor. Cartilla básica de extensión No 1. Convenio Colombo Alemán de biogás. Cali, Colombia, pp 4.

CVC., 1995. Monitoreo ambiental para el Municipio de Dagua. Corporación Autónoma para el Valle del Cauca. Sub dirección técnica, grupo de apoyo. Cali Colombia, pp 2-6.

FAO., 1995. Integrated plant nutrition systems. FAO fertilizer and plant nutrition. Bulletin 12. pp 35.

Fuentes, J. L. Y., 1987. La crianza de la lombriz roja. Ministerio de agricultura, pesca y alimentación. Madrid, España, pp 27-35.

Krishna, R. P., 1994. A time proven procedure of biogas plant installation in Nepal. Biogas and agricultural equipment development company.

Marchaim, U., 1992. Biogas process for sustainable development. FAO Agricultural Services, Bulletin 95. Rome, pp 94-120.

MARENA, 1995. Situación de la leña en Nicaragua. Ministerio de Recursos Naturales y del Ambiente (MARENA). Managua, Nicaragua, pp 2-10.

Ortiz, M., 1994. Reactivación de la producción de abono orgánico, Instituto Mayor Campesino. CIPAV. Cali, Colombia, pp 15.

Ramachandra, T. V., 1994. Efficient wood energy devices for cooking and water heating purpose. Centre of Ecological Sciences. Karnataka State, India.

Solarte, L., 1995. Biodigestor plástico de flujo contínuo, generador de gas y bioabono a partir de aguas servidas. Fundación Centro para la investigación en Sistemas sostenibles de Producción Agropecuaria (CIPAV). Cali, Colombia, pp 1-14.

Solarte, A., Dolberg, F. and Corrales, E., 1994. Descriptive study of the local resources used as animal feed and animal production systems in a rural community on the pacific coast of Colombia. M.Sc. Thesis. Swedish University of Agricultural Sciences. Uppsala, Sweden, pp 8-10.

Soto, R., 1993. Avaces proyecto de una fertilización orgánica de café (Coffea arabica) en una finca del municipio de Riofrío. Cali, Colombia, pp 5.

URPA, 1995. El Valle del Cauca en cifras. Unidad Regional de Planificación Agropecuaria   (URPA). Cali, Colombia, pp 4-15.

Valencia, C. E. and Villegas, M. C., 1995. Amor y sabor a diversidad. Cocina recreativa del pacífico Colombiano. CIPAV. Cali, Colombia, pp 5-8.

Van, M. M., 1994. Effect of biogas on the workload of women in Rupandehi-Distric, Nepal. University Leiden, The Netherlands.

Van Nes, W. J., 1994. The biogas support programme in Nepal. Kathmandu, Nepal.

Evaluation of Small Scale Biogas Digesters in Turiani, Nronga and Amani, Tanzania

Lotte Cortsen, Malene Lassen, Helle K.Nielsen

University of Aarhus,
8000 Aarhus C., Denmark
E-mail:
ifsk902004@ecostat.aau.dk, 891355@ps.aau.dk and katrine@stat.bio.aau.dk

Summary

The study was carried out in rural areas of Tanzania, where we were looking at the function and adoption of low cost small scale biogas digesters made of plastic. Our work was accomplished with the help from a Tanzanian non governmental organization, SURUDE, who has been responsible for installing the digesters.

In spite of a general satisfaction with the technology, we found that out of a total of twenty-six digesters installed by SURUDE we visited eleven were working, eleven were not working and four were new. Main problems were too low pressure and gas production, problems with burners and lack of water. It was mainly insufficient training of farmers and follow-up that was the reason for the relatively low rate of functioning digesters. The plastic quality did not seem to be a problem. In spite of firewood being free of cost in most places and thereby eliminating the direct economic benefit, the farmers expressed interest in the digester. The time saved collecting firewood is often counterbalanced by time spent getting water although human and animal urine can be used in its place. Household lighting is therefore, in many cases, the primary motivating factor for wanting a biogas digester in Tanzania, as kerosene is expensive and electricity mostly unreliable or not installed at all. It is concluded that in the future priority must be given to follow-up and maintenance. Equally important is optimizing gas production by correct retention time, research into effect of adding cattle urine, toilet waste, using other locally available material and recycling of slurry. A strong indicator of improvement in gas supply will be sufficient gas to cook a full main meal. Furthermore, the success of the biogas project is dependent on the farmers' payment of the digesters, which can be achieved by establishing an efficient credit system.

Key words: *biogas, digester, plastic, low cost, evaluation, on-farm, adoption, smallholder, Tanzania.*

Introduction

The world population is expected to reach 10 billion people towards the end of the next century (Preston and Murgueitio, 1992). This demands that future scenarios of resource utilization must be predicted on: (1) optimizing the capacity of the Earth's ecosystems to produce biomass, as the only renewable source of energy; and (2) minimizing waste through recycling, which reduces the need for raw materials and helps to protect the environment (Preston, 1995).

Work on integrated farming systems that embody these concepts are seen in Columbia (Preston and Murgueitio, 1992) and Vietnam (An and Preston, 1995) and recently also in Tanzania (CONDRIT Ltda, 1995). One of the components in the integrated farming system is the plastic biodigester where excreta are converted into methane and used as a source of energy. The digester has been widely accepted in especially Vietnam due to its low cost and flexibility (An et al., 1996).

The low cost plastic biodigester was developed by the non governmental organization (NGO) CIPAV (Centro para la Investigacion en Sistemas Sostenibles de Produccion Agropecuaria), located in Cali Columbia based on a design first promoted in Taiwan, known as the "Red Mud PVC" biodigester (CONDRIT Ltda, 1995). CIPAV made some modifications to the Taiwan model to lower the costs. They introduced polyethylene tubular film instead of PVC. This simple and low cost technology was first introduced in Colombia and later transferred to Vietnam and Tanzania. The technology was introduced in Tanzania in March 1993 through the technical Cooperation Programme of FAO, which assisted the government of Tanzania in the transfer and adaptation of technologies validated in other tropical, less developed countries (CONDRIT Ltda, 1995). Biogas technology was already introduced in Tanzania in 1975, but it was a very demanding and expensive type (cost up to 2000 US$) so it did not have a great impact.

The biogas project is expected to have several benefits:

- The women will spend less time collecting firewood (it can reduce the firewood requirements up to 50-70%), and cooking will be easier and cleaner, as there will be no smoke when using biogas.

- The reduction in firewood use will reduce deforestation.

- The increase of income, by saving money spent on charcoal, and by saving time from collecting firewood, which can be used for other income generating activities.

In Tanzania a number of biodigesters have been installed by a young NGO, SURUDE, within the last two years. The materials for one of the simple plastic biodigesters from SURUDE cost about Tanzanian Shilling 40.000 (70 US$), and they are expected to have more impact because they are cheaper. The plastic of the digester is expected to last for 3-4 years depending on management and maintenance.

The substrate for the digesters in Tanzania is cattle manure in contrast to Vietnam where the main substrate is manure from pigs. As cattle manure has a lower methane-producing potential than pig manure (The Biogas Technology in China, 1992) it implies special requirements to try to optimize gas production in Tanzania.

On-farm activities require evaluation and feedback to be successful. The objective of this study was to look at the implementation and adoption of the technology, and identify successful experiences and problems associated with the plastic biodigesters installed in Tanzania.

Materials and methods

Presentation of SURUDE

SURUDE is a young Tanzanian nonprofit NGO. The organization was established in 1993 as a cooperation between farmers and Sokoine University, Morogoro, but first registered in 1994, by Prof. Lekule and Dr. Sarwatt from Sokoine University. The organization is concerned with the promotion and use of renewable natural resources at village level. SURUDE's aim is to demonstrate the potential and development of decentralized farmer-driven technology development and transfer, as the means of improving the standards of living of rural people. Environmental protection is also a key element in SURUDE's programme. This is reflected in the name of the organization as "SURUDE" is an abbreviation of "Foundation for sustainable rural development". The farmers who get a biodigester from SURUDE become automatically members of the organization.

SURUDE has concentrated on especially four activities (Preston and Murgueitio, 1992):

- Installation of biodigesters
- Introduction of crossbred dairy cows
- Introduction of multipurpose fodder trees
- Introduction of sugarcane processing

The aim of the biodigester project is demonstration and training of farmers in installation and maintenance of simple low cost plastic biodigesters.

The more specific objectives are (according to SURUDE's project proposal 1994):

- to install low cost continuous-flow plastic biodigesters in the homesteads of small scale farmers
- to encourage incorporation of latrines as an integral part of the biodigester
- to protect the environment by not cutting trees for charcoal

The target group of the project are farmers with cattle. You need at least two cows to get enough manure to feed the biodigester. The only conditions for getting a biodigester are that the farmers use zero-grazing, and that they build a shed to the cows. SURUDE will distribute the digesters to all farmers who are interested in having one.

At the moment SURUDE has installed just under one hundred biodigesters in the three areas, Turiani, Nronga and Amani, where they operate. Most of them are placed in Turiani and there are three in Nronga and twelve in Amani.

The interviews

We interviewed Thirty-nine farmers from the three areas. Twenty-six of the farmers had a plastic biodigester from SURUDE, the rest would either like to have one soon, or had a more expensive and advanced type of biodigester.

The interviews were made in Turiani (17th-18th of Oct. and 13th-15th of Dec., 1995), in Nronga (24th-31st of Oct. and 13th-16th of Nov. 1995) and in Amani (21st-29th of Nov. 1995).

We managed to speak to all the farmers who had a biodigester from SURUDE in Nronga and Amani. This was unfortunately not possible in Turiani, as there were sixty biodigesters and we did not have time enough to visit all of them. In Turiani we made eleven interviews with farmers selected by SURUDE, meaning that we did not have any influence on the selection of farmers. Nevertheless, we saw both functioning and nonfunctioning digesters. The interviews were made at the farmers' home and usually it took from half an hour to one hour. We asked the same questions, making it possible to compare the answers from the different areas. Sometimes the farmers were expecting us, but in some cases it was not possible to let them know we were coming. Most of the people interviewed were women, as they are often responsible for the farm work - including looking after the biodigester.

Gas production and quality

Gas production was measured with a gas flowmeter in the three areas. The flowmeter used, was a Remus 3 G 1.6 from Compania De Contadores. Unfortunately we only managed to measure production from one digester in each area, meaning that the result is not representative for all the biogas digesters working. The flow meter was in all cases connected between the digester and the reservoir. In Nronga the production was measured in the house of Mrs Helen A. Usiri once a day for 21 days. In Amani production was measured once a day for 7 days in the house of Mr Alimasi Althumani. In Turiani, we measured production at the SURUDE Training Center once every hour for 24 hours. The measurements in Turiani indicate, that the measure is an estimate of consumption rather than production, as there only seemed to be a flow when the burner was on, meaning that gas otherwise was accumulating in the digester bag. To overcome the problem, it would be necessary to let out all the gas for at least 24 hours while measuring production. We did not find this appropriate to do.

The gas quality, i.e. the methane content (CH_4) was measured for the three plants mentioned above. One was measured with a portable gas detector; Digi Flam 2000, directly from the burner and the other two by reaction with base (10 gas samples from each), as the base reacts with carbon dioxide (CO_2) from the gas. The principle is that a low pressure will be created inside the glass vial with the gas sample, when CO_2 is reacting with base e.g. sodium hydroxide (NaOH). This low pressure will draw base into the vial equal to the amount of CO_2 that has reacted, and the percentage of CH_4 in the gas can be calculated. This is assuming that the gas contains only CO_2 and CH_4 and therefore the method might give an overestimation of CH_4 (Rockson, 1992).

Results and discussion

We visited twenty-six biogas digesters and interviewed the farmers in Nronga, Amani and Turiani. Eleven of the digesters were working, four were put up recently and eleven were not working.

We found it interesting to uncover the main reasons for the relatively high number of not functioning digesters. The interviews show a pattern of problems, and reasons for these, which appear to be general. The main problems are: too low pressure, low gas production, lack of water and problems with burners. The reasons are diverse, but they can all be related to three aspects. One aspect is a technical/practical one, concerning the actual management and maintenance of the digester. The second aspect has to do with the organizational design of the biogas project. The third aspect consider the financial sustainability of the biogas project, and is related to the other aspects by being a condition for a successful organizational performance of SURUDE. It is important to stress, that we do not look at the profitability of the digester for the individual farmer. We acknowledge, that profitability might have some influence on the farmers motivation for wanting and maintaining the digester. However, to assess the profitability it would be necessary to conduct a cost-benefit analysis. This was not possible on the basis of our data.

Benefits and problems listed by farmers

The benefits according to the people interviewed (table 1) are in agreement with the expected benefits in several respects. Cooking is easier, cleaner and healthier as there is no smoke in the kitchen and they can get up in the morning and start to cook straightaway. Some people also expressed an economic benefit as they saved charcoal and kerosene, but only in Nronga and at few places in Turiani, did people pay for firewood and got a direct

Table 1. *Benefits according to interview answers*

	Benefits listed by farmers
Turiani	easy to cook, save money for charcoal and kerosene, save labour time, no smoke
Nronga	easy to cook, save money for charcoal, kerosene and wood, save labour time, no smoke
Amani	easy to cook, save money for charcoal and kerosene, save labour time, no smoke, better fertilizer, forest conservation

money saving benefit from the digester. In Amani firewood can be collected without payment from the forest once a week. Several people said they were saving time collecting firewood, however in some places, especially Turiani and Amani, water was very distant and the time spent getting water counterbalanced the time saved from collecting firewood. In one case this problem was overcome by using cattle urine instead of water.

Only in Amani, did people mention the better quality of fertilizer and the forest conservation, as being benefits. The forest conservation issue might be due to a forest conservation project (IUCN) in the area. In spite of the general satisfaction with the technology the benefits do not fully comply with the expectations and a list of problems appeared.

Table 2. *Problems according to interview answers*

	Problems listed by farmers
Turiani	plastic melted, clay burners crack, lack of water, not enough gas for cooking heavy meals, pressure too low
Nronga	not enough gas for cooking heavy meals, pressure too low, a burner with too high capacity
Amani	not enough gas for cooking heavy meals, gas handle broken, cow destroyed the plastic, inlet blocked, pipe blocked, lack of water

The problems listed in table 2 are diverse, but general ones in the three areas are too little pressure and gas production. It was our impression, however, that many of the problems could be solved by small adjustments.

Technical problems

Main problems were too low gas production and pressure. Gas production measured in the three areas was; Nronga: 417.0 l/day, Amani: 539.7 l/day and Turiani: 921 l/day (Cortsen, Lassen and Nielsen, 1995). The gas production rate is the gas amount produced by each cubic metre of fermentation chamber per day. Gas production rates for Nronga, Amani and Turiani were respectively 0,104; 0,164; and 0,279 $m^3/m^3/day$. As comparison, results from Vietnam show a gas production rate of 0.354 $m^3/m^3/day$ for the same type of digester (An and Preston, 1995). The difference might be partly because cattle manure has lower biogas-producing potential than pig manure (The Biogas Technology in China, 1992). Pig manure is the main substrate in Vietnam, whereas cattle dung is used in Tanzania.

The low pressure is due to the design of the digester, but is partly over-come when the reservoir is full. There is a big wish to connect the digesters to biogas lamps, which we saw one successful attempt to do. In this par-ticular case there were two digesters in a series. One vertical digester, which had a stirring mechanism, produced gas for a lamp, and a horizontal digester with continuous flow produced gas for cooking. The effluent from the first digester was used in the second and therefore served as an inoculum. To get enough pressure for connecting lamps to a single digester it will be necessary to do some modifications to the design as it is now. However, if it is made possible to connect lamps it probably would pro-mote the dissemination of the technology as the supply of electricity is un-reliable and kerosene is very expensive.

The capacity of the digester size as promoted by SURUDE might be too small for the energy requirement according to household size (Cortsen, Lassen and Nielsen, 1995), but small adjustments can be done to increase gas production, and thereby also compensate for the pressure problem. Effort should be done to keep the temperature stable, for instance by dig-ging the digester further under the ground, and retention time should be adjusted according to the temperature in the given area. Care should be taken not to overload the digester as it can cause "wash out" of the methanogenic bacteria and souring of the slurry caused by accumulation of acids (Fulford 1988). A way to check if the digester is being overloaded, is to look at the colour of the effluent, dark colour indicating full digestion (Rodriguez, personnel communication). Alkalinity, pH or CH_4 content of the gas can also indicate overload (Fulford 1988), however, these measures are more difficult to monitor. The CH_4 content was measured for one di-gester in each area, values being 62%, 75.7% and 67,3% respectively for Nronga, Amani and Turiani. Recirculation is important for inoculating fresh material with active bacteria (The Biogas Technology in China, 1992) and its significance should not be underestimated. We only found three of twenty-six who actually did recirculate the slurry indicating that farmers need more information on this matter. Recirculating more of the effluent can also compensate for the lack of water as can the use of urine. Dung from cattle fed with poor feed tend to have high C:N ratios and gas pro-duction can then be enhanced by adding cattle urine or fitting latrines to the plant (Fulford, 1988). One family used entirely urine instead of water for the biodigester and had no problems so far, meaning that this might be another solution to the water problem in areas where water is very distant.

Organizational implications

In some cases the digester was not functioning due to simple maintenance problems like a broken gas handle or a cracked burner. At one house a cow had broken the plastic and in one case it was melted due to the cover (iron sheets), but generally the plastic quality did not appear to be a problem. It should be added that none of the digesters visited were older than two and a half years. Instead of solving the problems people were waiting for a man from SURUDE. This attitude needs to be changed trough better training of the farmers. This means, that the organizational performance of SURUDE must be improved. The training of farmers must comprise instruction in how to operate the digester as well as information about the type of problems that might occur. Even with this better training the farmers need some practical assistance that must be available nearby. This could be accomplished by the stationing of local technicians who must be supported in terms of payment, flow of information and education. Better communication is needed to insure the dissemination of innovations and knowledge. This can be done by having the locally stationed technicians meet regularly. Another way of strengthening the communication is continuously to let a man check up on the development in the different areas.

Financial sustainability

The organizational tasks require manpower and economic resources. As only a few of the farmers have started to pay for the digester, SURUDE has to establish an effective credit system that ensures the payment of the biodigesters to make the biogas project financially sustainable.

In Amani and Nronga it could be a possibility to deduct a monthly instalment from the milk sale. In Turiani it is necessary to think of another solution. Perhaps SURUDE could demand a deposit before they install the biodigester. Another way to obtain financial sustainability is to finance the biogas project through other income generating activities, but this will require more manpower and resources.

It is important that people are motivated to engage in the work of the biodigester and we think, that if people pay for it they will be more responsible in maintaining and taking care of it, but our data is inadequate to support this statement.

We looked for a coherence between the function of the digester and the profession of the person in the household, who is responsible for taking care of it as such a relationship is suggested in data from Vietnam (An, 1996). However, only few of them had another profession next to being a farmer, so it was not possible to conclude on this issue.

Conclusion

In the present study it was mainly insufficient training of farmers and insufficient follow-up by SURUDE that was the reason for the relatively low rate of functioning digesters. The plastic quality did not seem to be a problem. In spite of firewood being available free of cost, the farmers expressed interest in the digester. We think it will promote the dissemination of the technology if it can contribute to household lighting. Time saved from collecting firewood is often counterbalanced by the time spent getting water for the digester because water is very distant in many places. However, using urine instead of water, or recirculating more of the effluent might be a solution.

In the future priority must be given to follow-up and maintenance. Equally important is optimizing gas production by correct retention time, research into effect of adding cattle urine, toilet waste, using other locally available material and recycling of slurry. A strong indicator of improvement in gas supply will be sufficient gas to cook heavy meals. Furthermore, the success of the biogas project is dependent on the farmers payment of the digesters. This can be achieved by establishing an efficient credit system.

Acknowledgement

We would like to thank the people from SURUDE, Dr. S. Sarwatt, Prof. F.P. Lekule, Mr. Moses Temi, Mr. Peter Machibula and Mr. Alimasi Althumani and the people on the training center in Turiani, for introducing us to their work with the biodigesters in Tanzania. We are very grateful, that you took the time to make all the necessary arrangements for our field work. Also, we warmly thank the Nronga Women Dairy Cooperative for their great hospitality. Especially we would like to thank the members of the executive committee and the chairlady Mrs. Helen Usiri.

We send our warmest thanks to Mrs. Renalda Mrutu for taking care of us in Amani. We also want to thank Mr. Mody M. S. Nyimbile for taking time to help us with our interviews, and the people from the IUCN project for letting us stay at their hostel. A special thanks to all the farmers, who received us in their homes and took time for the interviews. Finally, we thank our teacher, Mr. Frands Dolberg, for supervision and for establishing the contact to SURUDE.

References

An B.X. and Preston T.R. (1995). Low-Cost Polyethylene Tube Biodigester Development in Vietnam. University of Agriculture and Forestry, Thu Duc, Ho Chi Minh City, Vietnam.

An B.X., Preston, T.R and Dolberg, F. (1996). The introduction of low-cost polyethylene tube biodigesters on small farms in Vietnam. Swedish University of Agricultural Sciences, Department of Animal Nutrition and Management, Uppsala.

CONDRIT Ltda. (1995). Improvement and Promotion of Turbular Biogas digesters, report to Coordinator Farming System Program (Dar es Salaam).

Cortsen L., Lassen M. and Nielsen H.K. (1995). Small Scale Biogas Digesters in Turiani, Nronga and Amani, Tanzania. A student report. Aarhus University, Denmark.

Fulford D. (1988). Running a Biogas Programme, A handbook. London, Intermediate Technology Publications.

Rockson J.K. (1992). Biogas production. Hurup, Denmark: Nordvestjysk Folkecenter for Vedvarende Energi

Preston T.R. (1995). Tropical animal feeding: a manual for research workers, FAO, Rome.

Preston T.R. and E. Murgueitio (1992). Strategy for Sustainable Livestock Production in the Tropics, Cali, Colombia: CONDRIT Ltda.

SURUDE's (1994). Project Proposal to DANCHURCHAID 1994, SURUDE, Morogoro.

The Biogas Technology in China (1992): Chengdu Biogas Research Institute of the Ministry of Agriculture, P.R.C., Beijing: Agricultural Publishing House.

Low Rainfall Agriculture

Use of Crossbred Dairy-Draught Cows for Optimal Use of On-Farm Resources and Sustainability in the Ethiopian Highlands Mixed Crop-Livestock System

E. Zerbini and C.E.S. Larsen

International Livestock Research Institute (ILRI),
P.O.Box 5689, Addis Ababa, Ethiopia
E-mail: e.zerbini@cgnet.com

Summary

On-station research results has showed that using crossbred dairy cows for both milk production and draught work is a feasible technology as long as the increased energy requirement is met. Working cows will have increased dry matter intake, a slightly lower milk yield but increased oestrous cycle length. Economic analysis based on three years of on-station data has showed that the value of work more than compensated for the small decline in milk production and longer productive cycle found in supplemented working cows. An anthropological survey conducted in the on-farm research area before implementation showed that although the technology was new to the farmers, 20% of them were willing to try it. The most likely group to adopt the technology are the younger and better educated, who have slightly smaller households, and considerably smaller land, grazing land holdings and herd size. The research is on-going, but indicators for cow traction technology adoption on-farm could be: Farmers continue to use cows for ploughing after the research project ends, farmers who have received crossbred cows are reducing the number of their local cattle, farmers in the research area who were against cow traction changed their mind, new farmers spontaneously adopt the technology.

Key words: Animal traction, milking cows, intensification, on-farm research.

Background

In most high potential agricultural areas of Africa the pressure on land resources from human and livestock is increasing rapidly. Traditional agricultural systems have been sustainable for centuries but they often require relatively large supporting land area. The cultivated land needs to be left fallow for years to recover its fertility since the input level is low. In addition, large herds need vast areas of natural pasture land in order to cover their nutritional requirements year-round. When the population and livestock pressure increases the delicate ecological balance could move towards increased soil degradation and accelerated erosion.

In the Ethiopian highlands draught oxen are used for ploughing and threshing. An estimated 7 - 8 million oxen cultivate around 90% of the arable land in Ethiopia. However, oxen work only for approximately 60 days a year but must be fed all year round competing for scare feed resources with other livestock. The use of dairy cows for traction could benefit total farm output and incomes through increased milk production, while alleviating the need to feed draught oxen year-round and to maintain a follower herd to supply replacement oxen (Gryseels and Anderson, 1985; Gryseels and Goe, 1984; Barton, 1991; Matthewman, 1987). Besides contributing to a better utilization of scarce feed resources, the use of dairy cows for draught would allow males to be fattened and sold younger, and could also lead to greater security of replacements. More productive animals on farm could result in a reduction of stocking rates and overgrazing, thus contributing to the establishment of a more productive and sustainable farming system.

The primary need of the working animal, especially where the main feed source is roughage, is to increase feed and ME intakes to meet energy requirements for work and avoid deleterious body weight losses. This becomes more critical in working cows requiring extra energy for lactation and reproduction.

Objective

- The first objective of this research was to determine the technical relationships between work, milk production, reproduction and feed utilisation in crossbred dairy cows used for draught.
- The long term objective is to develop cow traction technologies, including improved feed production and management systems to optimize farm productivity, and increase sustainability in the Ethiopian Highlands mixed crop-livestock farming system.

The hypothesis is that less but more productive cattle will increase the overall farm productivity and reduce the livestock pressure on the land. Increased revenue, from more productive animals could be used for improving productivity in other sectors of the crop/livestock system.

Research methodology

The approach used in testing the dairy-draught cow technology and its transfer on-farm has been interdisciplinary. A team of scientists within the disciplines of animal science, agricultural economy and anthropology has designed, implemented and evaluated research activities in a dynamic process of optimising the available physical and human resources.

Scientists implementing the project represent different institutions. The two main collaboration partners are the International Livestock Research Institute (ILRI) and the Ethiopian Institute of Agricultural Research (IAR).

IAR has provided the research site and facilities while ILRI has partly funded the research by providing international researchers, vehicles and other operational inputs. An additional sponsor of the project is the World Food Programme through its dairy development programme in Ethiopia.

For the on-farm research a collaboration agreement has been made with the Ministry of Agriculture (MoA) coordinating forage production and animal health activities. Collaborations is also made with anthropologist from Department of Sociology at Addis Ababa University and with graduate students from Alemaya University of Agriculture and the Veterinary Faculty of Debre Zeit.

The research is conducted at different levels: from strategic in the first on-station phase to more applied in the second on-farm phase. In the second phase emphasis is concentrated on verifying the technology feasibility and technology transfer methodology (Figure 1).

Figure 1. *Research set-up for the crossbred dairy-draught project*

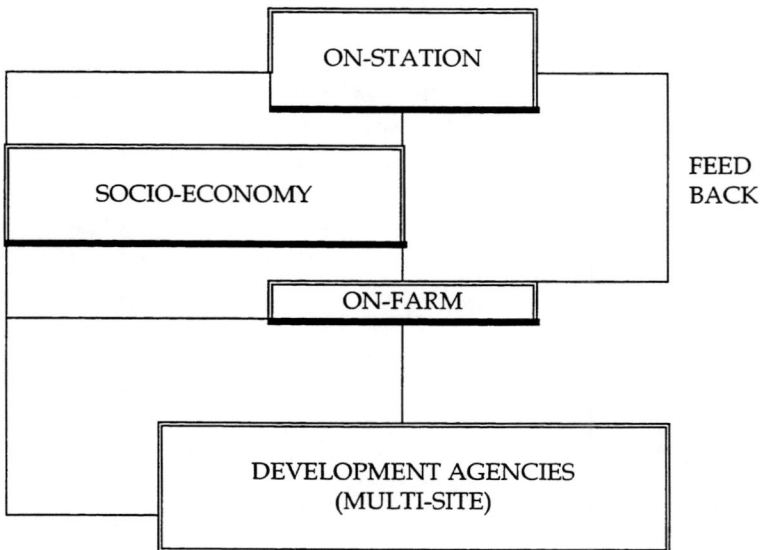

On-station research results

Production parameters

A study evaluated the technical relationship between milk production, reproduction, feed utilization and draught work in F_1 (Friesian x Boran and Simmental x Boran) crossbred dairy cows used for draught. The research was conducted at the Holetta Research Station of the Institute of Agricultural Research (IAR), located 45 Km west of Addis Ababa, at an altitude of 2400 m and with mean annual rainfall of 1100 mm. Cows were assigned to work 100 days a year for three years pulling sledges with a draught pull of about 400 N (10% live weight). Cows were either supplemented to meet their estimated requirements or fed a natural pasture hay diet which met only partially their requirements.

Over a period of two years, dry matter intake and digestibility were greater for working compared to non working cows (114.3 v. 99.5 g/kg live weight[.75] and 0.54 v. 0.51, respectively; p<0.05) (Zerbini, et al., 1995) (Table 1). Non supplemented cows lost weight throughout the first two years period, while supplemented cows tended to maintain or gain body weight over one and three years. Over two years, supplementation of working cows reduced liveweight loss by 73% and doubled the number of conceptions and parturitions (Gemeda, et, al. 1995). Milk yield was similar for working and non working cows, but was greater for supplemented compared to non supplemented cows (1986 v. 2206 kg and 3115 v. 1077 kg, respectively). Draught work significantly increased oestrous cycle length (20.5 to 32.5 days; p<0.5). This was due to the occurrence of ovulations without oestrus during the working period. However, once cows started ovulation they continued to ovulate regularly whether they were worked or not. Postpartum anestrous interval was extended and conception was delayed (Zerbini, et al., 1993a; Bamualim et al., 1987). The occurrence of ovulation without oestrus during the working period seems to decrease with time as cows adapt to draught work activity. Conception rate tended to be greater for non working compared to working cows (62.5 v. 40%, respectively). The diet effect was considerably larger than the work effect (73 v. 30%, respectively, for supplemented and non supplemented cows). A productivity index (PI) (output/input) was used to describe the overall productivity of cows in each treatment group. The productivity index over a period of two years of non working and working cows was similar (0.32 and 0.30, respectively).

$$PI = (ME\ 4\%\ FCM^* + ME\ completed\ pregnancy + ME\ work)/ME\ food\ intake.$$

*FCM = Fat corrected milk

Table 1.   *Performance parameters of crossbred dairy cows used for draught over a period of two years*

Parameter	Working cows (n=20)	Non working[a] cows (n=20)	SEM	Significance work
Dry matter intake g/Kg$^{0.75}$	114.3	99.5	3.5	**
Organic matter digestibility	0.54 (n=18)	0.51 (n=18)	0.017	*
Milk yield (Kg)	1986	2206	219.0	
Oestrus cycle length (days)[b]	32.5 (n=8)	20.5 (n=8)	4.0	*
Productivity index	0.32	0.30	0.03	

[a] Include supplemented and non supplemented cows in equal proportion

[b] Only supplemented cows

Over a period of two years, diet was the main factor which affected milk production, number of conceptions and number of calvings of crossbred cows, whether working or not. On the other hand, supplementation did not affect work output of cows. Work performed by supplemented cows had no adverse effects on lactation but delayed conception.

Economic evaluation

The ILRI Herd Model was modified and used to treat the effects of the introduction of supplementary feeds and forages and draught work, through changes in herd productivity. To generate the multi-year effects of working, 3 years of on-station experimental results were used to simulate the herd production parameters over ten years. The multi-year production effects of using crossbred cows for both milk and traction were compared to using supplemented crossbred cows for milk production alone. In addition introduction of supplemented crossbred cows for both milk and traction was compared to the traditional system of using local cows for milk production and local oxen for work.

The incremental internal rate of return for supplemented working cows over supplemented non-working cows on-station conditions is greater than 100% and the benefit/cost ratio is about 3.5. The value of work more than compensated for the small decline in milk production and longer productive cycle found in supplemented working cows.

Anthropological survey

In 1993 before starting on-farm research activities, an anthropological survey was carried out in the research area, by Dr. Alula Pankhurst from

Addis Ababa University. The survey included a total of 52 farmers (Pankhurst, 1993).

The survey showed that the cow traction technology was virtually unknown to farmers. When asked if they believed that ploughing with crossbred cows was possible 20% responded positively while 80% responded negatively. Cultural reasons included concern about the consensus of other members of the household or of the village community. Other included reasons such as: "It is morally wrong to use cows"; "God created cows to give calves and milk and the ox to plough". However, more than 50% of the reasons against cow traction were of technical nature such as: "The cow can not plough and give milk at the same time"; "Pregnant cows can not plough because they will then get weak calves"; "Cows are not strong enough to plough on stony, hilly, muddy fields".

The fact that on-station results had shown the technical feasibility of cow traction technology and that 20% of the interviewed farmers agreed to test the technology was encouraging and led to on-farm testing.

A follow-up anthropological survey was conducted in 1995 with in-depth open-ended interviews with 131 farmers in 13 sites in the Holetta-Addis Alem area (Pankhurst, 1995). This survey included farmers already participating in the project for the past two years, farmers selected in 1995, and non participating farmers. One of the important questions was whether there has been a change in attitude since the previous survey two years earlier. Many of the farmers reported that people were initially suspicious of the project but that much of the distrust has disappeared during the past 3 years. Fears related to the constant measuring of feed, yields etc, produce and the tagging of animals seem to have abated. Several of those who received crossbred cows reported that previously farmers who were visiting them considered the crossbred cows as unprofitable. However, presently their attitude has changed and they want to obtain crossbred cows themselves. Those who believe that cows can plough are younger and better educated, have slightly smaller households, and considerably smaller land, grazing land holdings and herd size.

"Ploughing with oxen was only introduced by man and not given from God". Ato Tolla was thus suggesting that it is cultural and not natural. A woman by the name of Sandafe from the neighbourhood appeared while we were discussing cow traction and reacted: "If the cows have to do two things they have to eat much". I asked her: "Are you convinced that they can plough if they eat enough?". She replied "No. However, we have to live according to the order of the day". She meant that circumstances are going to force people to plough using cows even if they may not want to do so.

On-farm research

Methods

The on-farm trail is carried out in the central highlands of Ethiopia, in Addis Alem and Wolmera woredas, 50 km west of Addis Ababa. These woredas are situated at an altitude of 2200 to 2400 m and receive an annual rainfall of 1060 mm. Mean maximum temperatures range from 18.7 to 24.0°C. This research is conducted by ILRI in close collaboration with scientists from the National Institute of Agriculture Research at Holetta Research Centre and with extension workers from the local Ministry of Agriculture offices in Holetta Genet and Addis Alem.

The on-farm trail started in the spring of 1993 with only 14 farmers and was gradually expanded from late 1994 to mid 1995 to a total of 66 project farmers and 20 control farmers. In the beginning of 1996 it will reach it's full scale with 70 project farmers and 60 control farmers.

The initial 14 project farmers were divided into two groups; seven farmers were only using cows for milk production and seven farmers used cows both for milk production and draught work. Each farmer was provided with a pair of crossbred cows (F_1 Friesian x Boran) at subsidised price. However, before they could buy the two crossbreed cows there were a number of conditions farmers had to fulfil. Each farmer had to plant a minimum of a half hectare of oats and vetch mixture, and plant napier grass, fodder beets and multi purpose trees in his backyard. Moreover, farmers were advised to sell three to four local animals, so that on-farm stocking rate would remain similar.

An enumerator was assigned for every two farmers to collect technical and socio-economic data. Enumerators reside and work six days at each farm.

The following biological data is being collected for all cattle:

- Feeding (behaviour, feed intake, grazing time)
- Milk production (yield)
- Calf rearing (suckling time, weaning, castration)
- Reproduction (heat detection, mating, calving date, sex)
- Work (days worked, hours worked per day, area of land ploughed per day, type of land ploughed) by cows and oxen
- Health (diseases, diagnose, treatment, drug, doses, cost)
- Body weight and condition score

Most of the initial 14 farmers were relatively rich farmers, with large landholding, large herd size and surplus of manpower. These farmers were mainly chosen because they had showed a firm interest in the project. When additional 60 farmers were included in the project it was decided that these farmers should be more representative. The quest for additional farmers was forwarded to different project assistants in the two woredas. Encouraged by the benefits they had observed that the first 14 farmers had obtained by participating in the research programme, a large number of farmers enlisted.

Conditions for participation and the research set-up for the additional 60 farmers was like the one for the first 14 except that data collection is presently monitored one week per month. For the remaining three weeks data collection is based on interviews. With the less intensified data collection scheme one enumerator can cover four farmers. The same type of data is collected from the control farmers.

The interested farmers were divided into three different categories according to their resource level (Table 1). Moreover, a group of farmers was selected to serve as control group. The group of resource rich farmers is the largest because most of the first 14 farmers was found to belong to this category.

Table 1. *Categories of farmers*

Categories of farmers	No.	Characteristics
Resource poor farmers	17	0-1 ox, less than 5 animals, 1.5 Ha under plough, limited labour
Medium income farmers	20	One pare of oxen, less than 10 animals, 2.5-3.0 Ha under plough, family labour
Resource rich farmers	33	Minimum two pare of oxen, more than 10 animals, above 3 Ha under plough (rent additional land), additional income from off farm activities.
Control farmers	60	Selected at random in the areas where the project is conducted. Except from participating in training and discussion session these farmers are following traditional farming practice without crossbred cows and use local oxen for land cultivation.

For implementation of the on-farm research trial the major activities have been divided into six packages, most of them are carried out in close collaboration with extension officers and veterinarians from the local Ministry of Agriculture offices. The packages are as follows:

- *Improved genotype:*
  A pair of cows (F_1 Friesian x Boran) with larger body frame and a higher production potential than local cows.

- *Forage package:*
  Farmers are advised to plant minimum ½ ha of oats and vetch for hay production each year. In addition a backyard forage package has been

developed recommending farmers to plant Napier grass, fodder trees (Tagasaste and Sesbania) and fodder beets on their compound.

- *Health package:*
  The project provides veterinary drugs. The health scheme consist of regularly vaccination, de-worming and spraying procedures as well as routine visits to all project farmers. Moreover, emphasis is put on advising farmers to improve hygiene procedures and practice restricted grazing.

- *Breeding package:*
  The scheme consist of heat detection, timely insemination, pregnancy testing (PD) and control of reproductive diseases. All project cow are serviced with 50% Friesian x Boran semen through AI. The offspring will be serviced either with crossbred bulls (50%) or with local bulls. The scope is to maintain a population close to 50% exotic blood on-farm.

- *Improved management of cows and calves:*
  This package includes areas such as: stalling, calf rearing, heifer rearing, crossbred cow management, milking, crossbred cow management, milking and milk handling, draught work, manure handling and herd size.

- *Training package:*
  The aim of the training package is to increase farmers' awareness of the advances and constrains of the introduced technology. A complementary objective is to get direct feedback from farmers on the technology adoption process.

Outputs which are quantified and which will be used to evaluate dairy-draught technologies:

1. Milk
2. Draft work
3. Calves
4. Manure
5. Herd size
6. Household income and food consumption

However, this production related justification can not stand alone. It is also necessary to look on qualitative aspects such as:

7. Family priorities
7. The gender issue (inter household power balance)
8. Farmers attitude toward the new technologies
9. Adaptation in the surrounding society.

Results

Preliminary on-farm research results on production (Table 2) have shown that milk production of working and non working F1 crossbred cows on-farm were similar (2620 vs 2980 kg), ranging from 2010 to 3,400 kg for working cows and from 2018 to 3907 kg for non working cows. Calving intervals for working and non working cows were 525 and 495 days, respectively. First lactation average milk yield and days in milk of working and non working cows were 1,864 and 2252 and 376 and 410 days, respectively. Average service per conception for working and non working cows were 2.1 and 1.9, respectively. Over a period of two years cows worked an average of 26 days/year.

Table 2. *Performance parameters of working and non-working crossbred cows under on-farm conditions*

Parameter	Working cows (n=14)	Non-working cows (n=14)
Milk Production two years (kg)	2,620	2,980
Calving Interval (days)	525	495
Milk Production First Lactation (kg)	1,864	2,252
Lactation Lengths (days)	376	410
Service per conception (No.)	2.1	1.9
Work days/Year	26	0

References

Bamualim, A., Ffoulkes, D. and Fletcher, I.C. (1987). Preliminary observations on the effect of work on intake, digestibility, growth and ovarian activity of swamp buffalo cows. Draught Animal Power Project, DAP Project Bulletin, No. 3, pp. 6-10. James Cook University, Townsville (Australia).

Barton, D. (1991). The use of cows for draught in Bangladesh. ACIAR. Draught Animal Bulletin, N 1, pp 14-26. James Cook University of North Queensland, Australia.

Gemeda, T., Zerbini, E., Alemu G/Wold and Dereje Demissie. (1995). Effect of draught work on performance and metabolism of crossbred cows. 1. Effect of work and diet on body weight change, body condition, lactation and productivity. Animal Science 60:361-367.

Gryseels, G. and Anderson, F.M. (1985). Use of crossbred dairy cows as draught animals: experiences from the Ethiopian highlands. In Research Methodology for Livestock on farm trials. Proceedings of a workshop held March 25-28 at Aleppo, Syria. p.237-258. International Development Research Centre, Ottawa.

Gryseels, G. and Goe, M.R. (1984). Energy flows on smallholder farms in the Ethiopian highlands. International Livestock Centre for Africa, Addis Ababa, Ethiopia. ILCA bulletin 17:2-9.

Matthewman, R.W. (1987). Role and potential of draught cows in tropical farming systems: a review. Tropical Animal Health and Production 19:215-222.

Pankhurst, A. (1993). Anthropological survey of crossbred cows for dairy production and draught work. Internal report. 30 pp.

Pankhurst, A. (1995). Crossbred cows for dairy production and traction. A report based on case studies of 131 farmers and their attitudes to the Holetta project. Internal report. 95 pp.

Zerbini, E., Gemeda T., Franceschini, R., Sherington, J. and Wold A.G. (1993a). Reproductive performance of F1 crossbred cows. Effect of work and diet supplementation. Animal Science 57:361-369.

Zerbini, E., Gemeda T., Wold, A.G. and Demissie D. (1995). Effect of draught work on performance and metabolism of crossbred cows. 2. Effect of work on roughage intake, digestion, digesta kinetics and plasma metabolites. Animal Science 60:369-378.

Livestock Systems in Semi-Arid Sub-Saharan Africa

Brian Ogle

Department of Animal Nutrition and Management, Swedish University of Agricultural Sciences, P.O.Box 7024, S-75007, Uppsala, Sweden. E-mail: brian.ogle@huv.slu.se

Summary

Sub-Saharan Africa's population is predicted to grow rapidly during the next 35 years to 1.3 billion. This calls - among other initiatives - for intensification of animal production, if Africa has to meet its increasing demand for livestock products from domestic resources. On this background the paper discusses climatic and environmental determinants of livestock development in the various production systems typical of Africa and constraints to increased production from livestock. Results from a case study of the Dodomo (in Tanzania) Soil Conservation Project's experience with introduction of milking crossbred and zebu cows are presented. It is concluded that due to growing population pressures on available agricultural land, farming systems are evolving towards intensification and integration, and it is unlikely that the basic nature of this evolutionary process can be substantially altered by development initiatives, which, however, may attempt to make crop-livestock systems more productive and sustainable.

Key words: Livestock systems, Sub-Saharan Africa, intensification.

Introduction: The role of livestock in feeding Africa's growing population

It is projected that in the next 35 years or so the population of Sub-Saharan Africa will increase by about 2.5 times to around 1.3 billion (Winrock International, 1992). Current population growth is around 3.1 % per year, the highest of any region in the world, and the most rapid population increases are occurring in the urban areas.

If livestock production continues to grow at the same rate as between 1962 and 1987 (2.6% for meat and 3.2% for milk) the region will face massive deficits in milk and meat supplies by 2025, and around 11% of milk is already imported. The increasing urbanization of the region is cause for alarm, but if coupled with income growth will provide a growing market for livestock products for rural as well as peri-urban farmers.

Table 1. *Population projections for Sub-Saharan Africa, 1990-2025*

Year	Population (millions)	Annual growth rate (%, 5 year period)
1990	498	-
1995	580	3.1
2000	676	3.1
2005	784	3.0
2010	902	2.8
2015	1,028	2.6
2020	1,159	2.4
2025	1,294	2.2

It has been estimated by the World Bank that around 10% of the population of Sub-Saharan Africa are primarily dependent on their animals, while another 58% depend to varying degrees on their livestock. Increasing population pressures will lead to increased intensification of agriculture, as growing competition between crop and livestock farmers leads to the evolution of mixed crop-livestock systems, which are the most efficient and sustainable means of increasing food production (Winrock International, 1992).

Climatic and environmental determinants of livestock development in Sub-Saharan Africa

Sub-Saharan Africa can be broadly divided into 6 agro-ecological zones on the basis of rainfall, altitude and length of the annual plant growing period: desert, arid, semi-arid, sub-humid, humid and highlands. Generally, the economic importance of livestock in African farming systems increases with decreasing rainfall, as table 2 illustrates.

Table 2. *Value of agriculture and livestock products in selected African countries,1988*

Country	Climate	Livestock value, ($, millions)	Livestock share of agric output,
Botswana	arid	107	88
Mauritania	arid	158	84
Ethiopia	semi-arid	1,299	40
Kenya	semi-arid	826	38
Uganda	sub-humid	404	14
Zaire	humid	143	5
Total*		11,839	25

* Mean of 41 countries in Sub-Saharan Africa. Source: US Dept. of Agric., 1990

However, the estimates shown above do not include the value of manure and animal traction, which in East Africa equals the value of meat (table 3) and for Sub-Saharan Africa as a whole would increase the total gross value of livestock products by about a third. As mixed crop-livestock systems expand the relative importance of animal traction and manure will grow.

Table 3. *Relative contributions of food and food-related products to the total value of livestock production, Sub-Saharan Africa*

Output	Percent of gross value of output				
	West Africa	Central Africa	East Africa	Southern Africa	Sub-Saharan Africa
Animal traction	21	3	39	26	31
Manure	4	1	3	2	3
Meat	56	79	38	58	47
Milk	11	12	17	9	15
Eggs	8	5	3	5	4
Total	1460	349	3747	930	6486

Extensive livestock-based systems

Traditional agriculture in Africa was generally based on mixed crop-livestock systems, with pure livestock systems predominating only when rainfall was too low and uncertain to support some form of crop production (table 4).

Table 4. *Types of indigenous stock keeping in Sub-Saharan Africa*

Rainfall (mm/year)	Predominant type of farming	Main species kept
Under 50	Occasional nomadic stock keeping	Camels
50-200	Nomadism with long migrations	Camels
200-400	All types of nomadism, transhumance with some arable farming	Cattle, goats, sheep
400-600	Semi-nomadism, transhumance, but more emphasis on arable farming	Cattle, goats, sheep
600-1000	Transhumance and partial nomadism, even more emphasis on crops	Cattle, goats
>1000	Sedentary agriculture (semi-nomadism only the result of tradition)	Cattle, goats Monogastrics

Similarly pure arable systems only developed where disease and parasites precluded most forms of livestock keeping. However, even in notionally pure livestock systems pastoralists usually practice some form of opportunistic seasonal cropping, and in primarily arable systems farmers normally keep monogastrics or small ruminants as scavengers around the homestead.

Pastoral systems

Africa has the largest area of permanent pasture of any continent, and the largest number of pastoralists, with between 15-25 million people being mainly dependent on livestock. The arid (< 500 mm annual rainfall) savannas in Sub-Saharan Africa amount to almost 800 million hectares, around four times the area of cultivated land, and the semi-arid zones (500-1000 mm annual rainfall) an additional 400 million hectares. These regions include a wide variety of agricultural ecosystems, but generally the importance of livestock increases with decreasing rainfall.

Although the number of pastoralists in Sub-Saharan Africa is rather small they control around 15% of the cattle and small ruminants and the vast majority of the camels. Pastoralists are opportunists, using their traditional knowledge and reacting to changes in climatic and ecological patterns to find on a day to day basis the best forage and water resources available. Relatively few are pure nomads, the majority practising some form of partial nomadism or transhumance. which depicts the annual movements of the Baggara arabs of Western Sudan who migrate southwards in the dry season to take advantage of the wetlands grazing, and then move northwards in the rainy season to escape the mud and biting flies and utilize the ephemeral forage in the semi-arid rangelands.

In fact, there is a complex and shifting continuum between pure nomadism, partial nomadism, transhumance and sedentary agriculture, as illustrated in Figure 1.

Fig.1. *Diagrammatic representation of the continuum between pastoralism and sedentary agriculture (after Payne, 1990)*

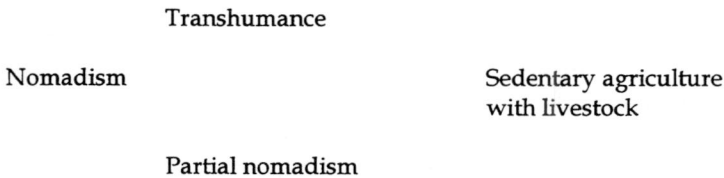

Transhumance

Nomadism   Sedentary agriculture
                                             with livestock

Partial nomadism

A common misconception concerning pastoral systems is that they are inefficient and unproductive, but the data presented in table 5 show that this is not the case.

Transhumance in particular is an extremely productive system, yielding between 50 and 600% more protein per ha than "modern" ranching in comparable ecological areas in the USA and Australia.

Table 5.   *Livestock production in the Sahel and two comparable regions*

Region	Livestock system	Protein (kg/ha/year)	Fossil energy input (Mcal/man-hour)
USA	"modern" ranching	0.3-0.5	25-35
Australia	"modern" ranching	0.4	150
Sahel	Nomadism	0.4	0
Sahel	Transhumance	0.6-3.2	0
Sahel	Sedentary	0.3	0

Source: (After Bremen and de Wit, 1983)

Constraints to the development of extensive systems

Population pressures are leading to increased encroachment of crop growing into rangelands, and this process is encouraged by traditional tenure systems, which are gained by cultivation, but not by migratory grazing. The result is the of spread of unsustainable cropping systems - which after only a decade or two often lead to irreversible land degradation - at the expense of sustainable pastoral systems, which depend on the ability to move herds to the dry season grazing areas for survival.

Another problem is the increasing tendency to regard livestock as capital investments. Wealthier farmers and urban dwellers invest their surplus capital in livestock, usually cattle, because of the lack of saving and other investment opportunities, and this tends to direct production strategies towards risk avoidance rather than improving productivity (Gilles, 1991). This usually involves attempting to increase numbers of animals and keeping a high proportion of males and older less productive cows, whose main advantage is their ability to survive.

The tsetse fly: scourge or saviour ?

The tsetse fly covers an area of around 10 million square km, making integrated crop/livestock systems (particularly those involving cattle) impossible. These regions include most of sub-humid and much of semi-arid Sub-Saharan Africa, and it has been estimated that the area could support 100-140 million extra cattle, and as many sheep and goats, and generating around 50 billion US$ in livestock products and additional food production from crops. In theory bush clearance eliminates the tsetse fly, but the cost is very high ($400-2,000 per square km), and it would take $4,000 billion to clear the whole tsetse area of Africa. Residual insecticides (eg DDT) are also extremely costly (and environmentally damaging) and biological control is almost as expensive. Regular dosing with drugs enables cattle to achieve milk and meat yields only around 20% lower than in tsetse-free areas, but this is not a viable solution for pastoralists and smallholders, so the best approach would seem to be to expand the stocks of resistant breeds of sheep, goats and cattle (such as the N'dama of West Africa).

However, in environmentally fragile areas rapid expansion in the number of grazing ruminants often results in overgrazing and land degradation, so unless some form of stall-feeding system is introduced the best long-term strategy could well be to leave things as they are.

In the more humid regions it has been estimated that sheep and goats are capable of providing around twice as much meat from a given input as cattle, and they could form the basis of some form of intensified mixed farming. For example they could be confined and fed diets based on leaves and pods from fodder trees grown in alley cropping systems as well as crop residues and agro-industrial by-products.

Do livestock create deserts ?

The concept of desertification being caused primarily by man-made activities is being increasingly questioned, and many experts now conclude that changing patterns of rainfall are manly responsible for the shifting boundaries of the Sahara. A United Nations report from 1991 concluded that " ....only under extreme densities of cattle is soil erosion serious..." and overgrazing has been greatly overestimated as an environmental problem. Similarly, Winrock International reported in 1992 that .... "heavy grazing has not seriously decreased the productivity of rangelands in Africa", ... and the greatest threat comes from human populations and the expansion of cultivation. Traditional land use systems encourage cropping, as tenure rights are usually secured by cultivation, not grazing.

However, there are certainly regions of semi-arid Sub-Saharan Africa that have been severely affected by land degradation in which livestock have been implicated. Often the process starts with the introduction of cash crops, which results in increases in the human population and expansion of shifting cultivation, leading to increased cropping areas at the expense of rangelands. Income from the sale of cash crops is usually invested in cattle, and the increased numbers and the reduced areas of rangelands result in overgrazing and eventually land degradation.

Intensification and integration of livestock systems

When population densities are low crop and livestock production tend to be separate enterprises, although examples of mutually beneficial cooperation are common - eg. in West Africa herd owners are encouraged to let their animals graze maize and sorghum stubble left in the fields after harvest, which through the provision of manure improves soil fertility. In spite of increased migration to the cities, the rural population of Sub-Saharan Africa increased by about 2% per year in the eighties (and this trend is likely to continue), resulting in an increasing scarcity of land and higher land prices relative to that of labour. This has been shown to lead to more intensive land use and increased integration of crops and livestock, such as the adoption of animal traction. If this process is combined with economic growth in the urban areas, creating markets for agricultural products, then there are strong incentives for farmers to invest in increased inputs to gain more output from a given unit of land. Several studies carried out recently have clearly shown that integration of livestock with crops results in improvements of 50% (Ethiopian highlands) to over 100% (Zimbabwe) or more in farm productivity and incomes compared to smallholders who only raise subsistence crops. This is not only due to the use of draught power and application of manure, although these are usually key elements, but also due to the sale of livestock to provide cash for purchase of inputs to increase crop yields. Mellor (1989) concluded that expansion of animal agriculture enables smallholder producers to intensify their agricultural efforts even on low-productivity resources, and it has vast potential for providing incomes and employment to the poorest farmers. It has been stated that when subsistence farmer incomes reach a certain threshold level, enough surplus cash becomes available to make investments that further enhance productivity and income, leading to an upward spiral of income generation and improved living standards.

McIntire et al. (1992) have characterized the evolution of crop-livestock interactions as follows:

i.   Livestock enters the farming system when expansion of crop lands and reduction in pasture and fallow reach the point that farmers seek substitutes to maintain fertility (fallow is normally cheaper than manure in maintaining soil fertility).

ii. Farmers paddock animals on cropland or otherwise collect and spread manure.

iii. To obtain even more manure they shift to systems of collection, processing, storage and application.

For herders the process is naturally different:

i.   As available pastures decline, they depend more and more on crop residues for feed and also begin to grow crops.

ii. Field grazing of crop residues gives way to confinement systems in which crop residues are collected and preserved, which results in more intensive use of these products and more efficient use of animal wastes.

iii. Finally hand labour is replaced by animal traction, that has become economic because of the high intensity of land use (hand tools are cheaper than animal traction in fallow based systems).

iv. In some cases a further step is the cultivation of forage legumes for feed, which in turn improves soil fertility and crop yields.

Intensive crop-livestock systems in semi-arid areas: a case study from Tanzania

A variety of highly productive small-scale, intensive mixed farming systems have evolved in the humid and sub-humid tropics, almost exclusively in areas of high population density where agricultural holdings had become too small to support traditional farming practices. Similar systems, based on stall-fed improved dairy cattle, have also developed in the high potential areas of Africa, such as the central highlands of Kenya and on the slopes of Mt.Kilimanjaro in Tanzania. However, such systems, virtually non-existent in semi-arid, low potential regions, are now also developing in central Tanzania, in low rainfall areas (500-700 mm) where agro-pastoralism had been the dominant agricultural form. Over the last few decades uncontrolled overgrazing and unsuitable cropping systems had resulted in severe land degradation and was leading to total ecological collapse. The start of the Dodoma Region Soil Conservation (HADO) project did not achieve the expected results, and so in 1979 and 1986 two of the most severely affected areas, in total over 2,000 sq.km, were closed to all grazing livestock, which involved the eviction of around 140,000 cattle, goats, sheep and donkeys. This resulted in problems of child malnutrition and declining crop yield due to reductions in soil fertility, and so in 1989 a decision was passed to allow back confined (stall-fed) dairy cows on a limited basis, and provided that a number of preconditions had been met. One of these was that only improved cows would be allowed, but due to shortages of suitable animals this rule was relaxed in 1993 and farmers were allowed to bring in their best milking zebus from adjacent areas. Today over 450 farmers have joined the scheme, with over 600 cows, around 75% of which are local zebus, which are progressively upgraded by crossing with F1 bulls. Average milk yields today are around 8 litres per day from improved cows, and around 3 litres from the zebus, much higher than expected (Ogle, 1995, unpublished data).

It was anticipated that dry season feed supply would be a major constraint, but studies have shown that dry season yields are only around 10% lower than for the rainy season. Farmers do not consider dry season feeding to be a real problem, and cited the following as their main feed sources (Larsson, 1993):

- Wild legumes, weeds and grasses
- Elephant grass (grows on river banks) and Makarikari grass (planted to reduce erosion on slopes).
- Maize stover, sugar cane tops, sweet potato and pigeon pea vines etc.
- Leaves of Leucaena leucocephela (introduced) and Sesbania spp (both wild and planted).
- Sunflower cake (by-product of artisan oil extraction) and maize bran.

Conclusions

The change from traditional extensive systems, to more intensive forms of mixed agriculture has in the past been a gradual long-term process, but deteriorating environmental conditions dictate a more rapid transition in the fragile semi-arid lands of Sub-Saharan Africa. Due to growing population pressures on available agricultural land, farming systems are evolving towards intensification and integration, and it is unlikely that the basic nature of this evolutionary process can be substantially altered by development initiatives. However the process can be supported and directed "by targeting components of these natural processes of evolution in ways that will accelerate intensification and make crop-livestock systems more productive and sustainable" (Winrock, 1992).

References

Breman H. and de Wit, C.T. (1983). Rangelands productivity and exploitation in the Sahel. Science 221: 1341-1347.

Gilles, J.L. (1991). Animal Agriculture in Sub-Saharan Africa: Socioeconomic issues. Columbia, University of Missouri.

Larsson, c. (1993). A study of smallholder zero-grazing dairy cow systems in the HADO areas of central Tanzania. Examensarbate 53, Swedish University of Agricultural Sciences, Department of Animal Nutrition and Management.

McIntire, J., Bourzat, D. and Pingali, P. (1992). Crop-livestock interactions in Sub-Saharan Africa. World Bank.

Mellor J.W. (1989). The political and economic context for development of animal agriculture in developing countries. USAID report, Washington DC.

Payne, W.A. (1990). Animal Husbandry in the Tropics.

Winrock International.1992. Animal Agriculture in Sub-Saharan Africa. Winrock International, Morrilton, Arkansas, USA.

A Study on the Multipurpose Sugar Palm Tree *(Borassus Flabellifer)* and Its Products for Animal Feeding in Cambodia

Borin Khieu, J.E. Lindberg and T.R. Preston****

*Department of Animal Health and Production,
Ministry of Agriculture, Forestry and Fisheries,
Phnom Penh City, Kingdom of Cambodia
E-mail: borin@forum.org.kh

**Department of Animal Nutrition and Management,
Swedish University of Agricultural Science,
P.O. Box 7024, S-750 07, Uppsala, Sweden

***La Finca Ecologica, University of Agriculture and Forestry,
Ho Chi Minh City, Vietnam
E-mail: thomas%preston%sarec%ifs.plants@ox.ac.uk

Summary

The Sugar palm tree plays an important role in the small farm systems in Cambodia. It provides different products for human as well as for animal production such as juice, sugar, leaves, timber, fruits and roots as traditional medicines. Therefore, sugar palm syrup production is an important product and source of seasonal income for the farmers. But at the same time the production of syrup requires a huge amount of energy to boil the juice for sugar. It was shown in this study that when firewood was purchased, some of the sugar producers lost money.

It was shown in a study of 12 farms that they got an average profit of 150 Riels per tree by feeding pigs with juice compared to 11 Riels for sugar syrup production. Also the collection of leaves for thatching was shown to be of interest in comparison to sugar production, as this gave a profit of 6 Riels per day per tree. Furthermore, the environment benefits when trees do not need to be used as energy source for sugar production. Feeding pigs with palm juice as the main dietary energy source was shown to be an alternative, sustainable production. In order to improve the production, pigs need to be reared in pens which makes it easy to collect manure for household gas production in biodigesters and decrease the disease risk. In addition, it will improve the pig production in the rural communities which provides meat in human nutrition and employment for farmers.

Key words: Cambodia, Borassus flabellifer, palm juice, palm sugar, fuel, profitability, environment, pigs, sustainable production.

Introduction

The sugar palm has been considered a valuable multipurpose tree for several centuries for the production of food and construction materials that will produce income for Cambodian rural communities. Romera (1968) reported that the government of Cambodia launched a project for increased production of palm juice and by-products on a producer-co-operative basis, covering 35,000 hectares. However, the political disruption which followed has prevented the realization of this programme. It should be noted that approximately 75% of the total area of 181,035 km² of Cambodia was covered by forest, and so fuel for sugar palm processing was not a problem. Farmers could collect fire wood around their households, or in a neighbouring forest. Today fire wood has become the major constraint for sugar production. This raised the following questions: How can the problem be solved? What modifications or diversification of palm juice can be applied in the rural communities? If sugar production is to continue what kind of energy sources could be used as a replacement for fire wood?

At present, with 85% of the population being small farmers, the Cambodian economy is mainly dependent on agriculture. Rice cultivation has been the main agricultural activity and one crop per year gives an average yield of 1.5 tonnes per ha (Ministry of Agriculture, 1995). Farmers also have other complementary enterprises such as vegetables, cassava, sweet potatoes, water melons and sugar palm production. However, the contribution made by animals is considerable in the small farm system providing food (meat and eggs), power and cash income. Devendra (1993) reported that in south-east Asia, the contribution of livestock is consistent with the relatively large numbers of different animal species found in the small farm sector. Furthermore, raising animals has other advantages such as providing fertilizer and fuel, in addition to utilisation of by-products and their social value.

The majority of the farmers own 2-3 cattle, which are commonly kept to provide draught power, organic fertilizer and as a form of savings and wealth to share at marriage to sons or daughters. Chickens are commonly raised for household consumption and gifts, and in addition a few are produced for the market. The rural farmers in Cambodia raise 1-2 pigs to provide seasonal cash income, but pigs are also raised to produce meat to be used for example for wedding feasts. Devendra (1993) reported a very high annual growth rate (16.6%) of the pig production in Cambodia. The main reasons are the availability and successful transfer of proven technologies in pig production, a large and growing market for the products, good credit facilities and the rapid turnover of the capital investment.

In the work of the first author Participatory Rural Appraisal (PRA) techniques (Chamberss, 1993) were applied in all the village studies. The goal of the studies were to better understand the traditional pig production systems in the rural communities, and to study the use of palm tree products for animal production. The aim also was to evaluate the economic profitability of the traditional sugar production practices compared to using sugar palm juice as the energy source for pig feeding.

Materials and methods

The studies were conducted in five villages of Takeo and Kadal provinces in South and North Phnom Penh City. The first PRA was conducted from September to October 1995 in Takeo province and the second from November 1995 to January 1996 in Kandal province. The studies were not focused on animal production only, but also on the farming systems as a whole as was requested by NGO (non-governmental organisations) supported projects in those villages.

The study sites

The first PRA was implemented at Or Pheasang and Kraing Sbauv villages, where the trial to evaluate pig feeding with palm juice was conducted (Borin et al., 1996). The villages are located approximately 25 and 50 Km south of Phnom Penh City, respectively. The pH of the soil is about 5.5 with an average rainfall of 1,200 mm per year and the humidity ranges from 55 to 75%. In these villages there was already an ongoing (integrated rural development) project supported by a non-governmental organization (Lutheran World Service). Pheasang village is populated with approximately 184 households (856 persons). The total population of Kraing Sbauv is 122 households and 502 persons.

The second PRA was conducted in three villages (Sre Tasek, Tropain Putrea and Komnop), Tumnop Thom commune, Punhea Leu district, Kandal Province, approximately 55 km north-west of Phnom Penh City. The area is completely deforested since 1979, and has poor sandy soils. There is a total of 109 families with 551 persons, and there are projects supported by a non-governmental organization GOAL also in these villages.

Participants

A total of 24 local staff from different projects of the two non-governmental organizations (Lutheran World Service and GOAL) were trained for a week before conducting the village studies (PRA). The team was composed of people with different professions, such as animal nutritionists, veterinarians, teachers, water engineers, foresters, economists, nurses, rural community workers, credit specialists and agronomists.

The villagers, including the 12 farmers who had taken part in the pig research project (Borin et al., 1996) and representatives of the local authorities participated in data collection, evaluation of the earlier pig trial and planning of future activities.

Data collection methods

The data collection was implemented through Participatory Rural Appraisal (PRA) methods as described by Chambers (1983 and 1993) and Kirsopp-Reed (1994). The secondary data were obtained from local authorities, the two non-governmental organizations and the earlier review contained in (Borin et al. 1996). PRA methods used included interviews, maps, transect walks, seasonal calendars, matrix, ranking and scoring and direct observation. The teams lived in the villages during the week of the village studies.

Measurements

In order to analyse the profitability of sugar palm production, 1.50 m^3 of fire wood (approximately 450 kg) from the same tree were purchased near the village and distributed to 7 farmers to boil palm juice. Sixty six kilogram of wood were distributed to each participating farmer in Or Pheasang village. Juice was also weighed and the sugar content measured before boiling. After the boiling procedure, the remaining fire wood was weighed again to determine fuel consumption. The sugar palm syrup obtained after boiling was also weighed.

Economic calculations

The economic calculations were based on the exchange rate during 1995 which was US$ 1=2,350 Riels in early 1995, and in September it was US$ 1=2,500 Riels.

Results

Farming systems

Rice cultivation is considered the priority activity in the local farming systems. Local varieties of rice are used which are characterised by long and medium growing periods (5-6 months). The villagers begin preparing their land for rice production after the onset of the rains. In July the seed beds are prepared and sown, transplanting is in August to September, and harvesting between November and January. Farmers use two sources of fertiliser in the rice fields, cattle manure and earth mounds (generally used by those without a source of animal manure). The average yield of rice was reported as being between 1 to 1.5 tonnes per hectare per year. Sweet potato is also a common crop that is grown in the middle of the rainy season and harvested before the rice. The farmers grow water melons in October and harvest them in December, and at the same time some farmers begin sugar palm production, this activity finishing in April-May.

Sugar production

Sugar production has been considered the second most important source of complementary income. At present, some of the sugar palm producers have stopped this activity in places where fire wood is scare. However, the sugar palm still gives a good income to farmers who have access to free fire wood. The price of imported refined sugar was 1,200 Riels and the local sugar 1,100 Riels per kg at the time of the studies. The price of sugar syrup varied from 400 to 600 Riels per kg. The price always increased after May, which is the time when sugar production finishes.

Animal production

The common animals kept by the farmers in these villages are cattle, pigs and poultry, mainly chickens but including a few ducks. Most of the animals are local breeds and are fed solely with locally available feed resources such as rice straw, rice bran, banana stems, water spinach, sweet potatoes, water hyacinth etc... (see table 1). Most of the animals are scavengers.

Table 1. *Different feed resources used for animal in the four villages of the study*

Feed	Chickens & ducks	Pigs	Cattle
Rice bran	yes	yes	some
Broken rice	yes	yes	
Paddy rice	yes		some
Rice straw			yes
Kitchen waste		yes	
Banana stem		yes	yes
Water spinach		yes	
Water hyacinth		yes	some
Water lylium (Prolit)		yes	
Duckweed	yes	yes*	
Palm juice		yes**	some
Scum		yes	
Palm fruits		yes	yes

*   Started from 1993 during a FAO/TCP project in Cambodia.
** Started in 1995 with small scale demonstration.

Cattle

Approximately 60 to 70% of the villagers own cattle (used for draught power). The breeds are commonly local cows (small yellow cows). There are also crossbreeds of Hariana, which are raised in regions where green feed is available around the year, such as along the lake shore or lagoons, while the local cows are kept in areas with the poor soil. During the wet season, cattle graze on the rice bunds and during the dry season their diet is mainly based on rice straw. Sometimes palm juice and scums from sugar palm production are poured on the rice straw and fed to cattle. The main diseases of cattle in these areas are Foot and Mouth disease (FMD), Black-leg and Haemorrhagic Septicaemia, which appear in different times of the year. The Haemorrhagic Septicaemia and FMD normally appear during the rainy season and Blackleg in the dry season. The price of beef is 4,000 to 6,000 Riels per kg, although cattle are not sold by weight but usually through an agreement between the owner and a middleman. A 200 kg cow may cost (1995) approximately 200,000 to 250,000 Riels.

Pigs

One to two pigs are commonly raised for sale or for ceremonial purposes e.g. wedding banquets. They are purchased after the rice harvest and reared for 8-10 months. The main feed resource is rice bran and kitchen waste, and some farmers feed their pigs with the residual solution from the pan after sugar production. Piglets are commonly purchased at a live weight of 4-6 kg, and the common breeds are crossbreds between York-shire or Landrace and local breeds, which were found at all the five study sites. The main health problems are parasites, while infectious diseases such as hog cholera, pasteurellosis and erysipelas also cause considerable losses and the mortality rate is over 50%. The live weight price was 2,600-3,000 Riels per kg in June-July, 1995 and it was increased to 3,300 Riels in January, 1996. The price of pork meat was approximately 6,000 to 8,000 Riels per kg.

Chickens

Each family raises 2-3 chickens, but some farmers raise up to 10 breeders, resulting in approximately 20 head for sale in the market per annum. Chickens are fed some rice in the morning and the rest of the day is spent scavenging for food. Chicken cost 3,800 to 4,500 Riels per kg liveweight (January, 1996). The most important diseases are Newcastle disease and Fowl Pox which usually appear after the first onset of rain. Approximately 80 to 90% of the chickens are affected in this period.

Marketing and market accessibility

There were a few private sugar palm syrup collectors in the villages, which meant that farmers could sell their daily sugar syrup production locally. The sugar syrup can also be stored in containers during the production period in order to sell later at a higher price. Sometimes farmers bring sugar syrup to the city to be sold in the market, and sugar can also be sold in the villages for immediate consumption.

Animals are generally sold through middlemen who frequently visit the villages. If possible farmers use their own transport to take animals to the slaughter house, as the price of pigs at the slaughter house is better than the price offered by the middlemen. A few years ago farmers were allowed to kill animals and sell the meat in the nearby villages, but this is not permitted today.

Introduced technology
Since 1993, about 30 cheap plastic biodigesters (An et al., 1994) have been installed in Kandoeung commune and one in Tumnop Thom commune. The digesters were connected to pig pens and latrines and there is one mixer in each digester for cow dung. The gas from the biodigester covers about 80 to 85% of the domestic fuel consumption per day.

Discussion

General aspects
The population of sugar palm trees in Cambodia has increased since 1901 (Dolbert, 1991), when the King of the Kingdom of Cambodia decreed that every Cambodian must plant a few sugar palm trees and the trees have been an important contributor to rural farmer livelihoods since that time. The product that has been considered to be the primary source of income is the juice. However, leaves, fruits, fibre and trunk also provide food, animal feed, materials for construction apart from other domestic uses. The trees have the potential to produce inflorescences all the year round or a total of 10-15 inflorescences during the production season. In this situation, the inflorescences can be preserved after the preparation procedures have been successfully completed. Normally, the palm juice is not commonly collected during the rainy season. There are many reasons that farmers do not practise juice collection in the rainy season such as the strong winds which make it dangerous to climb to the tops of the trees. Furthermore, it is difficult to protect the collectors from rain, and the tree trunks become slippery. However, some people still collect juice during the rainy season to sell as fresh drinks in the cities and towns. The price of sugar syrup varies depending on the location, and in 1995 the price per kg was 500-600 Riels in Bati district in Takeo province and 350-400 Riels in Kampong Chhnang province (Nol Paan 1995, personal communication). The price of fire wood is the major factor which influences the cost of production (table 2).

In 1995 the price of fire wood was 8,000 Riels per m^3 in Kompong Chhnang province and about 20,000 Riels per m^3 in Takeo province. However, transport distance may also influence the cost of sugar syrup, because the important market is located in the city (Phnom Penh). At present there is a fuel problem for boiling juice in the five sites studied. Fire wood has

Table 2. *Costs and return (Riel) of sugar palm syrup production per palm tree per day*

	Cost	Return
Fire wood	250	
Depreciation cost (ladders and collectors)	20	
Opportunity cost (palm juice)	20	
Opportunity cost (labour of boiling juice)	120	
Other costs	15	
Total	*430*	
Sugar palm syrup, kg		0.8
Market price, kg		550
Sum of return		*440*
Net return		*10*

- The opportunity cost of hired labour for the rice cultivation is used (3,000 Riels per day). An average of 6 hours are spend for boiling and steering syrup of 15 palm trees.
• Price per kg of sugar palm syrup 500-600 Riel.

become scarce and sugar production was found to be not profitable when fire wood was purchased. In the traditional sugar production system, fuel consumption has not been included in the cost of production because fire wood could be freely collected (Borin et al., 1995).

Leaves are the second most important product from the sugar palm tree, and they can be harvested two to three times a year, except for the trees from which juice is being collected. The leaves are mostly used for thatching, but also matting, baskets, hats, rice storage boxes, fans and fancy boxes are common products. Morton (1988) and Davis and Johnson (1987) reported that a well thatched roof will last for 2-3 years. The price of thatch depends on its quality, and varies from 70-300 Riels (1995) per thatch.

Animal production

Preston (1995) reported that the growing discrepancy between the expanding world population and its food producing capacity is primarily caused by competition between livestock and humans for cereal grains. The solution that is proposed is to develop alternative feeding systems for livestock using non-cereal, perennial, high-biomass producing crops that can be grown under rain-fed conditions on sloping lands not suitable for cereals.

Juice and scums from the sugar palm have been used for cattle feeding. The juice or scum is poured over rice straw and fed immediately to cattle. The green and mature fruits are considered a good supplement for cattle in the dry season, and the leaves from small trees are also eaten by cattle during the dry season when there is no green feed available. The waste fruit after human consumption is chopped and fed to cattle, especially to the draught animals. Sometimes the mature fruit is soaked in water to give a yellow pulp which can be fed both to cattle and pigs. The fresh yellow pulp has been reported to be rich in vitamins A and C (Morton 1988). Concerning cattle it is possible the effect is at rumen level as the pulp may stimulate early rumen microbial invasion of the roughage (Leng et al., 1995).

The remainder after cleaning the pans used for boiling juice has traditionally been fed to pigs in the rural communities. Farmers mentioned that pigs seem to have better performance during this period compared to periods when they are fed rice bran alone. The price per kg of rice bran varies with its quality from 150-250 Riels (1995), and the bran found on many farms and used as pig feed appeared to contain more than 50% of rice hulls, which makes the diet unpalatable. Payne (1990) suggested that rice bran should not be fed to piglets, and no more than 30-50% of the total ration for fatteners due to its high fibre content and laxative effect. The quantity and quality of feed varies according to the money available to the farmers. For example, it is not only feed deficiency which affects the performance of pigs, there are several other factors that should be taken into consideration, such as management, breeds and diseases (especially parasites). Farmers raise pigs until they reach 70-80 kg, and normally do not have any plans to sell them while they still are growing and look healthy.

At present, the production systems with free ranging pigs seem to cause many problems in the community. Since the human population is growing there is a need to use available land to produce food and this comes in conflict with free ranging pigs. During the dry season when the temperature is high, pigs do not scavenge in the paddy fields, but rather go inside the vegetable gardens and cause damage to the gardens. Sometimes they are killed by the pesticides that the farmers have recently sprayed over their vegetable. There is also a high incidence of internal parasites in this system. Payne (1990) reported that the humid tropics provide an almost perfect environment for many parasites. The high mortality in pig production in the rural areas in Cambodia are due to the inadequate veterinary service, poor and unbalanced diets and the incomplete management provided by the owners of the pigs.

Farmers have their own opinions, based on experience, as to the choice of pigs to raise. Some farmers were of the opinion that they could only rear white pigs, while the others preferred black and white pigs. Normally piglets of 4-6 kg live weight are purchased in the rural areas. They are cheap and if they die, the farmers would only loose a small investment, as at the time of the study the price of a small piglet (4-6 kg) was approximately 20,000 to 25,000 Riels and a weaner (15 to 20 kg) about 50,000 Riels (1995). Borin (1994) suggested that it is preferable to purchase pigs with an initial weight of at least 15 kg when feeding sugar cane juice as sole energy source as this weight performs better at small farm conditions and Cunha (1977) suggested that heavier pigs can use higher levels of molasses more effectively than small pigs.

Economic comparisons of different sugar palm products
The average profit from feeding palm juice to pigs was 150 Riels per tree per day (table 3).

Table 3. *Daily intake and cost of pig production (Riel)*

Ingredients	Amount (day)	Cost (kg)	Cost (day)	Return (day)
Palm juice	2 palm trees	-	40*	
Soya beans	400 g	625	250	
Water spinach	500 g	100	50	
Salt	5 g	150	1	
Lime	5 g	500	2	
Depreciation cost (ladders and collectors)			40	
Health care			26	
Depreciation cost (pen construction)			24	
Opportunity cost (labour)			100	
Cost of piglet			335	
Total			868	
LWG, kg/day				0.36
Market price, kg				2,800
Sum of return				*1,008*
Net return				140

* Cost of hiring 2 palm trees per day
Opportunity cost of hired labour for the rice cultivation is 3,000 Riels per day. Approximately 2 hours are spent to feed and clean 6 pigs per day.

The market price per kg pig live weight pig is 2,600-3,000 Riels compared to 11 Riels per tree per day for sugar syrup production. The profit from selling thatch was 6 Riels per day per tree, including the fruits. The comparison of juice feeding to pigs with sugar production is particularly interesting. Four of the seven families made no profit from making sugar when the cost of fuel was taken into account, and the highest profit from sugar production was still less than the lowest profit from pigs. Four out of the seven farmers (Yem Khnol, Thol Onn, Houy Kiel and Pauv Pauv) were in fact loosing 22, 7, 45 and 36 Riels per tree per day, respectively (see figure 1).

The quantity of fuel consumption depends mainly on the efficiency of the stove that individual farmers have built, and also the skill and experience of each tapper to produce good quality juice. When the inflorescence is managed in the proper way, the concentration of sugar (Brix) will be high, and with less water in the juice the time needed for boiling will be

Figure 1. *Comparative profit of different farmers from sugar palm trees used for sugar production or for pig rearing*

Riels/tree/day

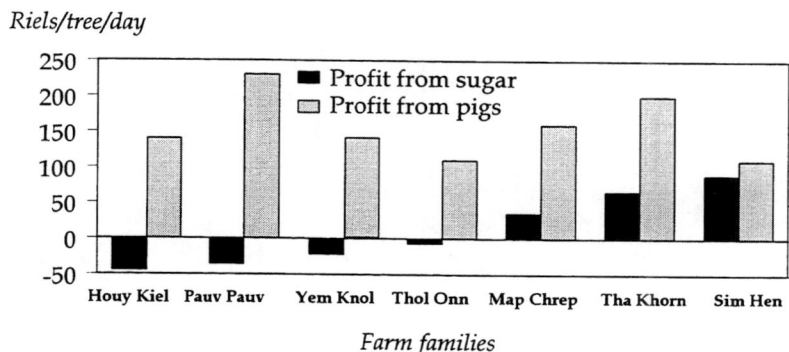

Farm families

Source: Borin Khieu et al
(1995)

shorter and less fuel will be needed. It was also found that the each sugar producer needs to buy 8 to 10 palm trees for fuel every year in addition to the fire wood that they have bought or collected earlier.

The mean net profit recorded was 22,500 Riels annually per tree by feeding the juice to pigs, compared to 1,600 Riels annually per tree for sugar production. When the purchased price of fire wood was not taken in consideration for sugar production, the profit was 49,000 Riels per palm tree per year. The average cost for fire wood is approximately 39,000 Riels annually per sugar palm tree. Labour costs directly related to the sugar production were not included in the calculation, but the opportunity cost of working on rice cultivation was used (2,500 to 3,000 Riels per day). In regions where fire wood is still available and free, farmers consider that sugar syrup production is one of their main sources of annual income. Moog et al (1994) reported that feeding pigs with sugar cane juice ad libitum (supplemented with soybean meal) was a profitable alternative and gave higher income per unit of land compared to processing for sugar. This alternative is more environmentally friendly because it avoids the use of fossil fuels and allows the use of manure for more sustainable agricultural production.

Feeding palm juice to pigs is not only profitable for people in the rural areas in Cambodia but also benefits biodiversity and the environment, as trees will not be destroyed as fire wood to boil the sugar palm juice. Farmers also save time when juice is fed to pigs, allowing them to concentrate on other important activities. It takes approximately only 2-3 hours per day to climb 8-10 palm trees and 1-2 hours per day to feed the pigs and clean the pig pen compared to sugar production which takes about 16 hours per day of work for the whole family. In addition, the farmers are able to collect the manure in the pig feeding system as substrate for the biodigester and can use the effluent as organic fertilizer, which is also more beneficial to the environment compared to the use of chemical fertilizers.

Dolbert (1991) in his survey on sugar palm syrup production in Cambodia reported that the whole family was occupied almost all the time with sugar production and spent about 16 hours per day in collecting fire wood, chopping it and monitoring the boiling process. In some places fire wood has been replaced to some extent by rice hulls using a new stove design. With such a stove 1 kg of sugar requires around 4-5 kg of rice hulls, at a cost of approximately 200-250 Riels per day per tree which is equivalent to 33,750 Riels per season per tree. This is cheaper than the equivalent amount of fire wood which costs 49,000 Riels. According to one farmer (Mr Em Pheap) it was much more convenient to boil with rice hulls than with fire wood and also cheaper, but the hulls, however, were more difficult to acquire.

There are several domestic options for using palm leaves, but the most common is for thatching and mats. Twenty five to thirty six leaves were collected 1-2 times a year from trees that were not being used for juice production. When leaves were used for thatching the profit was 2,250 Riels per tree per year, while the profit was somewhat higher for making mats, although labour costs (opportunity cost) were also included in the calculations. Using the leaves for thatching or mat making, is better from the environmental perspective than for sugar production.

Sugar palm juice for pigs in the rural areas, has great potential. Elliott and Kloren (1987) reported that the use of fibre free energy sources such as raw sugar or sugar cane juice permits greater use of cheaper vegetable protein sources which are not usually included to a great extent in conventional diets because of their high fibre content. Therefore, the use of sugar palm juice in pig feeding will allow farmers to use their own local available feed resources and the cost of pig production will be lower. Water plants such as water spinach, water hyacinth and duckweed will become potential protein resources when pigs are fed with sugar palm juice, as soya beans are expensive and are not considered as a common protein supplement for farm animals in Cambodia. Dry soya beans cost $US 265 per tonne in December 1994 and the price had increased to $US 520 six months later. Soya bean prices are usually lowest in December-February at the early stage of harvest, but the price varies from year to year according to demand at local and regional markets.

Low cost plastic biodigester

Biodigestion and gasification are complimentary technologies for generating combustible gas and substrates for the chemical industry. Biodigestion is the technology of choice for farmers for recycling excreta from animals and humans and converting it into biogas (methane). This provides an environmentally friendly system for fuel production and also organic fertilizer suitable for crops or fish ponds (Preston, 1994 and An et al., 1994). The low cost plastic digesters were installed in Or Pheasang village to collect faeces and residues from pig pens and from the human latrines. Also one mixer was installed in the entrance tube of the biodigester. This promises to be an effective way of making the use of latrines more popular by rural families who find the gas production an incentive to use latrines and keep the house clean (Than Soeurn 1994). In order to encourage farmers, to adopt the technology they were initially asked to pay only 40% of the total cost of installation, because it was new and unknown technology which they were uncertain of. Six pigs are able to produce manure sufficient for one digester which will provide gas enough to replace 80 to 85 % of fire wood consumption for cooking per day for one family of 5 members. In addition, the effluent from the digester can also be used as fertilizer for duckweed production, which provides a good source of protein for pigs (Leng et al.,1995) as illustrated in figure 2.

Figure 2. Integrating the sugar palm (Borassus flabellifer) in farming systems in Cambodia

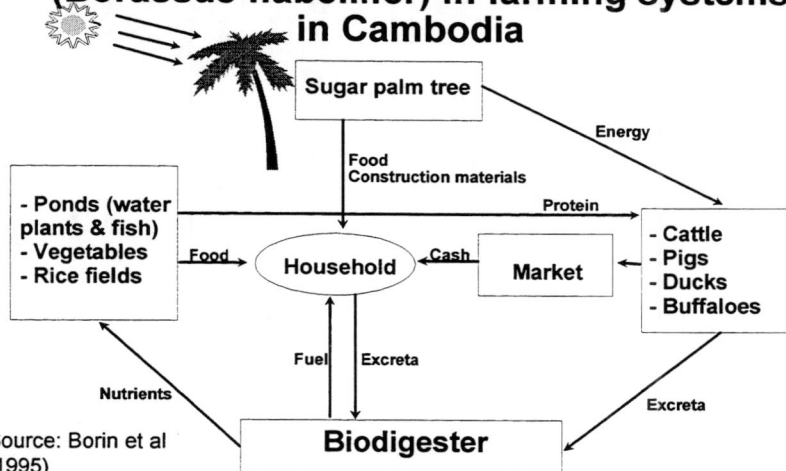

Source: Borin et al (1995)

Conclusion

The studies show clearly that feeding pigs with sugar palm juice gives the farmers more profit than sugar syrup production, when the cost of fire wood is included in the cost of sugar production. However, it is still a good source of seasonal income for the farmers who have access to free fire wood. Therefore, feeding pigs with palm juice is an alternative source of income in the rural areas where fire wood is expensive and furthermore it will contribute to reduction in deforestation. It is also suggested that pigs need to be reared in confinement pens as that will give less opportunity for diseases and more importantly in this system pigs can not destroy vegetable gardens.

Due to the fluctuation of the price of soya beans, other alternative protein sources should be used in order to ensure profitability. Sugar palm juice has a good potential for pig feeding in rural areas in Cambodia and can be complemented by the use of cheap protein sources which have a high fibre content.

It is important to install the digester for methane production, so that it can be connected directly to the pig pens. The system provides gas for cooking and keeps the house clean without any pollution from smoke and odours from the pigs' wastes. This system has additional advantages in preventing deforestation and can be considered to be environmentally friendly and healthy.

Acknowledgements

The financial support from Swedish Agency for Research Cooperation with Developing Countries (SAREC) and the Ecumenical Scholarships Programme in Germany (DIAKONISCHES WERK) is gratefully acknowledged.

Thanks are due to the farmers in Kandoeun commune in Takeo province and in Thom commune, Kandal province, who spent their valuable time to provide us with information on their experiences and indigenous practices and also provided accommodation during the Participatory Rural Appraisal exercise. Thanks also due to the PRA team from the LWS, FHI and GOAL organizations.

References

An Bui Xuan, Man Ngo Van, Khang Doung Nguyen, Anh Nguyen Duc and Preston T. R.,1994. Installation and performance of Low-Cost Polyethylene Tube Biodigesters on Small Scale Farms in Vietnam. In: Proceeding of National Seminar-workshop "Sustainable Livestock Production on Local Feed Resources" (Editors: T. R. Preston, B. Ogle, Le Viet Ly and Lu Trong Hieu), Ho Chi Minh City, November 22-27, 1993. pp 81-90.

Borin Khieu, 1994. Feeding Livestock on Local Resources in Cambodia. In: Proceedings of National Seminar-workshop "Sustainable Livestock Production on Local Feed Resources" (Editors T. R. Preston, B. Ogle, Le Viet Ly and Lu Trong Hieu), Ho Chi Minch City, November 22-27, 1993. pp 98-104.

Borin Khieu, Preston, T. R. and Ogle, B. 1995. Fattening pigs with the juice of the sugar palm tree (Borassus flabellifer). Livestock Research for Rural Development, 7(2). 25-29.

Borin Khieu, Preston, T. R. and Lindberg, J. E. 1996. Juice Production from the sugar palm tree (Borassus flabellifer) in Cambodia and performance of growing pigs fed sugar palm juice. pp 1-9. Part I of M.Sc. Thesis. Department of Animal

Nutrition and Management, Swedish University of Agricultural Sciences, Uppsala.

Chambers, R. 1993. Challenging the professionals: Frontiers for rural development. Intermediate Technology Publications Ltd. 103-105, Southampton Row, London WC1B 4HH, UK. pp 97-105.

Chambers, R. 1983. Rural Development: Putting the Last First. Longman Scientific & Technical. Copublished in the United State with John Willey &Sons, Inc., New York. pp 191-208.

Cunha, T. J. 1977. Swine feeding and Nutrition. pp 154 and 241.

Davis, T. A. and Johnson D. V. 1987. Current Utilization and Further Development of the Palmyra Palm (Borassus flabellifer L., Aracaceae) in Tamil Nadu State, India. 1988, by the New York Botanical Garden, Bronx, NY 10458. Economic Botany, 41(2). pp 23-44.

Devendra, C. 1993. Sustainable animal production from small farm systems in South-East Asia. Chapter 3 and 4. pp 23-44.

Dolbert, P. 1991. The Artisan Sugar Palm Syrup Production in Cambodia. Project of GRET (Agricultural Technique and development of Cambodia), translated from Cambodian language version. pp 1-9.

Elliott, R. and Kloren, W. R. L. 1987. The use of sugar in diet for monogastrics. Recent Advances in Anim. Nutr. in Australia. pp 290-296.

Kirsopp-Reed, K., 1994. A review of PRA Methods for Livestock Research and Development. In: RRA Notes, Number 20: Special Issue on Livestock (Edited by K. Kirsopp-Reed and F. Hinchcliffe). pp 11-35.

Leng, R. A., Stambolie, J. H. and Bell R., 1995. Duckweed-A Potential High-Protein Feed Resource for Domestic Animals and Fish. Livestock Research for Rural Development. 7(1). (16Kb).

Moog, F. A., Samiano A. M., Valenzuela, F. G., Raymundo, S. G. and Fementira, E. B. 1994. Utilization of Sugarcanes Juice for Swine in the Philippines. Proceedings of the 7th AAAP Congress in Bali, Indonesia. 129-135.

Morton, J. F. 1988. Notes on Distribution, Propagation, and Products of Borassus Palm (Arecaceae). 1988, by the New York Botanical Garden, Bronx, NY 10458. Economic Botany, 42(3). pp 420-441.

Payne, W. J. A. 1990. An Introduction to Animal Husbandry in the Tropics. Fourth edition. Tropical Agriculture Series. Longman Scientific & Technical. Copublished in the United State with John Wiley & Sons, Inc., New York. pp 660-666.

Preston T. R., 1995. Research, extension and training for sustainable farming systems in the Tropics. 7(2). 1-7.

Preston, T. R., 1994. Renewable Energy from Tropical Agriculture: The respective roles of Biodigestion and gasification. CIPAV, Cali, Colombia. In: Proceeding of National Seminar-workshop "Sustainable Livestock Production on Local Feed Resources" (Editors: T. R. Preston, B. Ogle, Le Viet Ly and Lu Trong Hieu), Ho Chi Minh City, November 22-27, 1993. pp 94-97.

Romera, J. P. 1968. Le Borassus et le sucre de palme au Cambodge. L' Agron. Trop. 8, 801-839.

Than Soeurn, 1994. Low cost biodigesters in Cambodia (FAO/TCP/CMB/2254). In: Proceedings of National Seminar-workshop "Sustainable Livestock Production on Local Feed Resources" (Editors: T. R. Preston, B. Ogle, Le Viet Ly and Lu Trong Hieu), Ho Chi Minh City, November 22-27, 1993. pp 94-97.

Effect of Management Practices on Yield and Quality of Sugar Cane and on Soil Fertility

Nguyen Thi Mui

Goat and Rabbit Research Centre,
SonTay, HaTay, Vietnam
E-mail: mui%bavi%hue%ifs.plants@ox.ac

Summary

This paper is focused on ways of increasing biomass yield and quality of sugar cane when it is used as a feed resource for animals. Two experiments were conducted at the Goat and Rabbit Research Centre, which is located in a hilly area in the north-west region of Vietnam and an evaluation was made with farmers in the neighbourhood close to the Centre.

The first experiment carried out from January 1993 to December 1995 aimed to study the effect on biomass yield and juice quality (Brix) of row width (75, 100 and 150cm), plant material (stem cuttings and tops), and removing or leaving the dead leaves on the soil surface (mulching). The results for the first three years showed that: when sugar cane was planted by using stem cuttings there was a higher yield in the first and the second year and the cost of seed was reduced by 9-10% compared with the traditional way used by the farmers in north Vietnam. Increasing plant density by shortening the inter-row spacing from 150 to 75cm doubled the yield of stalk and total edible biomass with no reduction in Brix or extraction rate of the juice. Mulching with the dead leaves increased yield, the difference being more marked in the second year and even more apparent in the third year (6.3, 20 and 30% increase in stalk yield for planting and 1st and 2nd ratoon crops, respectively). Mulching also improved soil fertility and increased the amount of carbon sequestered in soil, indicating that mulching is a most positive alternative to the traditional way of removing the dead leaves. Soil fertility increased steadily with increasing ratoons indicating that the growing of sugar cane does not exploit soil nutrients, but in contrast, has beneficial effects on the growth of subsequent crops.

The second experiment, also carried out at the Goat and Rabbit Research Centre in Bavi district, extended over two years (January 1994 to December 1995). The treatments in a Latin square split-plot design included four varieties (Hoanam, POJ 3016, F 156, My 55-14), two planting distances (90 or 150 cm between rows) and mulching or not with the dead leaves. After two years (planting and 1st ratoon crop) it was apparent that all the introduced varieties (Hoanam and My 55-14, F156) produced more edible biomass than the traditional variety POJ 30-16, which has been planted for a long time in

Bavi district. Returning the dead leaves to the soil increased yields of edible biomass, the amount of carbon sequestered in soil, and the growth rate of maize planted in soil samples from the experimental plots after harvest. Narrowing the row distance from 150 to 90cm led to increases in biomass yield without affecting Brix or extraction rate of the juice.

For both experiments there were no advantages in yield nor in feed value when sugar cane was harvested at 10 months

In the on farm evaluation, carried out on ten neighbouring family farms, the objectives were to substantiate the main findings from the on-station work. The interventions were spacing between rows of 100 or 150cm and return or not of the dead leaves. The results confirmed that returning the dead leaves to the soil and decreasing the row distance lead to higher yields of stalk and of total biomass, and improvements in soil fertility.

The general conclusions are that reducing row distance, returning dead leaves to the soil and selection of newly introduced sugar cane varieties are management practices which can be recommended to farmers in the Bavi area.

Keywords: Row width, Mulching, Plant material, Brix, Soil fertility, Sugar cane

Introduction

The sugar cane industry in Vietnam

Industrial system: During the past 30 years, world sugar production has more than doubled reaching in 1993 a total of 110 million tonnes. There are over 70 countries planting sugar cane for processing to make sugar, representing some 60% of total world sugar production. The Asian region produces 28% of the world total (Handbook of General Information of Sugar cane Industry in Vietnam, 1994).

Sugar cane is widely grown throughout Vietnam for chewing, for juice extraction by roadside drink sellers, for artisan production of raw sugar and for industrial centrifugal sugar production. The annual sugar cane production in Vietnam reached 6.34 million tonnes in 1993 resulting in a sugar production of some 500,000 tonnes (Handbook of Official Statistic, 1994), of which slightly less than one third (about 150,000 tonnes) was produced in 6 factories where the cane crushing capacity is from 700 to 1800 tonnes per day during a 4-5 month milling season.

Artisan system: Two thirds of the sugar in Vietnam is produced at cottage industry level, with simple 2-5 roll mills. The extracted juice is put into an open pan and the impurities are flocculated by heating. This flocculated material is skimmed off with a large spoon and is called "scum". The clarified juice is concentrated by heating in open pans with continuous stirring until it is at the point of crystallizing. The product is put into moulds where it forms brown-coloured sugar which is used for human consumption. This process of sugar extraction is done in thousands of small and medium scale units which have a capacity ranging from 50 to 1000 kg raw sugar per day.

Fractionation of sugar cane

1600 kg

```
                    ┌─────────────────────────┐
        ┌───────────│      SUGAR CANE         │───────────┐
        │           └─────────────────────────┘           │
        │ 300 kg            │ 1000 kg                      │ 300 kg
        ▼                   ▼                              ▼
   ┌─────────┐        ┌──────────┐                  ┌──────────┐
   │  TRASH  │        │  STALKS  │                  │   TOPS   │
   └─────────┘        └──────────┘                  └──────────┘
     Soil                  │
   ecosystem               ▼             Sheep/goats/buffaloes
                         (MILL)
              ┌────────────┴────────────┐
              ▼                         ▼
      ┌──────────┐              ┌──────────┐
500 kg│  JUICE   │              │ BAGASSE  │ 500 kg
      └──────────┘              └──────────┘
       Pigs/ducks          Sheep/goats/buffaloes
                                  Fuel
```

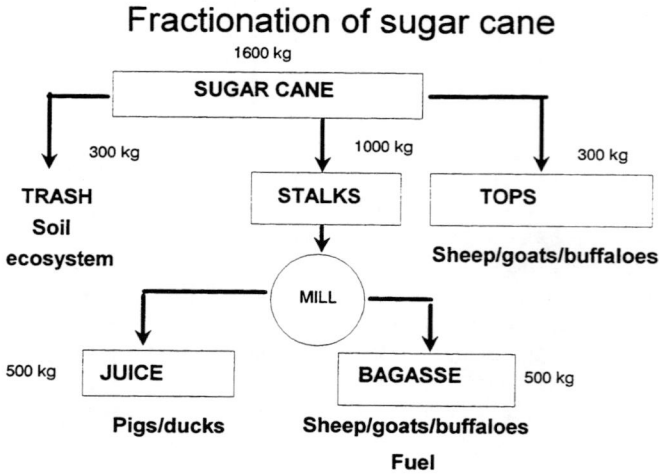

The scum may be fed directly to livestock or more usually it is concentrated to approximately 50% solids content to prevent rapid fermentation. It is obvious from the mode of its preparation that its feed value is superior even to cane juice, since it is enriched with protein and minerals (Preston and Murgueitio, 1992). In the northern region, low grade coal is used almost invariably as fuel at both factory and cottage industry level. Thus all the bagasse is available for other purposes.

Importance of sugar cane as livestock feed: The results of recent research show that sugar cane, which is probably the most productive crop in the tropics, can be used as a basis of intensive animal production systems (Preston, 1995a). The three possibilities for using this crop are: (i) in the form of by-products after extraction of the sugar; (ii) as integral whole sugar cane; and (iii) by fractionation into different end products without extraction of sugar. In this latter case, the juice is fed to pigs and the bagasse and tops to ruminants, giving a potential production capacity of over 5,000 kg animal liveweight/ha (Preston, 1995b).

The potential benefits from feeding the fresh sugar cane juice to pigs, and in some cases the raw syrup, will depend on the opportunity price of sugar cane, syrup and sugar. When markets for sugar cane are close to the sites of production, it is generally more profitable to sell the cane stalk for chewing, or to make sugar. The use of sugar cane for livestock is likely to be most appropriate in remote mountainous areas that are distant from markets and where the alternatives for using the sugar cane are limited.

In Vietnam, the concept of sugar cane fractionation was first introduced to farmers in Tuyen Quang province, in the mountainous region of north Vietnam (Nguyen Thi Oanh et al., 1993). In this province, as in other mountainous regions of Vietnam, there is a long tradition of artisan sugar production at family level. The farm households typically have from 500 to

2,000 m² planted in sugar cane. This is harvested in the winter period from December through to May. The cane stalks are crushed in simple 2- and 3-roll mills driven mainly by buffaloes and occasionally by cattle. The juice is evaporated in open pans using the bagasse, supplemented with wood, as the source of fuel. The juice may be concentrated to syrup for subsequent refining to white sugar or crystallized into blocks of brown sugar. Both sugar and/or syrup are sold in the market and are the main source of income for the families. The cane tops are fed to the buffaloes and cattle and the scums to the pigs. Most families keep pigs, which traditionally are fed on rice bran and fresh and waste vegetables.

In a pilot project with 10 households it was found that the economic benefit from pig production was equal to, or greater than, that from making sugar, and as expected the impact was greatest with those families living furthest from the road (Nguyen Thi Oanh et al 1993). The technology has now been taken up by many other farmers in Tuyen Quang and has spread to other mountainous regions in Bac Thai and Cao Bang provinces (NIAH, 1995).

Justification for the study and hypotheses

Justification

Recent research at the National Institute of Animal Husbandry and associated institutions, especially the Goat and Rabbit Research Centre at Bavi, has concentrated on increasing feed resources by growing plant species with high yields, good nutritive value and ready adaptation to hilly areas especially in the dry season. Sugar cane and multi-purpose trees have proved to be particularly appropriate for these conditions, and now form the basis of the livestock feeding programme at the Goat and Rabbit Research Centre. Pigs and kids are fed with sugar cane juice, the tops (growing points and leaves) are fed to goats, the peeled stem is the basal diet for rabbits and the pressed sugar cane stalk after partial extraction of juice is fed to buffaloes and cattle. These feeding systems are being introduced on family farms in Bavi district.

An evaluation of different cropping practices in Bavi district showed that sugar cane was the preferred crop for most farmers. Growing sugar cane was more profitable than cassava, elephant grass or guinea grass as assessed by net farm-income (Nguyen Thi Mui ,1993). Probably no other crop has the enormous latent potential to produce biomass that characterizes the Saccharum species. The potential of sugar cane, and its intrinsic advantages over other tropical grasses, as a converter of solar energy into biomass is the rationale for the concept of "energy cane" (Alexander, 1985). However, sugar cane has other characteristics which make it especially appropriate as a feed reserve for livestock in the tropics and superior to almost all other forage crops. In contrast with other tropical grasses, the nutritive value of sugar cane increases with maturity (Alvarez and Preston ,1976). Furthermore, the time of reaching maturity coincides with the dry season, which is when the sugar cane has its optimum nutritive value. At this time of the year it yields much more digestible biomass than grasses such as *Panicum maximum* (Rodriguez and Ruiz, 1983).

Hypotheses

The research to be described in this paper is part of a long term study: (i) to develop the most appropriate ways to plant and manage sugar cane destined to be used as animal feed; (ii) to monitor the effects of management practices on soil fertility. (iii) to compare the total biomass yields of different varieties available in north Vietnam; and (iv) to evaluate the effects of the most promising interventions on the nutritive value of the total biomass and its different components as animal feed.

The specific experiments that were carried out aimed to test the following hypotheses:

- Planting the sugar cane at narrow distances between rows will increase yield of total biomass and of soluble sugars.

- Planting stem cuttings rather than growing points would support higher plant populations and yields especially in the second and third ratoons.

- Leaving the dead leaves on the soil surface will have beneficial effects on soil organic matter, overall nutrient status and on yield of biomass and total sugars.

- Varieties not used traditionally for chewing, or for making sugar, because of low sucrose content, may produce higher yields of biomass and total sugars, including reducing sugars, and thus be more appropriate when the aim is to produce feed for livestock.

- Sugar cane harvested after 12 months can have a better nutritive value for livestock than when harvested after 10 months.

General discussion

Number of harvests

In the traditional management of sugar cane in Vietnam, only one or two crops are taken before replanting. This is in contrast with experience in most sugar cane growing countries where the crop is ratooned (the crop is re-grown without planting over successive years) for as many as five to seven harvests. Reasons advanced by farmers for the short life cycle of sugar cane in Vietnam are the reduction in yield with successive ratoons and the increasing incidence of pests and diseases. In Nguyen Thi Mui et al. (1996a) it was reported that by the time of the 2nd ratoon, the numbers of plants decreased by:

13.5 % for the 75cm row distance compared with 46% at 150cm;
26% for stem cuttings compared with tops (22%);
26% for both mulching and no mulching.

Decreases in stalk yield were:

45% for 75cm rows compared with 60% for 150cm rows;
54% for stem cuttings compared with 50% for tops;
47% for mulching compared with 57% for not mulching.

It can be concluded that the interventions (decreased row distance, use of stems as plant material and mulching with dead leaves) in general reduced the rate of decrease in population and yield with successive harvests.

Removal of dead leaves

A feature of traditional sugar cane management in Vietnam is the removal of the dry leaves at regular intervals throughout the growing period. These are used for fuel or for thatching of roofs. It is also believed by many farmers that removing the dry leaves helps to reduce incidence of disease.

In industrial and artisan sugar production in Colombia, dry leaves are allowed to accumulate on the soil surface. In the industrial sector these leaves are burned immediately prior to harvest so as to facilitate the work of the cane cutters. In the artisan sector, which is located mainly on sloping hillsides under rain-fed conditions, the leaves are not burned. It is also customary to harvest only the ripe, mature stalks, leaving the immature ones to develop further. It is rare for all the sugar cane area to be renewed. Instead, new plant material, usually the growing points, is planted in the spaces where the original stools (the root system of sugar cane) have died. In this way a green canopy is maintained perpetually in the sugar cane growing area. The farmers believe these practices - accumulation of dry leaves on the soil and their subsequent decomposition, and harvesting only of mature stems - control erosion and maintain soil fertility in what are otherwise ecologically fragile zones (Rodriguez, 1995 pers. comm.).

The burning of dead leaves causes pollution through release of partially burned carbon residues and contributes to global warming because of release of carbon dioxide (Preston and Leng, 1987; Kirchhoff et al., 1991). It has also been shown that the decaying of the fallen leaves supports a microbial ecosystem capable of fixing appreciable quantities of atmospheric nitrogen, with an average of 5 kg N being fixed per tonne of dead leaves (Patriquin, 1982). Most of the nitrogen fixed will cycle through the soil organic matter before being made available to the plants. This means that a large pool of organic nitrogen accumulates as humus, which has several other benefits. Sugar cane being a perennial plant can continuously take advantage of the release of this nitrogen.

Soil, water and nutrient conservation are improved with use of mulches or cover crops which also protect the soil from erosion, desiccation and excessive heating, thus promoting good conditions for composition and mineralization of organic matter (Khalak and Kumaraswamy, 1993; Pretty, 1993; Borthaker and Bhattacharyya, 1992). Mulching generally increased the yields and quality in potatoes (Kaniszewski, 1994; Pakyurek et al., 1994), soybean (Kitou and Yoshida, 1994), sweet pepper (Siwek et al., 1994), mango (Farre et al., 1993) and bananas (Sarad et al., 1994). In general, the cheapest approach is to use plant residues from previous crops (Pretty, 1993).

In sugar cane it has been shown in India that recycling leaf trash to the soil conserves soil organic carbon and helps to sustain yields of successive ratoon crops (Yadav et al., 1994). Ball Coelho et al., (1993) have demonstrated that mulching increased the N uptake and the apparent recovery of applied N by the crop by improving the soil organic carbon compared with no mulching. The concentration of NO_3-N in the upper soil layer was higher with trash mulching than with trash burning and no mulching. The leaf trash as it decomposes confers long-term benefits and yearly applications will slowly increase soil-N levels and reduce fertilizer requirements according to Chapman et al (1992). There are more recent reports showing reduced weed growth and the resultant competition for moisture and nutrients. These combined effects increased production and improved soil fertility when dead sugar cane leaves are left on the soil surface (Phan Gia Tan, 1993; Mahadevaswamy et al., 1994). In research reported by Mendoza (1988), leaving post harvest cane trash on the soil as a mulch instead of burning it increased sugar cane yield, supposedly through fixation of atmospheric nitrogen as described by Patriquin (1982), Segundo et al. (1992) and Boddey et al. (1991).

The dead leaves of the sugar cane (leaf trash) were used as mulching for sugar cane in our work. In two on-station trials and one on-farm trial there were consistent improvements in yield and soil fertility when the dead leaves were returned to the soil compared with removing them, thus supporting the original hypothesis and the findings in the literature.

Plant material

Another management factor which may be related to the short life cycle of sugar cane in Vietnam is the use of the growing point as plant material. This is also in marked contrast with conventional practice in sugar cane producing countries, where invariably stem cuttings are used selected from actively growing immature cane. This latter system is obviously only feasible when the life cycle extends to more than 4-5 ratoons, as the quantity of sugar cane needed for planting would seriously reduce the potential yield of saleable stalks if replanting was after one or two harvests. By using only the growing point the yield of millable stalk is not affected.

When sugar cane is still at the immature stage (6-9 months after planting) the process of sucrose formation is not complete and levels of fructose and glucose and non-sugar soluble substrates are high. This facilitates germination and vitality of the young shoots (Tran Van Soi, 1987). Also the use of stem cuttings allows planting to be done at any time of year and not only at the time of harvest. Good quality planting material of sugar cane plays an important role, and reduces the empty area in the sugar cane plantation and increases plant density. When tops are used, only one and at most two shoots are produced per unit of planting material; with stems as many as 3 to 4 shoots are used for the same weight of plant material. Thus there is a saving in plant material which can be as much as one third (Nguyen Huy Uoc, 1987b).

One of the aims of the study reported in Nguyen Thi Mui et al., (1996a) was to find out if there would be advantages in using stem cuttings rather than the traditional system of planting only the tops. The results showed little difference between these two treatments other than in the first year when yields were slightly higher for the stem cutting system

Plant population

The biomass yield of sugar cane is a function of the density of mature plants at harvest time. According to Tran Van Soi(1987), the number of mature stalks per ha should be in the range of 56,000 to 82,000 plants. The recorded densities of 43,000 and 35,000 for the 150cm distance in the 2nd and 3rd years in Nguyen Thi Mui et al., (1996a) are well below this indicative value.

Traditionally in Vietnam, sugar cane is planted with a distance between rows of 1.5m. Research has shown that by narrowing the distance between rows the yields of biomass and of sugar are increased (Irvine et al., 1984; Sharma, 1982; Phan Gia Tan, 1993). Farmers in Vietnam believe that high plant densities lead to lower sugar production due to an increased tendency for the canes to fall and break. Damaged stalks are not suitable for sugar production since partial inversion occurs leading to conversion of some of the sucrose to reducing sugars which cannot be crystallized, and broken canes can not be sold for chewing or juice extraction by roadside vendors. However, when sugar cane is used for animal feeding, the sucrose content is not a problem as reducing sugars (glucose and fructose) have a similar nutritive value for animals as sucrose (Figueroa and Ly, 1990).

The results of our work showed conclusively the advantages in yield of stalk and total biomass of narrowing the row distance from 150 to 75cm (Nguyen Thi Mui et al., 1996a), to 90cm (Nguyen Thi Mui et al., 1996b) or 100cm (Nguyen Thi Mui and Preston, 1996). The nutritive value of the cane stalk for livestock (extraction rate and Brix of juice) was not apparently affected by the higher plant densities.

Varieties

There are two kinds of sugar cane planted in Vietnam,. One is used for processing for sugar production and the other for human consumption by chewing or as a juice drink sold at the roadside (Duong Duc Thang, 1991). For the sugar processing industry many varieties of sugar cane have been imported from Cuba, India, Hawaii, Taiwan, Indonesia and China. Crosses were made in Vietnam among some of these varieties (eg. F134, F156, VD54/143, VD/59/264, NCO310) introduced into the agricultural system of Vietnam. The most promising varieties and crosses were F134, VD54/143, F156, NCO310, CO775, JA60-5, CO419, CO290, VD59/264, POJ2878 and POJ3016 (Tran Van Soi, 1987). In general, these varieties have hard peel, tough bagasse, a high sucrose content, high resistance to environmental stress and are early maturing. At the present time, most of these varieties and crosses are planted as mixtures and very few fields of sugar cane are over 80% pure (Nguyen Huy Uoc, 1987a). Often in any particular area there is only one and at most two varieties.

In the artisan sector, there appear to be few instances where attempts have been made to compare the new varieties with the ones used traditionally. Some varieties that are used for direct consumption have a thin peel and soft pith and are easily broken (Duong Duc Thang, 1991). The planting season for these varieties can be spread all the year around and the ripening period is only about 9 months. Some of these varieties can be grown continuously for three years without flowering, however, they have lower sucrose content and high glucose (Duong Duc Thang, 1991. In the sugar industry the requirement is for sucrose content and cane stalk yield. For animal feed the need is total biomass yield and total sugars, adaptation to year round planting and early maturity. Therefore, varieties or crosses which are not suitable for commercial processing to sugar may be appropriate for animal feeding.

It was shown in Nguyen Thi Mui et al., (1996b) that the introduced varieties (F156, Hoanam and My-55-14) produced significantly more stalk (77.6, 74.9, 78.9 tonnes/ha) than the variety POJ39-16 (67.5 tonnes/ha) which is traditionally used by farmers in the region where the study was conducted.

Conclusions and recommendations arising from the research

- The results of the research carried out show conclusively that simple management practices, requiring no external inputs, can significantly increase the sugar cane yields as well as increase the soil fertility.

- Decreasing the row distance to 75, 90 or 100cm instead of the traditional 150cm had the most dramatic effect on yield with increases of: 71% in 3 years (Nguyen Thi Mui et al., 1996a), 48% over 2 years (Nguyen Thi Mui et al., 1996b) and 29% in one year (Nguyen Thi Mui and Preston, 1996).

- Returning the dead leaves to the soil increased yields by 18.7% over 3 years, 12.5% over 2 years and 11% over 1 year (Nguyen Thi Mui et al., 1996a; Nguyen Thi Mui et al., 1996b and Nguyen Thi Mui and Preston, 1996) and improved soil fertility and the carbon content of the soil.

- New varieties of sugar cane, recently introduced into Vietnam (Hoanam, F 156, My 55-14) gave higher yields of 8%, 5%, 8% (first year) and 15.5%, 30%, 30 % (second year), respectively, than the original "Java" POJ 30 16, traditionally planted by farmers in the region.

- None of the above interventions appeared to affect the quality of the sugar cane for livestock feed as Brix in the juice and juice extraction rate were not changed.

Acknowledgements

The studies were carried out at The Goat and Rabbit Research Centre of the National Institute of Animal Husbandry, North Vietnam, during 1993 to 1995 as requirement of the Master of Science course "Sustainable Livestock Systems in the Tropics" administered by the Department of Animal Nutrition and Management of the Swedish University of Agricultural Sciences, and sponsored by the Swedish Agency for Research Cooperation with Developing Countries (SAREC) and the Swedish International Development Authority (SIDA). I would like to express my sincere gratitude to these institutions and especially to Professor Hans Wiktorson, Goran Bruhn, Dr. Brian Ogle, Dr.Nguyen Thien, Dr. Le Viet Ly, Dr. Inger Ledin and Dr. Pernilla Fajersson for their help, and for giving me the wonderful opportunity to carry out this study in Vietnam and in Sweden.

I wish to thank the Director of The Goat and Rabbit Research Centre, National Institute of Animal Husbandry in Hanoi for granting me permission to pursue my research and for allowing me to use the staff and facilities to accomplish my research.

My study was made possible through my supervisor Dr. Ingvar Ohlsson and supervisor in Vietnam Dr. T.R.Preston, who freely offered their time for valuable advice, guidance, discussion and always gave me encouragement. They provided an environment that made this period a most enjoyable learning experience and I am greatly indebted to them.

I would also like to thank to Mr. Frands Dolberg, Dr. Anil Gupta, Dr. Peter Uden, Dr. Jan-Erik Linberg, Dr. John Ohrvik, Dr. Nguyen Quang Suc and Howard Benson for their advice and guidance during my studies.

Thanks are also due to Ngo Tien Dung for his technical assistance.

Last, but certainly not least, I would like to thank my friend, Dinh Van Binh, who encouraged me throughout my research career and always showed great interest in my work.

References

Alexander A.G., 1985. Sugar cane as source of biomass; Sugar cane as feed. pp. 46-59.

Alvarez F.J. and Preston T.R., 1976. Performance of fattening cattle on immature or mature sugar cane. Tropical Animal Production 1:106

Ball Coelho, H. Tiessen, J.W. B. Stewart, I.H.Salcedo, and E.V.S.B. Sampaio,, 1993. Residue management effect on sugar cane yield and soil properties in Northeastern Brazil. Agron J. 85:1004-1008.

Bassham J. A., 1978 Photosynthetic productivity of tropical and temperate crops Caribbean Consultancy on Energy and Agriculture. Santo Domingo, Dominican Republic. November 29th.

Boddey, R.M., Urquiaga Reis V., Dobereiner, J., 1991. Biological nitrogen fixation associated with sugar cane. Plant and Soil, 111- 117.

Borthaker, P.K. and Bhattacharyya R.K., 1992. Effect of organic mulches on soil organic matter content and soil pH in guava plantations. South Indian Horticulture, 352-354.

Chapman, L.S., Haysom, M.B.C. and Saffigna, P.G., 1992. N cycling in cane fields from 15N labelled trash and residual fertilizer. Proceedings of the 14th Conference of the Australian Society of Sugar Cane Technologists Mackay, Queensland, 84-89.

Duong Duc Thang, 1991. Sugar cane book, Agricultural University of Hue. pp: 25

Farre J.M.,Hermoso J.M. and Schaffer B., 1993. Mulching and irrigation effects on growth, cropping and fruit quality of the mango cv. Sensation. In Fourth international mango symposium, Miami, Florida, USA, Acta-Horticulturae, 295-302.

Figueroa Vilda Y. and Ly, J., 1990. Alimentacion porcina no convencional. Coleccion GEPLACEA, Serie Di versificacion Mexico, 215.

Handbook of General Information of Sugar cane Industry in Vietnam, 1994. pp. 24-32

Handbook of Official Statistic in Vietnam, 1993.

Irvine, J.E., Richard, C.A., Carter, C.E. and Dunckelman, J.W., 1984. The effect of row spacing and sub-surface drainage on sugar cane yield. Sugar cane, 1984, 2:.3-5.

Kaniszewski, 1994. Response of tomatoes to drip irrigation and mulching with polyethylene and non-woven polyethylene; Instytut Warzynictwa, Skierniewice, Poland. Biuletyn-warzywniczy, 29-38.

Kirchohoff, V.W.J.H., Marinho, E.V.A., Dias, P.L.S., Pereia, E.B., Calheiros, R., Andre, R. and Volpe, C., 1991. Enhancement of CO and O3 from burning in sugar cane fields. Journal of Atmospheric Chemistry, 87-102.

Kitou, M. And Yoshida, S., 1994.Mulching effect of plant residues on soybean growth and soil chemical properties. Soil Science and Plant Nutrition, 211-220.

Khalak, A. and Kumaraswamy, A.S., 1993. Weed biomass in relation to irrigation and mulching, and economics of mulching potato crop under conditions of acute water scarcity. In Journal of the Indian Potato Association, 185-189.

Mahadevaswamy, M., Kailasam, C. and Srinivasan, R.T., 1994.Integrated weed management in sugar cane, Indian Journal of Agronomy, 83-86.

Mendoza, T. C., 1988. Development of organic farming practices for sugar cane based farms. Paper presented during the 7th. Conference of the International federation of Organic Agricultural Movement. Ouagadougou, Burkina Faso.

Mohsan Atta, Hamid Mahmood, Hussaini, K.H., Malik K.B., 1991.Evaluation of new sugar cane varieties under Faisalabad conditions. Pakistan Sugar Journal, 1-5.

Mondharan M.L., Duraisamy K. and Vijayaraghavan H., 1990. Effect of management practices to improve cane yield and quality under moisture stress conditions. Bharatiya-Sugar, 19-25.

NIAH (National Institute of Animal Husbandry), 1995. Contribution to the establishment of a sustainable animal production in Bacthai (unpublished)

Nguyen Huy Uoc, 1987a. Performance of some sugar cane varieties on hilly land in Vietnam. Journal of Agricultural Science.

Nguyen Huy Uoc, 1987b. Research results on sugar cane varieties in Southern Vietnam. Journal of Agricultural Science.

Nguyen Thi Oanh, Dao Thi Quy, Preston, T.R., Le Viet Ly and Bui Van Chinh, 1993. Pigs fed by sugar cane juice in Tuyenquang. Contribution of the establishment of a sustainable animal production in Vietnam. 2nd National SIDA/SAREC Workshop on Sustainable Livestock Production on Local Feed Resources. Hanoi-HCM City.

Nguyen Thi Mui, 1993. Economic evaluation of growing elephant grass, guinea grass, sugar cane and cassava as animal feed or as cash crops on Bavi high land in North Vietnam. 2nd National SIDA/SAREC Workshop on Sustainable Livestock Production on Local Feed Resources. Hanoi-HCM City, pp. 16-19

Nguyen Thi Mui and Preston, T.R , 1996. On-farm evaluation of planting distance and mulching of sugar cane. M.Sc. Thesis. Department of Animal Nutrition and Management, Swedish University of Agricultural Sciences. Part 3.

Nguyen Thi Mui, Ohlsson, I. and Preston, T.R., 1996. Effect of management practices on yield and quality of sugar cane and on soil fertility. M.Sc. Thesis. Department of Animal Nutrition and Management, Swedish University of Agricultural Sciences. Part 1.

Nguyen Thi Mui, Preston, T.R. and Ohlsson, I., 1996. Responses of four varieties of sugar cane to planting distance and mulching. M.Sc. Thesis. Department of Animal Nutrition and Management, Swedish University of Agricultural Sciences. Part 2.

Patriquin, D.G., 1982 Nitrogen fixation in sugar cane litter. Biological Agriculture and Horticulture, 39-64.

Pakyurek A.Y., Abak K., Sari N., Guler H.Y., Cockshull K.E., Tuze Y., Gul A., 1994. Influence of mulching on earliness and yield of some vegetable grown under high tunnels. In Second symposium on protected cultivation of Solanacea in mild winter climates, Adana, Turkey, 12-16 April 1993. Acta Horticulturae, 155-160.

Phan Gia Tan, 1993. Effect on production of sugar cane and on soil fertility of leaving the dead leaves on the soil or removing them, pp. 28-32 National seminar-workshop, Sustainable livestock production on local feed resources.

Preston T. R. and Leng, R. A., 1987. Matching ruminant production systems with available resources in the tropics and sub-tropics, Armidable, Australia, pp. 142-155

Preston T. R. and Murgueitio, E., 1992. Strategy for sustainable livestock production in tropics. CIPAV, CONDRIT, Cali, Colombia, pp. 49-69.

Preston T.R., 1995a Research, Extension and Training for Sustainable Farming Systems in the Tropics. Livestock Research for Rural Development, Volume 7, 2:1-8.

Preston T.R. 1995b Sugar cane for feed and fuel: recent developments. World Animal Review.

Pretty J., 1993.The Transition to Sustainable Agriculture: Linking Process to Impact, pp. 119-124.

Rodriguez V. and Ruiz Corvea E., 1983 The utilization of sugar cane as forage for cattle. Production and use of sugar cane for animal nutrition. Editor: MINAZ-MINAG, Centro de Información y Divulgación Agropecuario Havana. pp. 7-29.

Sharma, R. A., 1982. Effect of pre-monsoon irrigation and inter-row spacing on yield and quality of sugar cane. JNKVV-Research-Journal, 136-141.

Sarad Gurung, Chattopadhyay P.K. and Gurung S., 1994. Influence of soil cover on production and quality of banana. In Annals of Agricultural Research, 445-447.

Segundo U., Katia H.S., Cruz and Robert M. Boddey, 1992. Contribution of nitrogen fixation to sugar cane: Nitrogen-15 and nitrogen-balance estimates. The Soil Science Society of America Journal, 105-113.

Singh G. and Singh O.P., 1992. Performance of sugar cane (saccharaum oficinarun) varieties at various row spacings when grown under flood-prone condition. In Indian Journal of Agricultural Sciences, 818-820

Singh, R.P. and Singh, P.P.,1984. Quality of juice and commercial cane sugar as influenced by crop canopy in sugar cane. Indian-Sugar, 21-28.

Siwek P., Cebula S., Libik A., Mydlarz J., Kockshull K.E., Tuzel Y. and Gul A., 1994. The effect of mulching on changes in microclimate and on the growth and yield of sweet pepper grown in plastic tunnels. Second symposium on protected cultivation of Solanacea in mild winter climates, Adana, Turkey, 12-16 April 1993. Acta Horticulturae, 161-167.

Tran Van Soi, 1987. Questions and answers on sugar cane planting technology; Agricultural Publishing House, Hanoi, pp. 82- 124.

Yadav R.L., Prasad S.R., Ramphal Singh and Srivastava V.K., 1994. Recycling sugar cane trash to conserve soil organic carbon for sustaining yield of successive ratoon crops in sugar cane. Bioresource Technology, 231-235.

A Review of Animal Production Work in the Context of Integrated Farming with Examples from East Africa

F.P.Lekule and S.V.Sarwatt

Department of Animal Science and Production,
Sokoine University of Agriculture,
P.O.Box 3004 Chuo Kikuu, Morogoro, Tanzania.
E-mail: SVSARWATT@hnettan.gn.apc.org

Summary

On the background of the need to enhance livestock production a discussion is conducted describing the main agricultural systems such as agropastoral systems, agro-sylvo-pastoral systems and sylvopastoral systems in sub-Saharan Africa and the possibilities for intensification of animal production within them. It is concluded sustainable crop-livestock systems can be developed in most areas of East Africa subject to improvement and introduction of fodder crops, leguminous tree crops, improved methods of feed storage, increased use of livestock for draught power and control of tsetse flies.

Key words: *Animal production, intensification, integrated farming, East Africa.*

Introduction

The population of Sub-Saharan Africa is increasing by 3.1% a year, the highest of any region of the world. The population is estimated at 500 million, up from 250 million in 1965. It has been estimated that by the year 2025 nearly 55% of the region's people will live in towns and cities as contrasted to 30% today. Urbanization will force the commercialization of agriculture and increase the demand for foods of animal origin (Winrock International, 1992). Both incomes of the people and food production will have to increase substantially to avert economic disaster.

Livestock play vital economic and cultural roles in the everyday life of the people of Africa. They can play a critical role in the intensification of agriculture and in the development of sustainable agricultural production systems. There is no doubt that sub-Saharan Africa has a great potential for livestock production. If properly developed the livestock industry can contribute enormously towards enhancement of food production, economic development, and human welfare on an equitable, sustainable and environmentally sound basis. The sustainability of the livestock production systems will depend to a large extent on how they are integrated with crop production systems. The livestock provide inputs to cropping including manure and draught power while the crops provide feed for livestock.

Why integrated systems?

Currently sub-Saharan Africa is a net importer of livestock products as well as cereals for human consumption and livestock. This is paradoxical when one considers that the region has 162 million cattle, 127 million sheep, 147 million goats, 13 million camels, 11 million pigs, 8 million donkeys, 3 million horses, 1.5 million mules and 631 million chickens (FAO, 1989) and has sufficient arable land. Ethiopia, Sudan and Tanzania have the largest number of cattle in Africa.

Gollin (1991) has estimated that ruminants and equines will require 63 million tons of crude protein, but only 50 million tons are available. Forage available for livestock is about 1400 to 1500 Mcal x 10⁹. The arid zone has no potential for higher forage production. There is, however considerable potential in the semi-arid, semi-humid and humid zones.

Table 1. Projected feed requirements by 2025

Animal groups	1986-88	2025
	Metabolizable energy (Mcal x 109)	
Ruminants	793	1,405
Equines	44	65
Poultry	32	102
Pigs	10	32
Total	879	1,604
	Crude protein (million tons)	
Ruminants	40.3	60.1
Equines	2.2	3.3
Poultry	1.4	9.5
Pigs	1.0	3.0
Total	44.9	75.9

Source: Gollin (1991).

Poultry and pigs are in most countries fed on byproducts or allowed to scavenge, with little grain being fed. It has been estimated that by the year 2025, the requirements for feed grains will increase 10-fold; i.e.from 2.5 million tons to 25 million tons. This requires about 15 million hectares. The other alternative is to import this feed which is not economically feasible.

The same is true for importation of animal products. About 200 million hectares have arable potential. Almost 50% of the land lies fallow.

Thus more food will be required for the increasing human population and for feeding nonruminants. To increase production and productivity of both crops and livestock, inputs and technology are required. This can only be achieved sustainably by intensification through mixed crop-livestock farming systems with efficient use of crop residues and animal wastes. Animal traction and manure have been shown to increase profitability of farming systems. According to ILCA, in 1975 the value of animal traction and manure was 34% of total value of livestock production (ILCA, 1987). Gittinger *et al.*(1990) working in Zimbabwe found that smallholders who combine crop and livestock production have twice the income of small-holders who only raise subsistence crops

Major integrated crop/livestock systems

Three major integrated crop/livestock production systems can be identi-fied in Africa:

- Agropastoral systems
- Agro-sylvo-pastoral systems
- Sylvopastoral systems

Agropastoral systems

In this system crops are grown in association with livestock keeping. In the semi-humid areas of East Africa, farmers grow maize, sorghum, millet, wheat and beans for food. After these crops are harvested, the remaining stovers and haulms are fed to livestock either in *situ* or harvested and fed to zero-grazed animals. If fed in situ the animals will drop the manure in the field. In zero-grazed animals the manure can be collected and spread on the farm. In this latter system recycling is not ensured as the manure may be used in more productive land. This is the case with the highlands where crop residues harvested from the lowlands provide feed to animals in the highlands and the manure is used to fertilize coffee and bananas.

Other crops which are now providing compatibility with livestock are sisal and pineapples. In the sisal estates livestock can be grazed to reduce the weeds while the sisal waste is a useful feed which can be fed fresh or ensiled. Animals can also be grazed in fields where pineapples have been harvested. The waste from pineapple processing can provide feed to live-stock. In Kenya it is commonly fed to dairy cattle.

One notable examples can be found in Central Tanzania. In Kondoa, en-vironmental degradation was so serious under pastoralism that all the animals had to be evacuated. With assistance from SIDA improved animals with higher productivity were reintroduced and zero-grazing system adopted (Ogle, 1996). Farmers' incomes have improved appreciably and manure use is now a common practice. Further integration needs to be worked out in the use of the animals for draught power. Farmers have shown some reluctance in using the same animals for draught power. The delicate question will be: "To what extent can farmers accommodate more animals without again endangering the environment?"

Agro-sylvo-pastoral systems

In this system field crops, tree crops and pasture are associated with livestock keeping. This is a system which is now common in the wetter areas. The livestock are shaded from maximum solar radiation. The manure and urine from the livestock are returned to the soil to enrich it and nutrients are eventually recycled to the livestock through the pasture and tree foliage.

An interesting example is found in the humid highlands. There is a vertical stratification of the vegetation which ensures efficient use of solar energy. At the highest strata there is *Albizzia lebbek* which is a leguminous tree. The leaves provide bedding to livestock which improves the quality of the manure. The canopy reduces the intensity of solar radiation to coffee which is a C_3 crop. Due to the vertical stratification of root systems of different trees and plants, the roots of *A.lebbek* absorb nutrients in the subsoil and return them for recycling. These nutrients would have, otherwise been lost through leaching. At the same time they fix nitrogen in the soil and hence provide long-term improvement of the soil. Bananas provide the second strata. While the fruits are harvested for human food the pseudo-stems and leaves provide feed to livestock. The shallow root system enjoys the mulch from *A.lebbek*. The third strata is coffee which is a cash crop and leaf fall provides mulch. Sometimes there is a fourth strata of pulses or tubers (taro, yams, potatoes). Manure from the cattle provide nutrients to these plants.

Sylvo-pastoral systems

This system has emerged in the coast where livestock graze under the shade of coconut palms. The soil is enriched by the manure and urine from the cattle. The coconut palms shade the cattle from the intense solar radiation. The animals keep down the weeds. After oil extraction, copra cake is obtained which is a good source of protein for livestock.

Effect of demographic changes on farming systems

In 1990, 71% of the population of sub-Saharan Africa lived in rural areas. It has been projected that in spite of substantial migration to urban areas, the rural population will increase by more than 68%. There will be an increase in intensity of cultivation. Land devoted to fallow and pasture will decline considerably. McIntire *et al* (1992) observed that as population pressures cause animal agriculture systems to become more intensified, mixed crop-livestock systems become more efficient than specialized systems of crop and livestock production. As population pressures on land grow, cropping increasingly replaces pasture and fallow, and manure and crop residues both become more valuable. At higher levels of intensification the trend is reversed with efficiency gains from intensification eventually levelling off. To attain higher levels of productivity greater inputs and use of technology are required.

In Africa 80-90% of the poor people live in the rural areas. The expanding urban population will create a market for agricultural products. This will put more pressure on the land. Population pressure, by creating a scarcity of land and increasing the price of land relative to that of labour, results in more intensive use of land (Boserup, 1965, 1981). Intensification of livestock will make animals play important roles in *traction* (power), *manure* (fertilizer) *and income (cash)*. Delgado (1989) observed that mixed farming is a practice that permits higher labour inputs per unit of land in a profitable manner.

The role of different livestock species in integrated systems

Cattle

One of the most successful integrated systems is the smallholder dairy production. This has occurred mainly in the highlands and peri-urban areas. In East Africa, the number of smallholders keeping dairy cows increased from 300,000 in 1979 to 600,000 in 1989 (Wanyoike, 1991). Milk production stood at 2,200 million litres. These cattle are fed mainly on locally grown forages, grain byproducts (maize bran, cottonseed or sunflower cake), brewers wastes and crop residues. Availability of these feed supplements depends to a great extent on successful crop production. In years of low crop production these animals equally suffer.

Sheep and goats

In East Africa goats are more important than sheep and these are mainly meat goats. They fit well in integrated systems because of their popularity and small size. This is more so for dairy goats, as will be discussed in a paper to follow on the experience of Norwegian goats in Mgeta (Kifaro *et al*, 1996). Elsewhere goats in large herds are grazed extensively in pastoral systems.

Poultry

Traditional poultry keeping involves free ranging where the birds are hardly fed concentrates except kitchen wastes. Commercial poultry production relies entirely on purchased feeds and the birds are totally confined. High technology is used.

Pig production

Small scale pig production is practised where agricultural byproducts like maize bran, wheat, rice polishings, oil cakes, garbage are available for feeding. The pigs are kept to make use of the available byproducts. There are a few areas where pigs are kept to provide manure to vegetable gardens and in turn vegetable wastes are fed to the pigs. Like in the case of poultry, large scale pig production is hardly integrated with crop production. It does, however, depend on byproducts of food processing. Shortage of grains for human food implies also limited scope for pig production. Pig production enterprises must be located close to feed sources.

Biotechnology in integrated systems

Introduction of biogas technology into the system has meant better utilisation of the excreta from animals (Cortsen *et al*, 1995). So far most of the biogas units in E.A. utilize cow dung. It is perhaps the most easily obtainable manure. The system improves the quality of manure as it reduces losses of N and other nutrients. In this system the losses of carbon are reduced. So most of the nitrogen and carbon will be recycled back to the soil.

Conclusions

- Integration of crops and livestock produces better returns to labour
- As a result, the strain on natural resources is minimized.
- Further integration with wildlife in the drier areas will increase the carrying capacity of land and hence reduce environmental degradation.
- Sustainable Crop-livestock systems can be developed in most areas of East Africa provided the following elements are given due consideration:

  ◊ Improvement and introduction of fodder crops, leguminous tree crops that will provide feed all year round.
  ◊ Improved methods of feed storage.
  ◊ Increased use of livestock for draught power.
  ◊ Matching the available resources with livestock numbers.
  ◊ Control of tsetse flies in areas of high potential (subhumid) to create opportunities for crops and livestock.

Acknowledgements

The senior author acknowledges the financial support provided by DANIDA through the ENRECA project and DanChurchaid to participate in this meeting.

References

Boserup, E.(1965). The conditions of agricultural growth: The economics of agrarian change under population pressure. New York:Aldine.

Boserup, E.(1981). Population and technological change: A study in long term trends. Chicago:University of Chicago Press.

Cortsen L., Lassen M. and Nielsen H.K.(1995). Small Scale Biogas Digesters in Turiani, Nronga and Amani, Tanzania. A student report. Aarhus University, Denmark.

Delgado, C. L. (1989). The changing economic context of mixed farming in savanna West Africa. A conceptual framework applied to Burkina Faso. Quarterly Journal of International Agriculture, 28:3-4.

Gittinger, J.P.,S.Chernick, N.R.Horenstein, and K.Saito (1990). Household food security and the role of women. Discussion paper 96. Washington D.C.:World Bank.

Gollin, D.(1991). Status, trends, and directions of animal agriculture in sub-Saharan Africa. Morrilton, Arkansas: Winrock International.

International Livestock Centre for Africa (1987). ILCA's strategy and long term plan. Addis Ababa.

Kifaro, G.C; Mtenga, L.A and Kiango, S.M. (1996). The Mgeta integrated farming system with emphasis on the introduction of Norwegian dairy goats. These proceedings.

McIntire, J., D.Bourzat and P. Pingali (1992). Crop-livestock interactions in sub-Saharan Africa. Washington D.C.:World Bank.

Ogle, B. (1996). Livestock Systems in Semi-Arid Sub-Saharan Africa. These proceedings.

Wanyoike, M. M. (1991). Milk production systems in Kenya. University of Nairobi.

Winrock International (1992). Assessment of Animal Agriculture in Sub-Saharan Africa. Winrock International, Arkansas. USA.

Goats

The Mgeta Integrated Farming System with Emphasis on the Introduction of Norwegian Dairy Goats

G.C. Kifaro, L.A. Mtenga and S.M. Kiango

Department of Animal Science & Production,
Sokoine University of Agriculture, Morogoro, Tanzania
E-mail: SVSARWATT@hnettan.gn.apc.org

Summary

This article describes the Mgeta, Tanzania mixed farming system which comprises of food production, vegetable gardening, fruit production and animals husbandry. Pigs, goats and poultry are the main livestock species. Maize, beans and peas are the major food crops in the area. There are a variety of vegetables and temperate fruits being grown for sale though some are consumed locally.

Sokoine University has introduced Norwegian dairy goats in Mgeta in order to improve the living standards of people and establish a site for on-farm research. So far the project has been a success. Growth rates above 70g/day have been recorded. Mortality rate of kids before 9 months of age is around 15%. Age at first kidding and kidding intervals have averaged 527 and 339 days, respectively. Overall daily and lactation milk yields have been 0.87 and 176 litres, respectively. Milk yield has significantly increased with increase in Norwegian blood in goats. It is concluded that the project has improved the general welfare of farmers and the efficiency of the Mgeta farming system.

Key words: Goats, farming systems research, crossbreeding, villages, performance.

Introduction

Administratively, Mgeta is a division under Morogoro rural district of Morogoro region. It is situated about 50 km south-west of Morogoro regional town in the western slopes of Uluguru mountains. Mgeta is reached by 11/2-2 hours' drive on a coarse, stony road. The altitude ranges from 1400 to 2000 m a.s.l. Annual precipitation varies between 1000 and 2000 mm depending on location. Rainfall is distributed bimodally with rain falling in months of March-May and November-December. Average annual temperature varies from 15 to 20° C decreasing with altitude.

The area is densely populated with inhabitants ranging from 100 to 300/km² depending on locality. The ethnic group occupying this area are called the "Waluguru" who follow a matrilineal system of inheritance (Menetrier, 1989).

Characteristics of the farming system

General overview

Agriculture is the predominant activity in Mgeta with over 70% of the farmers engaged in mixed farming system (Menetrier, 1989; Dolobel *et al.*,1991). Due to high population density, almost all land is occupied leaving no room for a fallow system. The production system has been transformed from the slash and burn system as well as fallow system to a market economy. All work is manual and that includes ploughing (by the hand hoe), planting, weeding and harvesting. Drought animal power has never existed. Within Mgeta, transportation of farm produce is done on peoples' heads or shoulders.

Multiple cropping is practiced throughout the year. The average farm size is less than 2 ha and the land holding is very fragmented and scattered on the area. Plots which are closer to households have permanently developed narrow terraces and those which are close to irrigation canals are constantly irrigated.

Farmers grow both food and cash crops. The main food crop is maize (*Zea mays*) which occupies 85% of the cultivated plots (Menetrier,1989). It is planted together with red beans (*Phaseolus vulgaris*) or peas depending on altitude. Fertilizer is rarely used thus harvests vary between <1 to 4 tonnes/ha. Peas is a cash crop but is completely integrated in the traditional cropping system. Several root crops such as potatoes (*Solanum tuberosum*), yams, cocoyams (*Colocasia antiquorum*) and cassava are grown as security crops in case of food shortage. Coffee is found in some plots. It used to be a compulsory crop some years ago. Sugarcane is grown along the streams. Another important perennial crop in the Mgeta farming system is bananas which can be seen up to 1700 m a.s.l. Eucalyptus is planted as an alternative source of domestic firewood.

The vegetable gardening system

Menetrier (1989) reports that 94% of farmers in Mgeta are involved in vegetable production and cabbage (*Brassica oleracea*) is the main cash crop. Other most represented vegetables include cauliflower, lettuce, carrot (*Daucus carota*), leek, garlic (*Allium sativum*), celery, parsley, onions and a variety of pepper. Vegetables are grown in pure stand or intercropped with other annual and perennial crops (Delobel *et al.*,1991). Most of these vegetables are sold to distant markets mainly Morogoro town and Dar-es-Salaam. Very little is consumed locally.

The fruit production system

Dolobel *et al.* (1991) have made extensive description of the Mgeta fruit production system. The most predominant deciduous fruit trees are plums (*Prunus salicina*) with five common varieties (red, ball, big yellow, small yellow, and Santa Rosa), peaches (*Prunus persica* L.), pears (*P. communis*) and apples (*Malus communis*) are very common fruits in Mgeta. Passion fruits (*Passiflora* spp) and tomato trees (*Cyphomandra betacea*) are quite common around houses and are used for both home consumption and for sale. Other sub-tropical species occasionally found in the area include avocado, guava and fig, but these are more or less considered as wild trees.

About 95% of the farmers own fruit trees with an average of 30 trees per farmer (Menetrier, 1989). Fruits are used for home consumption but 75% of farmers have fruit trees as a commercial crop. It is estimated that 30% of farmers realize the importance of fruit trees in soil erosion control. Others use trees to demarcate their plots preventing neighbours from gradual expansion. Fruits are sometimes used to feed pigs especially the small peaches which have fallen down. Other uses include serving as wind break, preserving privacy, providing firewood for domestic use and improving attractiveness of surroundings.

Fruits are considered least demanding in terms of labour and other resources compared to vegetables. Further, fruit production is seen to be less developed compared to vegetables. Both fruits and vegetables as perishables have common marketing problems of prices, poor roads, no grading system and lack of government assistance.

Animal husbandry practices

Generally Mgeta farmers do not keep cattle. In 25% of the farms, sheep and goats are present with an average herd size of 3 heads. The average number of poultry per household is eight (Kiango, 1994). Pig keeping is performed by about 60% of the farmers with an average of 1.5 head per farm (Delobel *et al.*,1991). Sarwatt and Lekule (1987) and Kiango (1994) however report the average number of pigs to be around 2 and the number of farmers owning pigs in three surveyed villages in Mgeta to range from 82 to 98%.

Pigs are kept permanently in small stalls and fed crop residues, weeds and agricultural by-products mainly maize bran bought from the market. The crop residues thrown into pig stalls become part of the bedding. Pigs provide meat, but their main function is to produce manure which is applied on vegetable plots thus maintaining soil fertility. In the survey made by Sarwatt and Lekule (1987) they reported on-farm reproductive performance of sows (table 1).

Table 1.   Reproductive traits of pigs in Mgeta

Trait	Range (of village means)
Litter size at birth	7.5 - 8.1
Litter size at weaning	6.2 - 7.2
Pre-weaning piglet mortality (%)	11.2 - 16.6
Weaning age (weeks)	7.5 - 8.1
Mean weight of weaners (kg)	8 - 12
Weaning to mating period (days)	36 - 90

Introduction of dairy goats

Evolution and development of the project

In 1982, a project called "Improved feeding of dairy cattle in the tropics" funded by NORAD started at the Department of Animal Science at Sokoine University, Morogoro. In 1983, a component of dairy goats was included and 63 Norwegian goats kids were imported from Norway. The purpose was to develop crosses with local goats with fair potential for milk production for semi-arid conditions. These goats did well at the start, but later poor reproductive performance and high mortality rate was experienced. Since 1988 it was decided to produce half-breds and thereafter mate them *inter se*.

It was hypothesized that probably pure Norwegian goats were not adapted to the hot Morogoro environment. It was decided to start an extension project whereby Mgeta was selected as project site.

Mgeta was chosen for a number of reasons:

i)   It is close to Sokoine University

ii) Because of high altitude, it has a pleasant cool climate

iii) All year round availability of forages and vegetable wastes

iv) Farmers had shown interest in dairy goats and had previous experience with stall feeding of local goats and pigs.

At the start, five interested farmers were identified from three villages, trained on dairy goat husbandry practices at the University and were given time to prepare goat sheds. Each family was thereafter (in May, 1988) sold two pregnant halfbred goats at subsidized prices (20 - 30 US $). Further, ten pure Norwegian females that had remained at the University were distributed to these farmers. In December, 1988 a pure Norwegian buck was stationed in one of the three project villages for breeding. One farmer was chosen to be a buck keeper. In December, 1989 each village had a buck centre. Bucks were used on introduced does as well as on local goats to produce 50% crossbreds. Bucks are rotated between villages annually to minimize inbreeding. Five other families were selected, trained and each given two pregnant goats in May, 1990.

One to two weeks courses were offered to all farmers (husbands, wives and/or sons/daughters) in general husbandry practices, breeding, recording, animal health and milk processing. In addition, the research team comprising of a breeder, a veterinarian and a nutritionist visited the farmers on monthly basis to give advise and exchange views. Farmer-to-farmer exchange programmes have also been used to impart new skills to the farmers.

Goats are fed by grazing them and/or being stall fed with grasses, herbs, branches and vegetable wastes. Maize bran with and sometimes without cotton-seed cake and minerals are offered at varying rates. Water is provided *ad lib*. Improved forages of Guatemala and Setaria were established on edges of their fields and terraces for feeding goats. Sesbania (*S. sesban*) was also introduced, but was not preferred because it is not leafy.

Birth weights of kids are taken by farmers, thereafter they are weighed once monthly by a project extension worker. Milk yield is measured volumetrically using plastic jars. Between 1988 and 1990 milk was recorded daily, but we have reduced it to once per week. Kidding dates, dry off dates, weaning dates and disposals are recorded. Bucks with < 75% Norwegian blood are castrated while those with higher blood levels are sold outside Mgeta for breeding purposes. Demand for bucks is extremely high. An up-grading programme is practiced in Mgeta.

Achievements obtained so far

The on-farm dairy goat project had the following objectives:

- To test the performance of Norwegian dairy goats under smallholder conditions.
- To develop a dairy husbandry system suitable for small scale farmers.
- To improve living standards of smallholders.
- To provide facilities for on-farm research for staff and students.

The achievements can be enumerated as:

a) One of the key achievement the project has made is the creation of awareness among Tanzanians of the importance of dairy goats in their farming systems. This has been reflected by requests to purchase goats, requests to replicate the project elsewhere, interest by other institutions to have similar projects and by regular visits by various people.

b) The number of project farmers has increased from 10 to 22 and over 500 goats have been born in Mgeta.

c) Milk for home consumption has gone up from nil to 1.8 litres per household and about 40% of their total farm income comes from dairy goat enterprise (Kiango, 1994).

d) Quite a number of undergraduate special projects, one M.Sc. and one Ph.D. study has been undertaken in Mgeta.

e) Four extension leaflets and three pamphlets have been published and distributed to farmers in Mgeta.

f) The first 10 farmers have formed "A Dairy Goat Farmers' Association" which took care of their farm inputs such as acaricides, drugs, minerals, lime, disinfectants, etc. This association has been transformed into a "Livestock Keepers' Association" which also caters for the needs of pig keepers.

Constraints faced

In spite of the achievements mentioned above, it is felt that some aspects have had some problems:

i) The spread of goats outside project farmers has been slow. Farmers feel it takes too long to get dairy goats through upgrading of their local stock. Consequently farmers have anxiously been waiting for assistance from the project to get started. Public meetings have been conducted involving the project farmers to motivate others to use the available bucks to get crossbred goats.

ii) Project funds have been declining with years. Project activities including visits to farmers have been dwindling and training of new farmers has not been possible.

iii) For the sake of sustainability of the project, the issue of recruiting replacement bucks is becoming very crucial. It is becoming increasingly difficult to import new bucks from Norway. There is a need to broaden the genetic base in Mgeta, improve recording so as to select buck mothers from farmers flocks.

On-farm performance of Norwegian dairy goats
Growth

Mean live weights of goats at different ages are presented in table 2. At all ages male goats were heavier than females and single born goats outweighed those born as twins or triplets. The 75% Norwegian blood crossbreds were heavier than higher exotic blood level crosses at all ages, indicating trends known from cattle crossbreeding programmes as well.

Pre and post weaning growth rates were 92±2 and 76±6g/day. Birth type had significant effect on pre-weaning growth rate. Kids born singles grew faster by approximately 19g/day compared to those born more than one.

Table 2. *Weights[1] of goat kids at various ages (kg)*

Group	Age (in months)			
	Birth	Weaning (3)	6	9
All	2.6 (256)	11.1 (235)	14.7 (121)	17.0 (41)
75% N	2.6 (141)	12.1 (132)	16.2 (73)	18.9 (25)
87 - 94% N	2.6 (62)	11.0 (54)	16.1 (26)	18.7 (8)
100% N	2.5 (53)	10.5 (49)	14.8 (27)	15.0 (8)

1) Based on least squares means estimated from a model in which sex, birth type, year of birth and genetic group were included.
2) In brackets are numbers of observations.

Reproduction and mortality rates

The average age at first kidding in Mgeta is 527 days (or 17.3 months) and goats kid after an average of 339 days (table 3). Pure Norwegian does had lowest age at first kidding, but their kidding intervals were longest. Twinning rate of Norwegian goats and their crosses has been found to be 48% and mean litter size of 1.5 kids per doe per kidding.

Table 3. *Reproductive performance of Norwegian goats in Mgeta (days)*

	Trait	
	Age at first kidding	Kidding interval
Overall	527 ± 20 (59)	339 ± 10 (111)
Genetic group		
50%N	521 ± 88 (19)	339 ± 14 (63)
75-94%N	554 ± 40 (30)	312 ± 39 (22)
100%N	498 ± 80 (10)	385 ± 30 (26)

Mortality rate among goat kids before 9 months of age averaged 14.3%. It is higher among male kids than among female kids (table 4). Kids born single have lower mortality (10.9%) compared to twins (16.2%). Pure Norwegian kids have highest mortality (25.9%). The high mortality rate for this genetic group is not very surprising because exotic breeds suffer high death rates in the tropics due to poor disease resistance and poor adaptability to stressful environments. The main causes of mortality are pneumonia, strangulation by rope (when tethered) (Madsen *et al.*,1990), bloat and general weakness which is associated with worm burden (Kiango, 1994).

Table 4. Mortality rates of goat kids at Mgeta before 9 months of age

Factor	No. at risk	No. died	% mortality	Significance*
Overall	314	45	14.3	
Sex				
Males	156	25	15.8	NS
Females	158	20	12.8	
Birth type				
Singles	110	12	10.9	NS
Twins	204	33	16.2	
Genetic group				
75%N	161	17	10.6	*
87-94%N	76	8	10.5	
100%N	77	20	25.9	

* Significance by chi-square test

Lactation performance

The daily milk yield average 0.87 litres with a coefficient of variation of 44.8% indicating, as expected from tropics, great variability among animals. The overall lactation yield was 176 ± 8 litres. Both daily and lactation yields increased as the level of Norwegian blood increased. Mean lactation length and dry period were 207 ± 6 and 127 ± 9 days, respectively (table 5).

Table 5. Least squares means of lactation performance traits of Mgeta Norwegian dairy goats

Factor	Daily yield (litres)		Lact. yield (litres)		Lact. length (days)		Dry period (days)	
Overall	0.87	(958)	176±	(113)	207±6	(119)	127±9	(92)
Genetic group								
50% N	0.74	(513)	146±21	(60)	221±9	(63)	128±19	(53)
75 - 94% N	0.83	(245)	168±32	(29)	214±19	(30)	87±44	(19)
100% N	1.07	(200)	220±31	(24)	228±18	(26)	108±42	(20)
Parity								
1	0.93	(412)	188±24	(47)	235±11	(49)	147±18	(39)
2	0.93	(247)	178±28	(28)	217±14	(30)	97±24	(26)
3	0.95	(145)	208±32	(17)	223±19	(19)	74±34	(19)
4	0.91	(89)	180±35	(11)	208±19	(18)	109±52	(5)
5	0.74	(65)	141±37	(10)	-		111±69	(3)

Conclusion

The Mgeta farming system is market oriented. Farmers have accepted dairy goat husbandry because it benefits them from economic, health and general welfare points of view. The successful introduction of this innovation can be attributed to the firm commitment of first farmers, collaborative efforts of the research team and the anticipated benefits to both parties.

References

Delobel, T.C.; Evers, G.R. and Maerere, A.P. 1991. Position and functions of deciduous fruit trees in the farming systems at Upper Mgeta, Uluguru Mountains, Tanzania. Acta Horticulturae 290: 91 - 102.

Kiango, S.M. 1994. Studies on factors affecting performance of dairy goats and socio-economic aspects of dairy goat production in Tchenzema and Dareda wards in Tanzania. Unpublished M.Sc. thesis, Sokoine U
niversity of Agriculture, Morogoro. 200 pp.

Madsen, A.; Nkya, R.; Mtenga, L.A. and Kifaro, G.C. 1990. Dairy goats for small scale farmers: Experiences in Mgeta Highland. In: Proceedings of Tanzania Society of Animal Production Vol. 17: 48 - 58.

Menetrier, J.B. 1989. Agroclimatology/rural development Uluguru Mountains. Mimeograph, Cnearc, France. 51 pp + appendices.

Mtenga, L.A. and Kifaro, G.C. 1992. Dairy goat research and development at Sokoine University of Agriculture: Experiences and future outlook. In: Improved Dairy production from cattle and goats in Tanzania. NORAGRIC Norway Occasional Paper Series B No. 11 p. 28 - 40.

Sarwatt, S.V. and Lekule, F.P. 1987. Traditional pig production in some villages in Morogoro district. In: Proceedings of Tanzania Society of Animal Production Vol. 14: 162 - 172.

On-farm Performance of Dual Purpose Goats and Farmers' Attitudes towards Introduction of Goats in HADO Areas of Kondoa

E. H. Goromela, I. Ledin** and P. Udén***

*Livestock Production Research Institute,
P.O.Box 202, Mpwapwa, Tanzania
Phone/fax: +255 61 24526.

**Swedish University of Agricultural Sciences,
Department of Animal Nutrition and Management,
P.O. Box 7024, S-750 07, Uppsala, Sweden
E-mail: inger.ledin@slu.huv.se and peter.uden@slu.huv.se

Summary

An on-farm trial was conducted in two villages located in the Kondoa eroded area in Central Tanzania to evaluate the performance of Anglo-nubian x Blended goats. In the area there were no goats due to the eviction of livestock in 1979 made by the government to arrest land degradation. In the present study twenty-four pregnant goats were sold to a total of twelve farmers pre-selected during a base line survey. The goats were confined in local barns constructed by the farmers themselves and fed *ad libitum* of locally available feeds. The goats were hand milked and milk output was es-timated by measuring the volumes of milk on a daily basis during a period of twelve weeks. Average daily milk yield recorded per doe ranged from 0.2 to 1.1 litres. Mean daily milk yield increased from 0.6 litres at parturition to a peak of 0.7 litres at the 4th week of lactation after which there was a decline to about 0.3 litres at the 12th week. Milk output recorded was positively cor-related with age of the does even though the coefficient of correlation was not statistically significant. From birth to 4 weeks of age the kids had the highest weight gains after which there was a decrease. Male kids had higher growth rates than females. Single-born kids grew faster than twin-born. The mortality rate from birth to 12 weeks of age was 14%. Two surveys were also conducted in the study area to assess farmers' attitudes toward introduction of goats and the acceptability of goats into the farming systems. The farmers were interviewed using two sets of semi-structured questionnaires pre-tested by experienced enumerators. The first questionnaire was used for interviews of 80 farmers in Baura and Bolisa villages and covered the attitudes of the farmers before the introduction of goats. The results of the first questionnaire showed that the farmers regarded cattle as more important than goats due to their multipurpose uses such as for traction, high income from sales of milk and live animals and large amount of manure. However, they pointed out that goats had socio-economic, managerial and biological advantages over cattle. About 33% of the respondents in Baura had at least tasted or drunk goat milk, respectively. Unavailability of goat milk, strong smell and taste and lack of traditional use were the major reasons preventing consumption of goat milk in the study area. A milk taste panel after the introduction of goats showed that a large proportion of the respondents preferred goat milk. Only 3.8% of the respondents claimed that goat milk had a strong smell. Goat milk received higher taste scores than cow milk. From this study it may be concluded that Anglonubian x Blended goats have a high potential for milk and meat for rural families and goat milk is as acceptable as cow's milk.

Key words: Goats, on-farm research, milk yield, growth rate, acceptability.

Introduction

HADO is the Swahili acronym for Dodoma Soil Conservation project launched in 1973 to arrest the accelerating land degradation occurring in some parts of Dodoma region. In tackling the problem the project employed physical conservation measures such as the construction of terraces, contour bands etc. However, due to the increase in the number of grazing animals and also uncontrolled water run-off from higher slopes denuded by over-grazing, all the physical measures were unsatisfactory. Thus a decision was made by the government to remove all livestock such as cattle, sheep, goats and donkeys in the most severely affected areas of Kondoa in 1979 and some parts of southern Dodoma including Mvumi division in 1986. It was later learnt that human nutritional problems ensued shortly after the eviction of livestock in these areas. The preliminary results of a SAREC/HADO (1988) survey indicated that mothers and children were the most affected by the lack of animal protein and other animal products. There was also a notable decrease of crop production due to the decline in soil fertility resulting from the lack of manure and inorganic fertilizers. A decrease in income was the reason for using less inorganic fertilizers.

In addressing the negative aspects associated with the eviction of livestock in the eroded areas, the Ministry of Agriculture with the collaboration of SAREC introduced stall-fed improved dairy cows to smallholdings in the HADO areas in 1989. The philosophy of this program was to demonstrate the possibilities for an increased sustainable food and fuel production in these ecologically fragile zones of Central Tanzania through changes in the management of livestock. Its important feature was to develop livestock activities which were in harmony with the philosophy of the HADO project. However, after some years of experience it became clear that not all smallholder farmers could afford to buy dairy cows. Introduction of dual purpose goats commonly known as "Blended goats" and their crosses was then viewed as an alternative to improve the living standard of especially resource-poor farmers.

The Blended goat is a three-way cross between Kamori (55%), Boer (30%) and Indigenous Tanzanian goats (15%) (Das and Sendalo, 1991). Performance evaluation of Blended goats and their crosses under different feeding regimes and management systems has been limited to various research centres. Preliminary results from these centres have shown a high production potential of milk and meat and better reproductive performance than indigenous goats (Das and Sendalo, 1991). However, there is no documentation available concerning their performance under smallholder conditions.

Das and Sendalo (1991) have reported that a large proportion of animal protein supply in rural areas is derived from small ruminants due to their high prolificacy compared to cattle. Goat milk is becoming popular in some regions of Tanzania and is mainly used where cow's milk is scarce and also traditionally used for therapeutic purposes in children and allergic persons (Das and Sendalo, 1991). However, strong flavour and taste of goat milk, low social status and the belief that goat udders are too small to be milked are among the factors which prevent people from consuming goat milk in Tanzania.

The objective of the present study was to make a preliminary assessment of the performance of dual purpose goats under smallholder conditions in the HADO areas of Kondoa based on principles of utilisation of locally available feeds and to investigate the farmers' attitudes towards introduction of dual purpose goats and the acceptability of goat milk.

Materials and methods

Location and climate of the study area

The study was undertaken in two villages, Bolisa and Baura in the divisions of Kolo and Kalamba, respectively. Bolisa and Baura are located approximately 15 and 30 km, respectively from Kondoa town on latitude 35° 58' N and longitude 4° 45' S. The elevation of the area is between 1300 and 2000 m above sea level. Average annual rainfall is 650 mm, starting from December and ending in May. The mean temperature ranges from 25 to 30°C. The soil types found include thin layers of red or grey sandy soils which are moderately poor in nutrients. The more fertile black or grey soils are found in valley bottoms and flood plains. Natural vegetation consists of different types of grasses, trees and shrubs.

The study was conducted between February and June 1995 which means in the wet season and part of the dry season. The study area was selected as a suitable area for the introduction and management of dual purpose (milk and meat) goats due to the following reasons:

1. Its climate is drier than other villages in the Kondoa Eroded Area and appears to be well adapted for goat production.
2. Its year round availability of feeds for small stocks.
3. Experience of farmers with zero-grazing systems
4. Expressed interest by the farmers for goat milk production.

Preliminary village contacts and selection of participating farmers

Following the identification of villages, the village leaders were contacted in February 1995 and informed about the proposed on-farm trial. The following day the village leaders organised two village meetings in order to meet with the farmers. The first meeting was to present and inform the farmers about the purpose of the proposed on-farm trial. The second meeting involved the farmers in a wealth ranking exercise because those intended to be included in the study should be resource-poor farmers. During the wealth ranking exercise, the farmers categorised themselves into three groups namely high, medium and low income earners according to their wealth resources. Those identified to be low income earners were referred to as resource-poor farmers. From this group those farmers who understood the purpose of the study and showed interest and willingness to participate in the study were selected.

Distribution of the experimental animals among the selected farmers and their management

Twenty-four does (F1-Anglonubian x Blended goats) diagnosed to be pregnant were purchased from Kongwa Pasture Research Centre and transferred to the study area in early February, 1995. The does were sold to 12 participating farmers pre-selected during the village contacts. Each farmer purchased two does. The farmers were taught all the details about the experiment. The goats were kept in pens with slatted floor made of branches of trees locally available. The pens had good ventilation and protected the animals from rain, parasites and predators. The animals were confined and offered *ad libitum* natural or improved grasses such as *Cynodon dactylon*, *Heteropogon contortus*, sugarcane tops, *Cenchrus ciliaris*, *Bothriocloa insculpta*, *Chloris gayana*, *Panicum maximum*, maize leaves and pigeon pea haulms. The animals were also fed a variety of leaves and twigs from trees and shrubs such as *Acacia tortilis*, *Dicrostachyus glomerata*, *Commiphora aminii*, *Commiphora africana*, *Grewia similis*, *Adansonia digitata*, *Acacia albida*, *Microglossa ablongifolia* and *Premna senensis*. Each doe was supposed to be supplemented with approximately 500g concentrates per day. However, the supply of the concentrates was not regular. The concentrate was made mainly from maize bran and sunflower seed cake that were available in the study area. The animals were provided with water all the time and also supplemented with Magadi soda which the farmers were recommended to dissolve in water rather than feeding it as licks. According to Urio (1981), the Magadi soda contains the following chemical composition: Phosphorous (P) 0.23%, Calcium (Ca) 0.08%, Sodium (Na) 30.4%, Chlorine (Cl) 0.01%, Fluorine (F) 0.18%, Carbon (C) 25.7%, Bicarbonate 22.9%, pH = 9.7 and its solubility at 28.7°C is 19.6 g/100ml of water. Before the experiment, the goats were treated with *Supa dip* against ectoparasites using hand sprayers and also dewormed with *Nilizan*.

On-farm measurements of the performance

Milk output: The assessment of milk production started after the first five days of lactation to enable the kids to suckle colostrum. The does were milked in the morning and evening and the amount of milk obtained was recorded daily for 12 weeks. Kids were kept separately from their dams and only brought to their dams after the milking periods. The milk was collected into calibrated cups with a capacity of 1 litre designed for estimating milk production. At each milking, the milk output was recorded in two note books. The records were entered by one of the household members, normally the farmer or his wife. The data recorded by the farmers were then checked and adjusted by the enumerators. This procedure served the dual purpose of saving important research resources and at the same time enhancing the level of farmer participation in the experiment. The milk output recorded did not represent the total milk yield because the farmers often milked a quarter or half of the total milk yield, leaving the rest for the kids.

Birth and subsequent kid weights: Kiddings, litter sizes, sex and birth weights were recorded within 12 hours after birth. Also stillbirths or deaths were recorded and examined. Kids were ear-tagged at birth for identification. The subsequent kid weights were recorded at weekly intervals using gunny bags with strings for hitching on a spring scale (Salter type) until the 12th week of lactation. The spring scales were calibrated into half kilograms and with a range of up to 25 kilograms.

Farmers' attitudes on the introduction of dual purpose goats and acceptability of goat milk

The farmers were interviewed using semi-structured questionnaires. Two sets of questionnaires were designed and pre-tested by the enumerators before being used in order to check if they are well understood. The questionnaires were written in simple English which was easy to translate to farmers. The first set of questionnaires covered the general attitudes of farmers towards goats and goat milk consumption. The questionnaires were used to interview 40 farmers in Bolisa village and 40 farmers in Baura village.

The second questionnaire aimed to capture information on goat milk acceptability relative to cow's milk through a simple milk taste exercise involving 53 farmers from Baura and Bolisa villages. In each village fresh goat and cow milk were purchased from the farmers. Cow milk and goat milk were randomly assigned **X** and **Y** labels, respectively. The farmers were asked to identify the source of the milk samples either by appearance (colour), odour, taste or texture. After identification they were asked to score the taste for each milk sample on a 5-Point scale score as used by Boor et al. (1987).The order of the scores were as follows:

5. Like very much

6. Like slightly

7. Neither like nor dislike

8. Dislike

9. Dislike very much

After giving their scores each participating farmer was requested to comment on the reasons for his or her scores.

Data analysis

Following the completion of the survey all the data were coded and stored in a spread sheet for later analysis. All data collected were analysed using simple data analyses based on descriptive statistics applying Minitab software release 10.2 (1994).

Results

Milk production

The two year old goats ceased producing milk earlier than the three year olds. Therefore the data on milk yields reported for 10 of the does were collected within a period of 12 weeks. Total and mean daily milk production by the goats excluding what was suckled by the kids is presented in Figure 1.

The data indicate that there was an increase of daily milk production from 0.6 litres at parturition to a peak of 0.7 litres at 4 weeks and thereafter a gradual decrease to 0.3 litres at 12 weeks of lactation. Total milk recorded and average milk production per doe in 84 days is presented in Table 1.

Figure 1. Average daily millk yield (litres) by Anglonubian x Blended
croosses recorded for twelve weeks of lactation periods

Daily milk recorded/doe

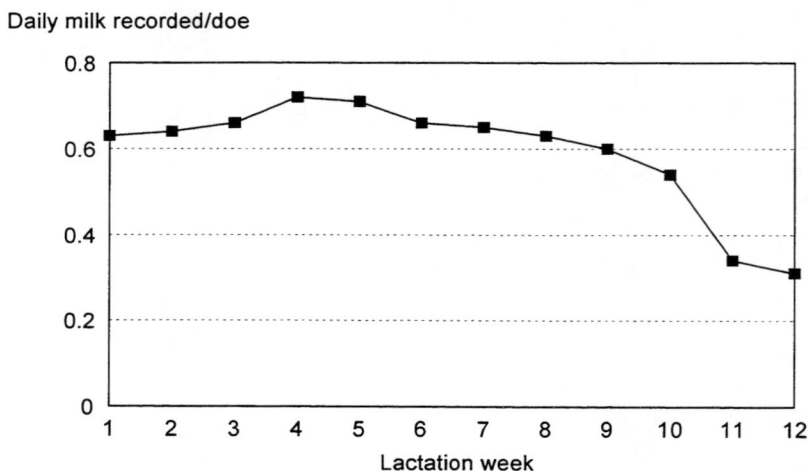

Table 1. Total and average daily milk production(litres) by Anglonubian X blended
crosses during 3 months of lactation

No. of does	Age of does (years)	Total milk recorded (lts)	Average daily milk recorded (lts)
1	3	88.0	1.1
2	3	36.6	0.4
3	3	31.7	0.4
4	3	55.3	0.7
5	3	34.9	0.4
6	2	73.8	0.9
7	2	73.4	0.9
8	2	49.0	0.6
9	2	15.3	0.2
10	2	38.9	0.5

The total milk recorded per doe ranged from 14.2 to 81.0 litres corresponding to approximately 0.2 to 1.1 litres per. Milk yield recorded was
positively correlated with the age of the does although the coefficient of
correlation was not statistically significant.

Birth weights and growth rates

Kid growth rates and live weights by sex and birth-type at various ages are
presented in Tables 2 and 3, respectively.

Table 2. *Mean kid weight gain (g/day) and their standard deviation (± Sd) at various age*

Variable	Birth - 3 weeks	4 - 6 weeks	7 - 9 weeks	10 - 12 weeks
Total growth rate	89.7±8.5	94.4±7.6	92.4±6.9	90.0±.6.3
Male	93.3±7.1	97.6±6.8	95.7±5.3	93.14±4.1
Female	83.6±7.6	89.2±6.8	86.9±6.3	84.9±6.2
Singles	94.6±8.8	98.7±6.9	96.1±6.3	93.8±5.7
Twins	81.3±6.2	84.6±2.6	84.2±13.1	83.1±7.3

Table 3. *Mean kid body weights gain (kg) and their standard deviation (± Sd) at various age of weeks*

Variable	Birth	3 weeks	6 weeks	9 weeks	12 weeks
Total growth rate	2.5±0.5	4.5±0.5	6.5±0.6	8.3±0.7	10.0±0.7
Male	2.7±0.5	4.7±0.5	6.8±0.6	8.7±0.5	10.5±0.4
Female	2.2±0.4	4.1±0.2	5.9±0.1	7.6±0.2	9.3±0.3
Singles	2.4±0.4	4.5±0.4	6.5±0.6	8.4±0.7	10.2±0.8
Twins	2.7±10	4.4±0.3	6.3±0.4	8.0±0.7	9.7±0.9

The overall mean growth rates indicate that there was a small increase in weight gains of the kids from birth to 3 weeks to a period of 4 to 6 weeks and thereafter a gradual decline until a period of 10-12 weeks. From birth to 12 weeks, male kids had higher growth rates than female kids. Single-born kids had higher growth rates than the twin-born kids. At parturition the twin-born had higher birth weights than single-born kids, but from the 3rd to 12th week of age the single-born kids had higher live weights (Table 3) whereas male kids had higher live weights than females from birth to 12 weeks of age. Two kids from twin mothers died and the mortality rate from birth to 12 weeks of age was 14%. The kids died due to general weakness probably because of their low birth weights.

Consumption and general attitudes of farmers towards goat milk

Frequencies of drinking cow and goat milk in the villages are presented in Table 4.

Table 4. *Frequencies and percentages of drinking cow and goat milk in the two villages*

	Cow milk	%	Goat milk	%
None	10	12	54	67
Once per week	3	4	6	8
Twice per week	15	19	6	8
Trice per week	3	4	4	5
Over three per week	49	61	10	12
Total	80	100	80	100

Sixty percent of the respondents claimed to drink or at least having tasted milk on a regular basis. However the sources of milk tended to vary from one household to another. In both villages about 88% of the respondents claimed to drink cow milk and about 33% goat milk and the reasons preventing farmers from consuming goat milk are presented in Table 5.

Table 5. *Reasons for not drinking goat milk in the two villages*

	Total frequencies	Total %
Not available	20	37
Strong smell	16	30
No tradition	18	33
Total	54	100

Tables 6 and 7 demonstrate the farmers responses towards the different milk sources and frequencies of taste scores of the two samples of milk, respectively.

Table 6. *Total frequencies and percentages of the taste panel on milk sources in both villages*

	Cow	%	Goat	%
Pass	24	45	27	51
Failure	29	55	26	49
Total	53	100	53	100

Table 7. Frequencies and percentages of milk taste scores for cow and goat milk in both villages

Score order	Cow	%	Goat	%
1	19	35	29	55
2	27	51	15	28
3	3	6	4	7
4	2	4	3	6
5	2	4	2	4
Total	53	100	53	100

About 52% on average of the taste panel failed to distinguish goat milk from cow milk despite the general belief that goat milk has a strong smell and taste (Table 5). Goat milk received higher scores of 1 (55%) than cow milk. Regarding the organoleptic qualities of the milk samples, it was observed that 66% of the respondents claimed that goat milk tasted sweeter than cow milk. About 17% of the respondents claimed that cow milk tasted salty while only 7.5% of the respondents claimed that goat milk had a salty taste. About 58.5% of the respondents claimed that goat milk was heavier and creamier in taste than cow milk (30.4%). In other words cow milk was lighter (35.8%) in taste than goat milk (7.5%). At the end of the experiment it was found that only 3.8% of the respondents claimed that goat milk had a strong smell and taste compared to approximately 30% before the experiment. Also after the milk taste exercise it was observed that over 95% claimed that goat was better than cow milk for the reason that it is heavier and creamier and only small amounts are required for home consumption.

Discussion

Milk production

Mean daily milk production by the goats increased from 0.6 litres at parturition to a peak of 0.7 litres at the 4th week of lactation. Then after the 4th week there was a gradual decrease in milk production until .03 litres at the 12th week of lactation. The patterns of daily milk recorded in the present study are similar to those reported by Banda et al. (1990) from non-dairy goats in Malawi. However, in their study the average daily milk yield increased to a maximum of 1.1 kg at the 2nd week post-partum and then decreased steadily until the end of lactation at the 12th week when production averaged 0.8 kg/day. The milk yields of goats in the present study were lower than an average of 1.0 kg/day reported by Sidahmed et al. (1985) from dual purpose goats held on-farms in 145 days. Also the milk yield in the present study was lower than that reported for on-station work by Das and Sendalo (1991) of 0.7 kg in 143 days of lactation for Anglonubian crosses, but was higher than 0.5 kg for blended goats in a133 days lactation. The low milk yield obtained in the study could be due to the low quality of feeds offered as in most cases a large part of the diets consisted of fibrous feeds. Concentrate feeds were not fed regularly. Also management practices may have played a part because some of the farmers were reluctant to milk the goats and they often milked a quarter or a half of the total milk leaving the rest for the kids. This can be clearly confirmed by the large variations in the milk records between animals although to some extent this may be attributed to the differences in age of the does which was positively correlated with milk recorded, but not statistically significant. Lyatuu et al. (1992) observed a significant (P<0.05) effect of the age of the does on milk yields and the lactation lengths were longer in older than in younger does which agrees with the present study.

Body weights and growth rates of kids

The overall mean live weights and kid growth rates in the present study were similar to those reported by Das and Sendalo (1991) for Anglonubian cross kids at Malya and Kongwa in a study undertaken in 1988. However the growth rates were higher than those reported for Anglonubian crosses for the 1971 and 1984 periods (Das and Sendalo, 1991). Male kids were heavier and grew faster than female kids all the time which agrees with the results reported by Das (1990) for improved meat goats in Tanzania and Inyangala et al. (1990) for dorper sheep in Kenya and Karua and Banda (1990) for Saanen crosses in Malawi. This superiority of male over female kids could be attributed to the hormonal differences between sexes and their resultant effects on growth (Bell et al., 1970). Single-born kids had higher growth rates than twin-born kids which agrees with the results reported by Das (1990) and Lyatuu et al. (1992), but at the parturition the twin-born kids had higher birth weights than the single-born kids.

Farmers' attitudes towards introduction and acceptability of goat milk into their farming systems

The farmers regarded cattle as more important than goats due to their multipurpose uses such as for traction, high income accrued from sales of milk and live animals and large amount of manure produced by cattle. However the farmers pointed out that the opportunities for goat production in the farming systems of HADO areas may be justified by their environmental, socio-economic, managerial and biological advantages over cattle. They indicated that the area lies in the semi-arid region of Central Tanzania which is characterised by low rainfall and shortage of feeds and water during the dry season. Goats appeared to be well adapted to and withstand such environments. The farmers indicated that goats have socio-economic advantages such as high profitability and faster turnover by attaining early maturity and shorter generation interval, high popularity and acceptability of meat among households and religions. The farmers reported that goats are not difficult to manage and that family labour is enough to manage goats even during the growing, weeding and harvesting periods which were identified to be labour demanding. The farmers also pointed out that goats are versatile animals and have high ability to adapt to various ecological environments, high disease resistance and require little of water. In addition the farmers viewed goats as a complement to cattle which would result in an efficient use of locally available feed resources.

Regarding the farmers' attitudes toward goat milk, about 60% of the respondents interviewed claimed to drink or at least to have tasted milk. In both villages 88% of the respondents claimed to drink cow milk. The underlying factors for not drinking cow milk were low income and unavailability of cow milk. Over 60% of the respondents claimed not to consume goat milk. The reasons mentioned for low consumption of goat milk included unavailability, strong smell or taste and lack of traditional use except for therapeutic purposes. These results are in agreement with those reported by Das and Sendalo (1991) that goat milk in Tanzania is mainly used where cow milk is scarce and also traditionally used for therapeutic purposes in children and allergic persons. Banda (1990) observed that over 90% of the respondents did not consume goat milk in Lilongwe because of tradition and 38% because of the unavailability of goat milk. Citing the work of Boor et al. (1987) in Kenya, Banda (1990) pointed out that unavail-ability may be strongly associated with the non-milking of the goats as a tradition which agrees with the present study. In the present study 30% of the respondents claimed that goat milk has strong smell and taste. These results correspond with those of Banda (1990) who reported that 19.2% claimed that goat milk has strong smell and taste. These results support the findings of French (1970) and Skjevdal (1979) who pointed out that the content of short and medium-chain fatty acids (especially C6-C10) and potassium chloride may be responsible for the strong flavour and taste of goat milk. In addition Banda (1990) demonstrated that strong smell and taste of milk from small ruminants such as goats and sheep may be due to unhygienic procedures and dirty utensils and probably not to the presence of the buck. This argument was confirmed by the results obtained from the milk taste experiment which indicated that 52% of the taste panel failed to associate the coded milk samples with the actual sources of milk tasted (Table 5). The implication of these observations are that goat milk is as ac-ceptable as cow's milk. Results from the taste panel indicate goat milk ranked the first by receiving higher scores of 55% for the first score com-pared to cow's milk (Table 7). The heavy and creamier taste of goat milk claimed by the respondents (58.5%) in the present study may be due to higher fat content (4.6-6%), which is higher than that reported for blended goats (Lyatuu et al., 1992; Paper III) and for Norwegian goats and Norwe-gian x indigenous crosses (Das and Sendalo, 1991). However the fat content was less than that reported by Banda et al.(1990) of 6-7% fat content for non-dairy goats in Malawi. The sweeter taste claimed by the taste panel (66%) of the goat milk compared to cow's milk may be attributed to the presence of a high lactose content (Devendra and McLeroy, 1982).

Conclusion

The present results show that Anglonubian x Blended crosses have a high potential to produce both milk and meat for rural families. The milk yields observed and high growth rates of the kids are good indicators of their potential under semi-arid conditions in Central Tanzania. Introduction of the Anglonubian x Blended crosses into the farming systems in HADO areas in Kondoa could also in the future solve the problems of insufficient milk and improve the economic status of the farmers. Acceptability of goat milk was high and the claim that fresh goat milk has a strong smell or taste could not be confirmed. The major problem influencing consumption of goat milk is its availability. Improving feeding and general management practices is required for optimum goat productivity.

Acknowledgements

The authors wish to thank Messrs Festo Msaka, Robert Dulle Nkungu, Robert Luambano, Agrey Kundya and Kidole Michael for their assistance in conducting all the survey work at the research sites. Mr Sianga (HADO Manager-Kondoa) and Dr. Antaro (DALDO-Kondoa) are highly thanked for allowing us to carry out this research work in their areas and for the assistance given in connection to the success of the study. Village leaders and farmers in Bolisa and Baura villages are also thanked for allowing us to carry out this study in their homes and for their full support and cooperation during the experiment.

References

Banda, J.W., 1990. Comparison of consumer attitudes towards and acceptance of goat, sheep and cow milk in Malawi. Small Ruminant Research and Development in Africa. In: Proceedings of the First Biennial Conference of The African Small Ruminant Research Network. ILRAD, Nairobi-Kenya.10-14 December, 1990. pp 105-114.

Banda, J.W. and Phiri, C.D., 1990. Investigations into the factors influencing the choice and consumption of milk and milk products in Malawi. Journal of Consumer Studies and Home Economics 14:123-131.

Banda, J.W., Steinbach, J. and Zerfas, H.P., 1990. The composition and yield of milk from non-dairy goats and sheep in Malawi. Small Ruminant research and Development in Africa. In: Proceedings of the First Biennial Conference of The African Small Ruminant Research Network. ILRAD, Nairobi-Kenya.10-14 December, 1990. pp 461-483.

Bell, G.H., Davidson, J.N. and Scarborough, H., 1970. Text book of physiology and biochemistry. Longman Group Limited, Edinburgh-UK.

Boor, K. J., Brown, D.L and Fitzhugh. H.A., 1987. Western Kenya: The potential for goat milk production. World Animal Review 62:31-40.

Das, S.M. and Sendalo, D.S., 1990. Comparative performance of improved meat goats in Malya, Tanzania. Small Ruminant research and Development in Africa. In: Proceedings of the First Biennial Conference of The African Small Ruminant Research Network. ILRAD, Nairobi-Kenya.10-14 December, 1990. pp. 445-452.

Das, S.M. and Sendalo, D.S., 1991. Small Ruminant Research Highlights in Tanzania (1960-1989). Ministry of Agriculture, Livestock Development and Co-operatives. Department of Research and Training, Dar es Salaam, Tanzania. pp. 10-16.

Devendra, C. and McLeroy, G.B.,1982. Goat production in the tropics. Commonwealth Agricultural Bureaux, Farnham Royal, Slough-UK, 1982.

French, M.H., 1970. Observation on the goat. FAO, Rome Italy. Agricultural Studies No.80. pp 54-67

Inyangala, B.A.O., Rege, J. E.O. and Itulya, S., 1990. Growth traits of the Dorper sheep. 1. Factors influencing growth traits. Small Ruminant research and Development in Africa. In: Proceedings of the First Biennial Conference of The African Small Ruminant Research Network. ILRAD, Nairobi-Kenya.10-14 December, 1990. pp. 505-515.

Karua, S. K. and Banda, J.W., 1990. Dairy goat breeding in Malawi: Gestation length, birth weights and growth of the indigenous Malawi goats and their Saanen crosses. Small Ruminant Research and Development in Africa. In: Proceedings of the First Biennial Conference of The African Small Ruminant Research Network. ILRAD, Nairobi-Kenya.10-14 December, 1990. pp.453-459.

Lyatuu, E.T.R., Das, S.M. and Mkonyi, J.I., 1992. Some production parameters of Blended goat in semi-arid regions of Tanzania. Small Ruminant research and Development in Africa. In: Proceedings of the 2nd Biennial Conference of The African Small Ruminant Research Network. AICC, Arusha-Tanzania. 7-11 December, 1992. pp. 241-245

Minitab Software Release (10.2) for Windows., 1994. Minitab Inc. 3081 Enterprise Drive State College, PA 16801-3008 814-238-3280

SAREC/HADO project., 1988. Preliminary informal survey in HADO areas of Kondoa and Mvumi.

Sidahmed, A.E., Onim, J.F.M., Mukhebi, A.W., Shavulimo, R.S., De Boer, A.J. and Fitzhugh, H.A., 1985. On-farm trial with Dual-Purpose goats on small farms in Western Kenya. Research Methodology for Livestock on-Farm Trials. In: Proceedings of a workshop held at Aleppo, Syria, 25-28 March 1985. ICARDA & IDRC. pp.101-132

Skjevdal, T., 1979. Flavour of goat's milk. A review of the studies on the sources of its variations. Livestock Production Science. 6: 397-405.

Experiences with Goat Project as a Tool in Human Development: Goats for Poor Women in Bangladesh

M. Saadullah , M. M. Hossain and Shajeda Akhter

Dept. of General Animal Science
Bangladesh Agricultural University, Mymensingh, 2202, Bangladesh
E-mail: saad@drik.bgd.toolnet.org

Summary

Income sources are few for land less, rural women in Bangladesh pushed into problems caused by loss of husband. After reviewing some research results concerning women's roles in farming in Bangladesh, the paper describes results of a project which used Black Bengal goats to assist poor women with positive results. It is concluded that goat rearing is an appropriate intervention in a capital scarce situation and that it can contribute significantly to household income without interfering with the main occupation of the women. Credit facilities are a major constraint to further spread of poor women's goat raising. Research and development can increases its contribution by attending to points like the roles of components of integrated farming in poverty alleviation and employment and by applying a gender analysis to its problem identification.

Key words: Goats, feeding systems, poverty alleviation, rural women.

Introduction

Rural women in Bangladesh are closely involved in various agricultural activities. They participate in the production process and perform many distinct activities which have great bearing on the production process in Agriculture. One estimate suggests that women's labor account for 25% of value addition in post harvest processing of rice alone (Scott and Carr, 1985). In addition to take care of children, preparing and serving food to the members of the family, rural women are also involved in other important sub sectors like fishery, livestock and poultry. On the whole, women's productive hours of work in Bangladesh is between 10-14 hours a day compared to 9-11 hours for men (Faruk, 1980). Employment opportunities in these sub sectors are increasing. However, one constraint often faced by poor women is lack of credit. Rearing of small ruminants like goats and sheep particularly by the weaker sections of the community in rural areas would go a long way to bring about social change by improving the incomes of these people.

Employment of rural poor in the livestock sector

The total work force in 1986 was estimated to be 34.91 million based on an assumed participation rate of 33.9% of the total population. The estimated agricultural labor force in 1986 was 27.23 million, or 78% of the total work force (Agricultural Employment in Bangladesh, April 1977 and BBS, 1986). Expansion of non-crop employment usually occurs first in livestock activities. The goat and poultry backyard based production systems are less capital-intensive than larger enterprises and can often be financed by dormant rural savings. In Bangladesh there are about 10.20 million small, marginal and land less farmer families. These farmers largely depend on livestock for their existence. Traditionally, they are bulk producers of milk, eggs and meat and most of them are unable to meet the requirements set by financial institutions and other loan giving agencies for agricultural loans. Thus, the small and land less farmers can get neither the opportunity to generate sufficient income to support the family nor to extend the livestock activities. So, it is important to review the present approach to not only to increase livestock productivity, but also to increase cash income and self employment and thereby improve the lives of these farm families.

Small ruminant and poultry are of economic importance for small holder farmers, but achieving maximum income is not their primary objective. The total income share of small ruminant tends to be inversely related to size of land holding, suggesting that small ruminants are of particular importance for land less people. Since agriculture provides only seasonal employment, rearing of this animal would provide employment and income as a subsidiary occupation. Goat raising is a very effective means of poverty alleviation in Bangladesh. It has been observed that 7-8 goats given to a poverty stricken farm family under a combined grazing and cut and carry feeding system can easily alleviate poverty.

An overview of goat production systems

Population & growth rate

In the largely traditional subsistence economy of rural Bangladesh, livestock plays a secondary, but crucial role. Besides providing draught power, animal protein to a poor diet, cash income, and fertilizer for crops; the export of animal products fetches about 13 % of total Bangladesh merchandise export earnings. Small ruminants like goats and sheep play an integral part of the livestock production systems in Bangladesh with a population and growth rate from 1990 to 1994 as shown in figures I and II.

Figure I. Ruminant population in Bangladesh

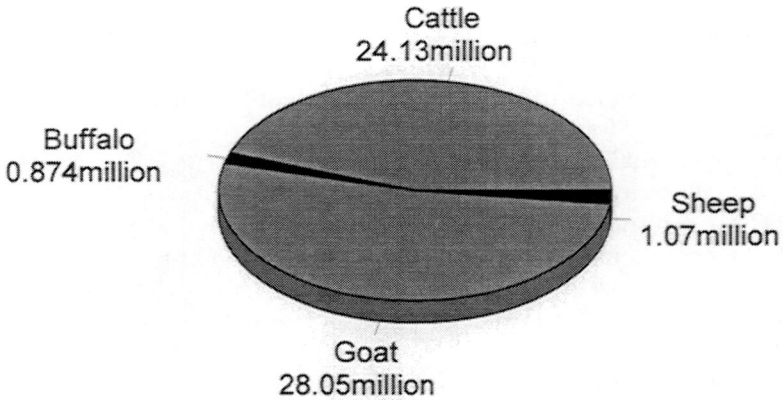

Cattle
24.13million

Buffalo
0.874million

Sheep
1.07million

Goat
28.05million

FAO Year Book vol 8,1995

**Figure II. Growth rate of livestock in Banladesh
from 1990 to 1994**

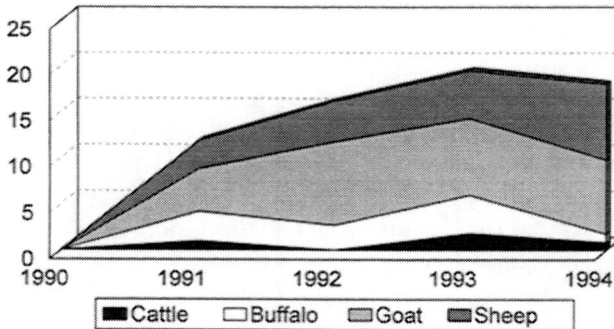

■Cattle   □Buffalo   ▨Goat   ▨Sheep

Ref: FAO Year Bool, Vol. 8, 1995

Based on the estimate of livestock population, it was observed that the average growth rate of cattle, buffalo, goat and sheep is 0.75, 2.15, 5.98 and 4.18, respectively during the period 1990-1994. A higher growth rate in small ruminants might be due to increasing human population leading to more land less households to keep goats. Increased production efficiency can be expected from goats because they have higher reproductive efficiency with the potential for increased litter size, shorter gestation intervals and relatively higher fertility compared to large ruminants. For these reasons, small-scale farmers are more inclined to raise goats when feed is a major constraint.

Goats are multipurpose animals producing meat, milk, hair, coarse wool and fine quality skin, which foreign exchange to the country. Goat milk contributes 28% of the total milk produced in the country (118,000 ton per year). It is estimated that goats produce 70,000 tones of meat per annum which is equivalent to 19% of the total meat production of Bangladesh (BBS, 1986). They provide a source of security during food insecure and other difficult periods.

Ownership:. It appears from Figure II that 86.6% of the total farm households have animals, crops and poultry. 30% of them have goats. It has been reported that 66% of the goats are kept by 0-0.5 ha holdings (Saadullah, 1986). The number of goats owned ranged from 3-29 with an average of 7 (BAU-FSRDP,1985-86). Another study found that more than three fourth of goat rearing households keep 3-8 goats, 27% between 9-11 and 7% between 15-29 goats (Huq et al., 1990).

Figure III. Distribution of Farms in the Farming Systems of Bangladesh

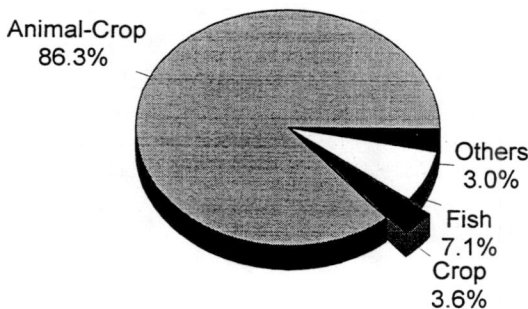

Animal-Crop
86.3%

Others
3.0%

Fish
7.1%

Crop
3.6%

Ref: BAU-FSRD (1986)

Production practices

Goats are grazed, fed on low quality crop residues and household waste. They have, due to their smaller body size, obvious advantages over cattle as sources of meat and cash income. Although cutting grass and grazing animals, are male dominated activities, women play a very vital role in small ruminant activities, especially older women. Children play a role in raising small ruminants and some about 4 hours a day . Women clean goat houses and feed and water the animals.

Feeding
Generally, the majority of the goats are grazed on the communal grazing land. They are allowed for grazing during the day on land such as roadsides, homesteads, in the forest, and on fallow land. Sometimes, mother does with small kids are kept tethered besides the house. In terms of feeding most farmers practice mixed feeding systems i.e. cut and carry, some grazing, and some also give rice bran, rice gruel and kitchen wastes as supplements.

Management
Goats are mostly herded by children and women and graze approximately 5-7 hours a day. Some grasses and cut twigs are provided during night.

Housing
Very few farmers provide separate houses for sheep and goat. They are housed on the verandah, corridor, cow shed, kitchen and in the open yard of the homestead. It has been found that 47% of the goats are housed in an open shed and 30 % in the cow shed , while the remainder are kept in the house (Saadullah, 1991).

Breeds
There are small and large Black Bengal Goats and they make up the majority of the country's goat population. The Black Bengal Goat is well adapted, prolific and known for its good quality skin and meat. There are a number of pure or crossbred Jamnapari goats in certain regions of Bangladesh.

Diseases
Infectious diseases (diarrhea, pneumonia etc.) are common in kids. From on-farm studies it has been observed that predators are also major cause of mortality in kids.

Breeding management
Practically all does are naturally mated. In most of the cases, availability of fully matured breeding bucks in the villages is rare since most of them are castrated at an early age as people prefer meat from castrated goat.

Women participation in homestead activities
The problems of rural women cannot be separated from the problems of rural Bangladesh. Their poverty, illiteracy and ignorance are merely a part of the total society. As a part of society women have an important role to play in the productive activities. In our country, especially in rural society, all women educated and uneducated alike generally remain involved in their domestic duties. The role of women in homestead and family life

needs to be assessed for future research oriented development activities of the nation. Women in rural Bangladesh are major, but largely unrecognized contributors to agricultural and other economic productivity. The role of distressed women (35%) having no land, even homestead have not been well recognized, but it needs to be appreciated for them to improve their lives.

The activities of rural women is shown tables 1 an 2. It appears from the tables that the rural women are mostly involved in washing utensils, clothes and cleaning the house, raising poultry and goats, post harvest activities on crops, cooking and fuel collection. Buying and selling of poultry, goats and eggs are also managed by women, besides vegetable production in the homestead. Women were responsible in

Table 1. *Work done by rural women*

Activities	In per cent				
	Women	Daughter	Son	Husband	Others
Washing,cleaning of utensils & Homestead	87	8	1	1	3
Raising goat & poultry	55	20	13	5	7
Post harvest activities	58	3	6	25	8
Cooking	95	1	0	0	4
Fuel collection within house	60	8	12	15	5
Purchase & Selling of goat & Chicken	70	0	3	1	1
Vegetable Production at homestead	65	10	5	5	5
Child care	90	3	1	2	3

Others: Labor, relatives etc. Ref.: Paul and Saadullah (1991)

75%, 69%, 50%, 90%, 97% and 91% of the following work activities: washing utensils and cleaning house compound, releasing poultry and its feeding, post harvest activities (crops), crop preservation, cooking and fuel collection (table 1). In the case of homestead work, daughters assisted more than sons. Others, like mother-in-law, grandmother, parents, sister in law and servants may help in the work. The husbands spent less time on homestead activities. From the study (Paul and Saadullah, 1991) it was estimated that 40% of house making, 48% of social festivals, 32% of selling vegetables, 25% of selling other crops, 17% of cattle sales, 21% of goat sales and 70% of selling and purchasing of poultry and its byproducts were done by women. From 50% to above 50% of the decisions were taken by women. However, 70 to 90 % of the decisions were executed by the husband's approval (table 2).

Table 2. Role of women in decision making

Activities	In per cent					
	Women	Husband	Both	Daughter	Son	Others*
Homestead cultivation (vegtbles.& spices)	49	56	95	2	2	1
House making	40	78	80	5	7	8
Particip. in social acts / festival	48	53	96	3	5	6
Selling vegetables	32	65	75	4	25	12
Selling crops	25	80	82	2	10	4
Selling cattle	17	85	70	1	22	5
Selling goats	21	83	90	5	15	3
Selling poultry	70	48	85	33	38	8

Note : *Others (Parents, servants and relatives). , Ref.: Paul and Saadullah (1991)

Examples of women raising goats

In this part examples are mentioned from a study that was conducted in Kanthal Union of Mymensingh District in Bangladesh during 1988-1992.

Objectives

To promote goat raising as a source of income generation and self employment to improve life style of the rural resource poor women.

- To evolve a practical model for replication elsewhere.
- To identify the potentials and constraints of wider involvement of women in goat raising.

Methodology

Location and environment: The project was located at Kanthal union of Trishal thana, about 10 km from Mymensingh District head quarter on the Dhaka-Mymensingh highway. It is a predominantly rice and jute growing area. About 76% of the people are involved in agriculture. There are about 5270 cattle and 7240 goats in Kanthal union.

Description of the activities: A sample survey was made in order to have an idea about the pattern of livestock and poultry rearing by the women. Based on a sample survey on the distressed women and widows, who were interested in goat rearing, 25 does were distributed to 25 women as credit, to assist in poverty alleviation under certain set condition as described below. Does were bought from the local market for an average price of Taka 500. The average live weight was 10 kg. Goats were grazed on roadside grass and tree leaves were also fed. A few breeding bucks were provided for breeding purposes of these goats. The bucks were managed by the farmers in the project area.

The condition laid down by the Project Director to the women when distributing goats (doe) were as follows:

1. They should rear with care and responsibility.
2. The goats should be marked with ear tags.
3. They should supply balanced feed using local feed resources as advised by the project.
4. Goats were not allowed to be sold or slaughtered before kidding.
5. They should report to the field supervisor of the project in case of any illness of the goats.
6. After first kidding, one kid should be returned to the project for distribution to another woman and subsequently the women will own of the remaining kids and the mother doe.
7. The women are to bear the risk in case of goats being lost or stolen due to carelessness or irresponsibility.
8. The field worker will regularly supervise the activities and the women should collaborate with them.

Results

Farm family characteristics

Characteristics of a few families are briefly presented below.

Farmer 1. Mrs. Jahanara Khatoon is a widow. Her husband died 4 years ago. She has only one child, but no land except her homestead. She has her own thatched house to live in. Before starting goat raising, her only source of income had been to work as maid servant in her neighbors' houses. She was provided with a young goat in December, 1988. Besides working in her neighbors' houses as maid servant, she also earns a substantial amount from rearing goats. During a four year period she got five male and six female kids including the one she returned to the project office. At this moment she has one doe. Her statement shows she earned Taka 5200 by selling goats. This amount was quite helpful for the widow in order for her to look after her family. With regards to cost of rearing, she said that it did not cost anything except her labor. She grazed her goats on the roadside and supplied tree leaves.

Farmer 2. Mrs. Hayatan Nessa is a widow. Her husband died long ago without leaving any land property for his only child and wife. She worked as maid servant in her neighbors' houses. She had also the skill of knitting "*katha*" (blanket from old cloths), but she could not get time for knitting, since she had to work the whole day as maid servant. In 1989 she received one she goat from the project. She reared the goat by grazing on public land. Now, she is no longer maid servant, but has engaged herself in knitting "katha" and in rearing goats. She earned about Taka 3000 by selling seven kids. She has one doe worth to Taka 650. She feels quite secure now as she has enough time for knitting katha . Moreover, by employing a minimum of labor she is to raise goats, which provide her financial security.

Farmer 3. Mrs. Rabya Khatoon a widow with 3 dependents. Her husband did not leave any property, not even a homestead. She had to work in her neighbors' houses for maintaining her family. In 1988 she received a goat from the project. She has since got nine kids excluding the one returned to the project. She earned about Taka 3000 by selling eight goat. She has an extra important source of income from rearing goats for which she does not require to invest any resources except her labor. She raised the goats on public land, and also fed them with collected tree leaves.

The characteristics of other families are not very different from those described above.

The income generated through goat raising by women
The income generated by the goat raiser during the period is shown in table 3.

Time for rearing goats
It was observed that on an average they spent 1.5 hours of a total 10 hours' working day for feeding and management of goats. The rest of the working hours are spent in working in neighbor's houses or their own.

Table 3 *The income generated through goat rearing by some of the women*

Name of the woman	Village	Amount earned (Taka)
1. Mrs. Sahera Begum	Bilboka	4720
2. Mrs. Meherun Nessa	Bilboka	3800
3. Mrs. Khairun Nessa	Bilboka	6000
4. Mrs. Rabeya Khatun	Bilboka	3000
5. Mrs. Bakul Jan	Bilboka	1725
6. Mrs. Hanufa Begum	Aynakhet	2000
7. Mrs. Zahura Khatun	Aynakhet	2115
8. Mrs. Kakar Jan	Tetolia para	800
9. Mrs. Hayatun Nessa	Singrail	3000
10. Mrs. Jahanara Khatoon	Singrail	5200
11. Mrs. Mallika Khatun	Singrail	6000

Feeding and management practices
Feeding systems practiced by the women are grazing on the public land, bunds, tethering, tree leaves, shrubs and weed. Ninety percent of the goats of the contact women are housed in the living room or kitchens. Kid was

higher (12%) than adult (4%) mortality. Higher mortality in kids is mainly due to predators and pneumonia. Housing is considered to be the major problem to increase the number of goats maintained by a woman .

Constraints faced

Women are unable to buy goats initially due to lack of capital. There is a problem of parasitic and bacterial diseases. Availability of feeds and fodder can be a serious constraint during periods with lack of rain. During the monsoon proper housing is considered a great problem especially for kids.

Important lessons learned

Support either in the form of funding or stock animals are good tools in starting the extension program. In addition basic knowledge of goat keeping should be provided directly. Through goat rearing women increased their income, improved nutrition of the family, stability of the households and their self-reliance. Rearing of goats by the women for improving their economic status created an immense interest among other people.

Some specific recommendations emerging from the Project are:

1. Training institutions may get involved in organising training for rural poor women in small ruminant production technologies and its importance for income generation.
2. Some efforts should be made to link up rural women with funding agencies.
3. Development of routine vaccination systems for livestock and poultry.

Conclusion

A. Self employment and income generation from raising goat:

- *Rearing goats requires less capital and is appropriate where capital is scarce.*
- *It may provide part time self employment without affecting the main occupation.*
- *Further, this type of enterprise will not demand very special skills compared to other agricultural enterprises.*
- *Credit facilities for poor women and widows might help to encourage them to rear goats.*

B. Research and Development directed towards human resources involved in small ruminant production

Small ruminants like goats play an important role in small holder farming systems. These roles include liquidity aspects (cash to meet short term needs), income generation, supply of manure for crops and insurance against risk. Research and development can increases its contribution by attending to points like the rolesof components of integrated farming in poverty alleviation and employment and by applying a gender analysis to its problem identification.

Acknowledgments

Livestock and Poultry Development Project, Rotary Club of Mymensingh, District 3280 and the International Rotary Club, RI 3H-87-9 Project, U.S.A. The first author wishes to express his sincere gratitude to the Danish Ministry of Foreign Affairs for providing a travel grant to contribute this paper.

References

Agricultural Employment in Bangladesh (1977). Working paper XI, Govt. of Bangladesh Dhaka, April 1977

BBS (1986). Report on Bangladesh Livestock Survey 1983-84. Bangladesh Census of Agriculture, Govt. of Bangladesh, Dhaka.

BAU-FSADP 1986. Farming Systems Research, Annual report, 1986. Bangladesh Agricultural University, Mymensingh, Bangladesh.

Devendra, C. 1992. Goats and rural prosperity. In proceeding of the Fifth International conference on goats. 2-8 March, 1992. New Delhi, India.

Faruk, A. (1980). "Use of time by the individual: A survey in Bangladesh". Rural Household Studies in Asia , Singapore University Press, King Ridge, Singapore 0511

FAO (1995). FAO Year Book, Volume 8, 1995, Rome, Italy

Huq, E. Md., M. Rahman and M.A. Miah (1990). A study on the relationship between management practices followed by the goat raisers with some of the selected characteristics in the selected area of Sathkhira Upazilla. Bangladesh J. Anim. Sci. 19:1-7

Islam S (1977) . Women Education and Development in Bangladesh; A few reflections, Role of women in Socio-economic development in Bangladesh-Proceedings of a Seminar, Bangladesh Economic Association Dhaka, March, 1977, pp121-131

Kabir K, Abed, A. and Chen, M. (1977). Rural women in Bangladesh: Exploding some myths. Role of Women in Socio-economic development in Bangladesh. Proceedings of a Seminar, Bangladesh Economic Association Dhaka, March, 1977, pp72-79

Paul D.C. and Saadullah (1991). Role of women in homestead of small farm category in an area of Jessore, Bangladesh, Livestock Research for Rural Development 1991(3)2:23-29

Saadullah, M. (1986). Integrated Crop and Small Ruminant System in Bangladesh. In: Small Ruminant Production Systems in South and Southeast Asia. Ed: C. Devendra. IDRC, Canada

Saadullah, M. (1991). Research and Development Activities and Needs of Small Ruminants in Bangladesh. In: Research and Development Needs of Small Ruminants in Asia. Small Ruminant Production Systems Network in Asia (SRUPNA). Ed. Andi Djajianegara and C. Devendra. Indonesia

Scot, Gloria and Marilyn Carr (1985). The impact of technology Choice on Rural Women in Bangladesh: Problems and opportunities, Washington, World Bank 1985.

Methodology, Impact Assessment, Credit etc.

Impact Assessment for Participatory Research

Janet Riley

Statistics Department, IACR-Rothamsted
Harpenden, AL5 2JQ, UK
E-mail: janet.riley@bbsrc.ac.uk

Summary

The author proposes approaches to the assessment of impact of participatory studies in natural resources research but argues that, for them to be successful, they must be designed into the study from its beginning. The project-implementation-impact spectrum is emphasised as an essential structure for the maintenance of efficient design, monitoring, evaluation and impact assessment. Good design and analysis techniques need to be used to correspond to clearly defined objectives. A summary is given of essential features of statistical design for participatory studies. The types of data which will be generated are discussed and indicators of relevance to impact assessment are highlighted. Further issues for impact assessment programmes are discussed.

Key words: Impact studies, participatory, natural resources, statistics, design, on-farm, data analysis.

Introduction

A multidisciplinary approach to the study of agricultural production systems is now practiced in all parts of the world, but particularly in the tropics and semi-arid tropics where climatic conditions may be extreme and the sustainability of agricultural productivity poor. Cooper *et al* (1996) note that many small-scale farmers cannot afford to replenish the loss of soil nutrients through the use of expensive fertilisers or other external inputs. They recommend the inclusion of agroforestry in farming systems as it has the potential to enhance soil fertility prevent soil erosion, and provide fodder and income.

The use of such low input - high output multi-species systems is desirable to try to provide a sustainable food supply yet maintain an economic balance at little cost to the environment. Because of the complex structure of these systems and the need to test realistic interventions upon them, the most appropriate place for such research is on the farm with inputs made by the farmers themselves. Additional research value is gained from this approach as the research results must be assessed by seeking farmers' opinions and by monitoring change to the farmers' lifestyles and to the neighbouring communities.

The very idea of collecting information on farmers' opinions and changes in the community for lengthy times to come, evokes a picture of mountains of data and exhausted analysts coming to few definite conclusions! Indeed this will almost certainly be the situation unless a very clearly defined strategy for data collection is defined at the concept stage of each study and maintained throughout all of its phases. The project cycle was defined by Gapasin in 1993 as

- specification of project objectives
- preparation of research proposals
- review of research proposals
- approval of research proposals
- implementation and monitoring of research
- evaluation of completed projects and impact.

This project-implementation-impact spectrum must be used as the basis of an information collection system in relation to the project objectives. Riley and Alexander (1996) argue that impact now requires a whole category of its own and make clear recommendations with regard to assessment techniques and needs for research in this area.   They also propose that impact assessment involves a long-term management approach based upon proper project design incorporating an impact assessment component and that the generated information requires skilful summary throughout the whole project cycle to assess impact in an unbiased way.

To ensure an adequate relationship between the choice of indicators and project objectives and structure, the skilful design of studies is necessary. Poor design will provide data which can support only poor estimates of variation, thus providing an incomplete picture of the true situation. Particular features of the design of participatory, on-farm trials are discussed in the next section, followed by a discussion on the choice and use of specific indicators. Finally a broad approach is proposed for the timing and location of collection of indicators for impact assessment and specific needs for data handling are addressed.

Design of participatory studies

Types of study

What is a participatory trial? Riley and Alexander (1997) proposed four groupings of such trials:

- on-station work, planned and activated by researchers but with a large farmer influence in the choice of design and comment upon outcomes

- on-farm work, planned and activated by the researcher but taking account of farmer needs and preferences
- on-farm work, planned by the researcher but activated by the farmer
- on-farm work, planned and activated by the farmer such that the researcher merely observes

Some simple examples of participatory studies might be:

- a survey of soil fertility management practices which may need both rapid rural appraisal techniques or a formal survey approach to investigate farmers' perceptions of change and policy response. The survey would be done on-farm by scientists in multidisciplinary teams but there would be a large component of farmer input. This would represent a study of the second type above.
- the improvement of nutrient use efficiency through the use of on-farm agroforestry trials designed by scientists and managed by farmers, together with on-farm trials designed and managed by farmers with concurrent monitoring by scientists for impact, sustainability and patterns of change. These studies would represent the third and fourth categories above.

Different sets of objectives may exist for on-farm work such as:

- the provision of a demonstration of a new technique to farmers
- the need to test interventions on-farm that have been successful on-station
- the testing of interventions across a large number of farms
- the testing of interventions across known different types of farms, thus generating a measure of interaction or the need for separate recommendations for different farm groups
- the testing of interventions across a wide range of farms, where sub-groups of farms may be suspected and the generated data must be used to determine the groups

Common features of all on-farm trials, acknowledging the multidisciplinary nature of typical systems, are listed in table 1. These indicate that a wide variety of inputs may be needed and a wide variety of outputs may ensue. Whatever the nature of the trial, the degree of complexity is likely to be great and to control the research strategy to achieve adequate and useful results, a clearly defined strategy is necessary at the study outset. Far from the textbook, single species design of one-season, on-station crop experiments, the statistical implications here are enormous and daunting. The statistical approach required for such studies is quite different from that needed for balanced on-station studies and it is necessary to approach

Table 1. *Common features of participatory on-farm trials*

- the community helps
- not particularly well-controlled
- involves a gathering of information
- information may be explanatory or comparative
- information may be numeric or textual
- information will be multidisciplinary
- may involve surveys of greater or lesser formality
- may involve comparative experiments
- may involve many locations
- may involve a long time period
- will be subject to large variability
- information may be difficult to interpret
- always benefit from a statistical strategy

participatory work with a new statistical *strategy* rather than the set of rigid statistical tools often gained in undergraduate agriculture courses.

Design backdrops

A study of an improved feeding strategy compared to farmers' practice using two cows on each of 30 farms may appear simple and straightforward to summarise. However, when factors such as wealth groups, farmer differences in management, labour availability, animal shelter ... are considered, the 30 farms may all differ from each other so much that they cannot be used as true replicates of the simple 'improved versus local practice' comparison. Some priority setting is necessary to choose the factors of most importance for estimation. When discussing variation in crop outputs, Scoones (1996) notes that soils types and rainfall are greatly influential, whilst field size, cattle ownership, labour availability, access to cash income, and ownership of tools or carts can all be expected to influence directly the crop output and variability in Zimbabwean studies. Such priorities can best be determined through observation of farmer needs or their determination may be the objective of the study. Whichever, it is certain that to achieve precise estimates of effects, essential adequate replication of farms within the groupings of factor combinations will be difficult to achieve and will mean that the study design is almost invariably unbalanced. Often researchers assume that the non-balanced structure of complex participatory studies renders them unsuitable for statistical treatment. However, statistical principles do apply, although not always in their traditional balanced form, this merely representing a specific application of statistical methodology to a specific type of problem.

Consider again the different priorities, niches, or backdrops, representing the environments in which we have specific interest. A simple example is illustrated in figure 1 where soil types overlap with different villages. If the specific soil type of interest is to be studied in villages 1 and 2

Figure 1. *Soil type by village combinations as different backdrops or niches*

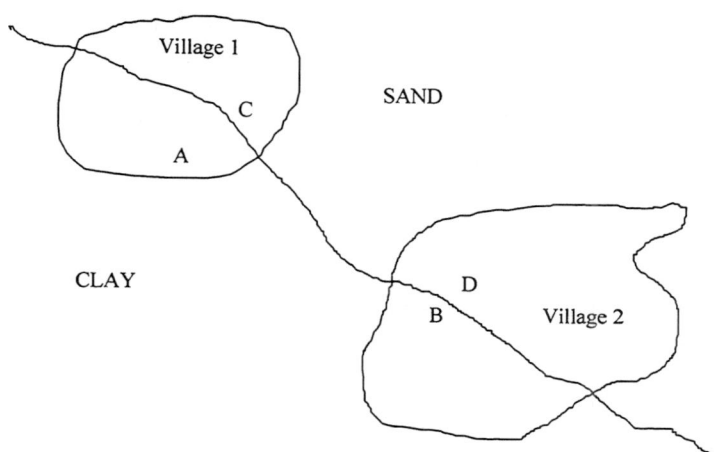

then data should be collected from, and recommendations made for, the backdrops or niches C and D only. If soil type is not to be studied however, then data can be collected and recommendations made for the backdrop, or niches, found in A, B, C and D. Niches can be formed from combinations of different sources of information such as different wealth groups, animal breeds, farmer gender, farmer perceptions, proximity to markets; what is important is to acknowledge that variability in the data will be caused by all of them and a successful and appropriate design strategy will aim to use appropriate subgroups from which to sample to provide data corresponding as best as possible to the study's objectives.

Where, for example, subgroups are suspected within areas or within the total available set of farms, a two-stage process may be most appropriate: participatory rural appraisal or formal survey techniques may be required to assess the existence of subgroups and then trials with different design features can be established appropriately for each group. Lamers *et al* (1996) demonstrate the potential for achieving useful information about fodder weeds from local experts and which could then be used as a starting point for detailed research programmes.

Choice of unit

Since on-farm, participatory trials are likely to be multidisciplinary and continue for a long time, the implications of volume of data must be considered early and plans made for their summary. Again, overall objectives are essential to be adhered to for the appropriate choice of unit to be measured is crucial to ensure both that the objectives can be achieved and to save time and possible loss of information if data have to be reduced to a smaller set through an aggregation process.

The issue of variability is crucial to design, the sampling strategy, the analysis interpretation and the recommendations. Scoones (1996) explores how biophysical variation (from soil and rainfall) and socio-economic variation (from assets and claims) and the interactions of their causes influence variation in household crop outputs. Shepherd and Roger (1991) showed that for on-farm agroforestry trials, variation could be found between plants, between plots, between blocks, between farms and between farm groups. In addition, each of these levels of variability could be ascribed to genetic, environmental or management sources and the last two, when caused by differences between farms and between groups of farms, were likely to be the largest if the trials were managed by farmers. Thus variability increases with greater non-researcher input, as less control can be achieved. However, as the purpose of on-farm studies is to assess systems under the management of farmers or local communities, this greater variation is to be expected and should be assessed: certainly it should not be ignored!

Where milk yields are needed after a regional programme of improved food supply has been initiated, herd data on a per farm basis may be adequate. But when live weight gain is required after the introduction of an immunisation procedure to small farms of few cattle, measurements on a per animal basis may be more informative. Similarly, measurements of soil fertility for a village may depend upon the average of single samples taken upon a wide range of farms within the village. But if farms are large and include areas of different farming systems, fertility will range throughout the farm and a greater number of soil samples within all systems in each farm may be more representative of the fertility picture.

Data collected on a relatively fine scale, say per animal, may be upscaled to a coarser scale, say per herd, and then to a coarser scale again of, say, the village. By forming mean values at each stage the appropriate dataset may be achieved. At each averaging stage, the loss of information about the variability may be great and if it is not recorded, it will be irretrievable. Steel and Holt (1996) provide examples of work where analyses based upon area level means give conclusions very different from those achieved from analyses at a finer scale. They propose a statistical model to describe the differences between the two analyses and which suggest to which areas units may belong, if this information is not available. However, the method depends heavily upon grouping variables, similar to the backdrop factors above, and clearly further work is needed to substantiate this approach in developing country on-farm work.

However, the need for a large number of datasets at different scales and the work to achieve them should be considered by all in the team. Datasets required by one team member, say the animal specialist may be on a different scale from those needed by the agronomist or the socioeconomist. A joint concerted effort is needed to achieve the most appropriate sets which can be used by all interested team members.

Samples and randomisation

When the most appropriate scale of measurement has been decided upon, such as the animal, the farm, the tree or the forest, how many of them within each backdrop factor are needed? Standard statistical formulae exist (Cochran and Cox, 1957) to determine this based upon the expected variability of the data within that backdrop factor. This implies that as data from farms are likely to be very variable, the required number of samples is likely to be large. Fielding and Riley (1996) showed that for adequate estimation of fertiliser effects in vegetable trials from 110 farms in Jamaica, plot sizes needed to be at least three times as large as those (large) plots already used on farms and for some vegetables, this implied plot sizes of 300 square m.! Riley *et al* (1996), examining plot size and sample size and shape within on-farm agroforestry plots in Brazil, showed that the number of maize plants to sample between rows of leguminous trees depended upon proximity to the trees and could be up to twice as many as are required from neighbouring sole maize plots. For animal studies, where the number of animals per farm may be small, the necessary sample size per farm may not be achievable. Where a survey of farmer opinion is involved, sufficient samples of farmers from, say, each wealth category within each farming region of interest will be required. In general, greater replication across farms is more desirable than replication within them in order to obtain efficient estimation of interventions across a region. Two replicates at each farm permit variability within the farms to be estimated, if this is required, and provides a buffer should one of them be lost or damaged. There is a need to balance the representation of all combinations of desired niches versus the achievement of good quality data from available facilities.

Randomisation of choice of units during the sampling process is also essential to ensure that the chosen individuals (animals, fish, people....) are not all of one size, one breed or from one specially treated category. True representation will be achieved through the numbering of all available individuals or units and the random selection of these numbers.

Thus, major design issues to be considered are the determination of clear objectives, the choice of a clearly defined target environment with appropriate niches or backdrops, the allowance for constraints caused by the multidisciplinary and long-term complexity of the study, the need to use unbalanced design structures and consequent powerful statistical analysis procedures, the determination of experimental unit (eg the cow or the herd), the determination of the response variables, the assessment of expected variability before the study is started, the acceptance that the level of available precision may be low, and the need to collect data appropriate to address the objectives. Participatory, on-farm studies can be compared to consumer testing stages of manufacturing processes. A whole range of statistical methods have been developed for this work, such as market testing techniques, survey analysis and qualitative response summary and which will be valuable for farmer participatory work. An introduction to such methods is in Barnett (1991)

Indicators for impact assessment

Choice of indicators

Indicators are data. They are named thus because they often are used to explore change, causes of change or impact of research interventions. A data value or an indicator can be quantitative, or numeric, or it can be qualitative, or non-numeric. Quantitative values can be discrete or continuous. Discrete values are numbers with distinct values such that no intermediate value is possible. Thus the number, 5, of chickens on a farm is discrete. A continuous value is a number which can take any value in a continuum such as fodder yield, 29.3 kg/plot or 17.26 kg/plot. A qualitative value can be nominal or ordinal. A nominal value is an unranked category such as a colour or an opinion e.g. the farmer likes a new animal feed or he does not like it. An ordinal value is also a category but has a clear rank, such as farmer preference for best, second best or worst.

In participatory research, indicators abound, yet the statistical and mathematical properties for their summary have received little attention. Riley and Alexander (1996) discussed typical indicators currently used in a wide range of research areas, including health, economics, industry, education, politics, medicine, social development, ecology and environmental sustainability. They concluded that reliable data analysis guidelines had been given only by Cox *et al* (1992) who stated that the use of weighting schemes for combination of indicators is often unrealistic and that established schemes may not be relevant for current studies. They recommended the avoidance of a single combined indicator, preferring simple weighting schemes for project components or disciplines with final combination of the weighted results. A great deal of further work in this area is clearly needed.

A refinement of the potential plethora of indicators can be made by listing essential and meaningful data variables in the following categories:

- changes in economic status
- equity issues (gender and age discrimination)
- environmental changes
- socioeconomic changes
- nutritional change
- performance values (of animals)
- sustainability
- changes in custom or beliefs
- practices or traditions

However, external factors, which will affect variability of the indicators above, must also be monitored. These can include climate, politics, the effects of other research projects. Project activities or monitoring and assessment activities may also influence the outcome of the research, and result in the need to change the choice of indicators as the project progresses.

Collection of impact assessment data

Broadly, research impact can be defined as the positive or negative changes brought upon a community or environment by a research project. Impact needs to be related to the project's environmental backdrop, but also to neighbouring environments upon which its results or activities may impinge and it needs to be assessed for a lengthy period of time. Some researchers depend upon the use of preset indicators to monitor research throughout projects and beyond their end. Rigid tabular monitoring structures, such as the logical framework or the ZOPPS approach, are often used by international donors (Uribe and Horton, 1993; GTZ, 1992) to monitor progress of projects during their lifespans. These use fixed indicators throughout the project. Few have yet had the opportunity to use them to assess sustainability beyond the life of the project. Others (eg Davies, 1995) propose that more realistic monitoring and impact assessment can be done through the use of indicators which are chosen at different stages of the project to correspond to outcomes at each stage. Thus the results of each project stage influence the direction and choice of indicators for the next stage. Such a process would imply that the original study objectives may be lost, but a more realistic and flexible approach to assessment may have been achieved.

Riley and Alexander (1996) proposed a combination of these two approaches, recommending the use of pre-set or core indicators for collection throughout and beyond the end of the project with sets of subsidiary indicators to be chosen as the project progresses, collected from neighbouring areas or at time points away from the project location and period. The quality of research can then be monitored throughout the project and change beyond the project can be assessed. Equally, impact upon other research projects or other communities can be determined as the project results diffuse throughout neighbouring environments. The flexibility of subsidiary indicator choice for this impact assessment will permit extensive coverage of other environments or niches as time progresses and which may originally have been expected to be impervious to its effects.

It is clear that the choice of indicators must be made for each specific project and they must be relevant to the objectives. Consistency of collection of data is essential so that change and sustainability can be estimated precisely. Once the indicators are collected, good structured databases are required to store them, whether qualitative or quantitative, on corresponding scales, such as the farm level or the regional level. The effects of data storage on different scales, their transformation to other scales through filters for large scale modelling procedures have been little explored (Gaunt *et al*, 1996).

Conclusions

A rigorous statistical approach is necessary in multi-disciplinary on-farm work, not only for analysis of collected data but for project design, study design and analysis, database formation and indicator determination and assessment. Familiar statistical methods typically are used for well-behaved experiments or surveys. On-farm work is likely to be more complex, generate farmer opinions as data and thus need powerful methodology for design and summary and which is now available in good statistical software packages. The choice of methodology and the interpretation of the computer output, however, requires professional statistical input to ensure that appropriate data are used and interpreted to give reliable recommendations.

Impact assessment data depends upon the collection of data from well-designed and appropriately-managed studies. The data need to be collected during the project and for some time after its termination both in the experimental area and its neighbourhood. The statistical properties of impact assessment indicators need assessment similar to experimental or survey data. The choice of indicators and the need to transform them depends upon the objectives of the study and the statistical methods to be used in their summary.

And finally, a recommendation for team collaboration: true multidisciplinarity is not always achieved in participatory on-farm projects. Yet it is essential for planning, to avoid duplication of data collection and to provide a wide perspective on farmer needs, design and interpretation of data summaries.

Acknowledgement

IACR-Rothamsted receives grant-aided support from the Biotechnology and Biological Sciences Research Council of the United Kingdom. This contribution to the workshop leading to this publication was achieved with funding from the UK Overseas Development Administration

References

Barnett, V (1991) *Sample Survey Principles and Methods*. London, Edward Arnold.
Cochran, W G and Cox, G M (1957) *Experimental Designs* (2nd edition). London, Wiley

Cooper, P J M, Leakey, R R B, Rao, M R and Reynolds, L (1996) Agroforestry and the mitigation of land degradation in the humid and sub-humid tropics of Africa. *Experimental Agriculture*, 32, 235-290.

Cox, D, Fitzpatrick, R, Fletcher, A E, Gore, S M, Spiegelhalter, D J and Jones, D R (1992) Quality-of-life assessment: can we keep it simple? (with discussion). *Journal of the Royal Statistical Society*, A, 353-394.

Davies, R (1995) An evolutionary approach to facilitating organisational learning:an experiment by the Christian Commission for Development in Bangladesh(CCDB) with the peoples' Participatory Rural development programme (PPRDP). In *Proceedings of ODI-CDS Workshop:the Potential for Process Monitoring in Project Management and Organisation Change.* 6-7 April, 1995.

Fielding, W and Riley, J (1996) Lessons from Jamaican on-farm vegetable trials. In *Contributed Proceedings of the XVIIIth International Biometrics Conference, Amsterdam, July 1 to 5, 1996.* p 303.

Gapasin, D P (1993) Research project management. In *Monitoring and Evaluating Agricultural Research. A Sourcebook (eds D Horton, P Ballantyne, W Peterson, B Uribe, D Gapasin & K Sheridan)* CAB International, Wallingford, UK, p 155-161.

Gaunt J, Riley, J, Stein, A and Penning de Vries (1996) Guidelines for effective modelling strategies. *Agicultural Systems* (to appear)

GTZ (1992) PFK, Guidelines for Project progress Review, Section 1002, Quality Assurance. Deutsche Gesellschaft fur Technische Zusammenarbeit (GTZ) GmbH.

Lamers, J, Buerkert, A, Makkar, H P S, von Open, M and Becker, K (1996) Biomass production, and feed and economic value of fodder weeds as by-products of millet cropping in a Sahelian farming system. *Experimental Agriculture*, 32, 317-326.

Riley J and Alexander C J (1996) 'Guidelines for an assessment method for the optimum uptake of research' Paper presented at the *Socio-economic Methodologies* Workshop, ODI, London, 29-30 April, 1996.

Riley J and Alexander C J (1997) Statistical literature for participatory on-farm research. *Experimental Agriculture*, 33, to appear

Riley, J, Oyejola, B and Smyth, S (1996) Variation in on-farm agroforestry data from Northern Brazil. In *Contributed Proceedings of the XVIIIth International Biometrics Conference, Amsterdam, July 1 to 5, 1996.* p 315.

Scoones, I (1996) Crop production in a variable environment: a case study from Southern Zimbabwe. *Experimental Agriculture*, 32, 291-304.

Shepherd, K D and Roger, J H (1991) Approaches to on-farm testing and evaluation of agroforestry technology. Working paper No 67, *Nairobi, International Council for Research in Agroforestry.*

Steel, D G and Holt, D (1996) Analysing and adjusting aggregation effects: the ecological fallacy revisited. *International Statistical Review*. 64, 39-60.

Uribe, B and Horton, D (1993) Logical frameworks. In *Monitoring and Evaluating Agricultural Research, A Sourcebook.(eds D Horton, P Ballantyne, W Peterson, B Uribe, D Gapasin & K Sheridan)* CAB International, Wallingford, p 113-119.

Managing Environments Sustainably through Understanding and Assimilating Local Ecological Knowledge: The Case of Honey Bee

Anil K Gupta

*Centre For Management In Agriculture, Indian Institute of Management,
Ahemdabad 380015, India,
Coordinator of SRISTI (Society for Research and Initiatives for Sustainable
Technologies and Institutions) and Editor, Honey Bee network
E-mail: anilg@iimahd.ernet.in*

Summary

Sustainable management of natural resources requires, it is argued widening of decision making choices and extending the time frame. The utilitarian logic by itself is unlikely to provide the long time frame necessary for the purpose. Similarly, decision making options cannot be widened without bringing in the tools and techniques that are available in modern science. Thus, while choice can be widened by modern science when blended with informal science, the time frame can be extended by granting the right of future generations and the non-human sentient beings.

Key words: sustainably, management of resources, local knowledge, modern science, institutions.

Introduction

Conceptually the developmental models can be arrayed on two dimensions: The time frame and decision making options or horizon. I have defined development as a process of widening the decision making horizon and extending the time frame of the households as well as the institutions (Gupta 1981, Gupta et al, 1995). The time frame refers to the period in which we appraise a technological or investment choice. The decision making horizon refers to the range of options that a decision maker is aware of and can access or avail of, in the given resource situation.

We can see the four possible developmental scenarios through combinations of these dimensions (fig.1).

Figure 1. *Development models*

		Time frame	
		Short	Long
Range of choices	Narrow	Non sustainable	Vulnerable
	Wide	Opportunistic, no sustainability	Sustainable

The implication of the above figure is that sustainability requires both the longer time frame as well as a wide range of choices.

The next question is: How do we widen the range of choices and extend the time frame? If a household does not have certainty of tenure or clarity of property rights vis-a-vis given resources, it is unlikely that the person may have a long time frame. Alternatively, in the absence of clear property rights, customary rights and informal institutions may exist and these could help extend the time frame. The cultural context, spiritual values and ethical basis of local knowledge systems also contributes to extending the time frame. That is why we notice some of the poorest households growing some of the slowest growing tree species in the homestead land. The widening of choices depends upon the

a) Access the households have to resources;

b) Assurances they have about others' behaviour vis-a-vis their own as well as about future returns from present investments;

c) Ability or skills people have to use available choices and

d) Attitudes towards nature, resource use and towards the concern for future generations.

To what extent the choices will be widened without impairing the ecological balance depends upon several factors which are summarized in Part One. Seven principles of Sustainability are discussed in Part Two.

Part one

The concerned choices: The Society for Research and Initiatives for Sustainable Technologies and Institutions (SRISTI) maintains an international network and publishes a newsletter named Honey Bee. The lessons of this experience is:

1. The erosion of knowledge is a greater and more urgent risk than the erosion of physical resources.

There has been a widespread concern and rightly so that natural resources are getting eroded very fast. However, this concern has often been converted into activities and investments which are focused only on physical resources. The knowledge that people have accumulated for conserving these resources over many generations has not been given adequate attention. The Honey Bee Network was started about six years ago primarily to arrest the erosion of ecological and technological knowledge, but also to document and disseminate the contemporary innovations produced by people for sustainable natural resource management. We have been trying to ensure two basic values while pursuing these goals.

a. That we acknowledge people we collect knowledge from. We must respect they are not like flowers, which do not complain just as flowers don't do, when a honey bee collects their pollen.

b. We connect farmer to farmer which is only possible in local languages just like a honey bee does by pollinating flowers.

The Honey Bee Network, which is now extended to seventy five countries is today one of the world's largest networks of indigenous innovators. We realize that knowledge of people whether traditional in nature or contemporary in origin will be lost, if not helped to grow through experimentation, value addition and dissemination. The inter generational gap makes the erosion of this knowledge an even more serious problem.

Conservation through competition

We set up a developmental voluntary organization viz., SRISTI (Society for Research and Initiatives for Sustainable Technologies and Institutions) to strengthen the Honey Bee Network and pursue research, action and advocacy around the issues of knowledge and resource rights of people. We have been organizing several kinds of competitions among young students, adults, scholars, public administrators, grassroots functionaries, farmers, etc. I describe below each of these competitions briefly and invite you to not only join hands, but also participate in the ones that are applicable to your interest and category. Let me add that while we respect the spirit of cooperation a great deal, we also feel that the spirit of competition to produce excellence is compatible with the spirit of cooperation. I also invite fellow scientists to consider other ways in which the competitive excellence can be generated at all levels to bring out the best in our society so far as strategies for natural resource management are concerned.

Biodiversity contests

The idea in this contest is to encourage children in the primary education classes (from 1st to 7th) as well as at other levels to look around themselves and collect knowledge about plant biodiversity to begin with along with its uses. On a given day, in the presence of a jury comprising local teachers, voluntary workers, herabilists, etc., each child brings the list of plants that they know about with or without uses in addition to the samples of the plants that they can identify. They are evaluated on the basis of five parameters:

- (a) number of plants listed, (b) number of plants brought (c) awareness about habitat (d) familiarity with uses and (e) style of presentation.

Those who excel are given prizes apart from a certificate of honour. Many children would ask their parents and more often their grandparents. In the process knowledge transfer would take place. We have noted that children do not compete only to win. Many times, some children bring only one or two leaves. Obviously they have no hope of winning the competition and yet the spirit of participation seems more dominant. In another study we noted that the ecological knowledge and academic excellence were not necessarily correlated (Shukla, Chand and Gupta, 1995). This study also indicated that children from backward and scheduled castes knew twice as much about the plant diversity as children from higher castes. Further, the girls did not have any disadvantage vis-a-vis boys upto class 4th, but after that there was a marked decline in their knowledge apparently due to increased household chores. The reason why lower caste children knew more could be because of their greater dependence on natural resources for their survival.

These competitions can also be organized among students at higher levels to understand the potential of embedded ecological knowledge. What has struck us most is the irony inherent in this process of learning. On the one hand the children identified through these competitions having excellence in ecological knowledge have to learn `a' for apple and `b' for bird. Otherwise, they become part of a large pool of so called `unskilled' labourers. On the other hand, we train students having no background in natural resource management in the field of ecology, botany, taxonomy, etc. Why couldn't we help these young ecological geniuses to grow as naturalists?

Competitions for scouting innovations by farmers, pastoralists, horticulturists, fishermen and women, etc.

We have been documenting indigenous innovations by farmers using local biodiversity for developing non-chemical sustainable technologies for agriculture, livestock, fisheries, agriculture and food processing, etc. Today we have one of the largest data bases in the world on farmers' innovations with names and addresses of the innovators and communicators of the ideas where these are drawn from traditional knowledge systems. These innovations have been collected from different parts of the World, but mostly from India and within India from Gujarat. More than 2000 villages have been surveyed with the help of under graduate students in summer vacations. Other nodes of the network are active in Tamil Nadu, Karnataka, Andhra Pradesh, Gujarat, Bhutan and Colombia.

In addition to documentation of innovations through the students, we also organized statewide competitions among grassroots functionaries as well as farmers in Rajasthan and Gujarat. The initial results have been extremely encouraging. What we have done is to circulate forms through state governments to various districts, talukas and villages seeking information about traditional ecological knowledge as well as contemporary innovations. A large number of examples have come out where farmers have developed herbal pesticides, veterinary medicines and growth regulators, etc. It is true that in most cases the scientists in formal labs have not given adequate attention to these innovations. But, farmers have taken up experimentation in informal labs at their farms, nevertheless. We have also been collaborating with Gujarat Agric. University, Jai Research Foundation, LM College of Pharmacy, Indian Institute of Science, MS University, Baroda, etc., in order to add value to peoples' knowledge systems. Our goal is to develop such products which can be either commercialized or disseminated directly among the farmers to reduce the costs and move towards non-chemical sustainable agriculture and resource use. We realize that this transition is not going to be easy given the massive influence chemical pesticide and other input industries have on the public administrators, policy makers and the scientific establishment. And yet, we are confident that through the coalition among public spirited scientists, grassroots innovators, and conscientious entrepreneurs, value added products can not only be developed and commercialized.

A part of the profits so generated will need to be shared with the innovators and a part with the research institutions and networks so that this coalition can evolve and survive through its internal dynamics. The fate of the coalition will not depend upon the benevolence of bureaucracy or aid agencies. The science and technology establishment in India has not even started thinking on these lines. One evidence of this indifference is the absence of a venture capital fund for small innovators in our country.

The competition for scouting innovations also helps in reorienting the thinking of the grassroots functionaries. Instead of focusing on the *lab to land* approach, they begin to see the importance of *land to lab to land*. Further, the developmental alternatives are explored in terms of what people have and not what they do not. As a consequence, the process generates humility and respect for indigenous innovators.

Involving innovators as researchers

Several artisans as well as innovators who have developed innovative uses of local biodiversity and other materials were requested to scout around for people of their kind. This process has generated a very participative way of learning from local innovations. In some cases, innovators so discovered would perhaps remain inaccessible by any alternative method.

Institutions for conserving, regenerating and diversifying biodiversity

The Honey Bee network has been building three kinds of data bases. One on indigenous technological innovations, a second on institutional innovations and a third on literature on indigenous ecological knowledge systems. The institutional innovations are no less important than the technological innovations. In fact, one could argue that without understanding the institutional context of technological innovations, we cannot speculate on the scope of sustainability.

To illustrate, a recent case of Thakuva village of Banaskantha District could be cited. In this village the common land of about 200 acres was quite degraded. Many people in the village had encroached upon it. In the neighbouring village on the Rajasthan side, there was an excellent auran land, i.e., land left for God and Goddesses. The local institutions required that nobody would take the wood from the auran land unless somebody is destitute. Even in such cases, the dead wood is taken for the funeral pyre purposes of those whose kith and kin cannot afford to buy the wood. The result is that one could see the dead wood lying around with grass growing on it. Balvantsingh, the leader of Thakuva felt that the common land of their village should also be regenerated like the land in Rajasthan village. He requested those who had encroached the common land to vacate it. He pleaded with them for several weeks with no result. Finally one day, he decided to declare to the entire village that he had encroached all the remaining land of the village. He refused to allow anybody to take their animals for grazing or drinking water in the common land. By evening the animals were thirsty and hungry. All the villagers then decided to pressure on the encroachers to vacate the encroachments. Balvantsingh took everybody to a temple and they had a meeting to evolve norms. They decided that henceforth nobody would graze their animals for two months after the rain and in the remaining part of the year they would graze, but not cut trees. Anybody observing anybody else cutting a tree, would be considered equally guilty. Accordingly, the common property land in the last eight years has regenerated in a very big way. People go for a collective patrol periodically and monitor any attempt to encroach by people even from neighbouring villages. They have decided that anybody caught poaching would have to contribute 30 Kg of grains to the common bird feeding centre. In case somebody persists with the offense, he or she would be outcasted and nobody would keep social relations with that person.

There are many other examples where contemporary institutions have evolved rules to manage common property resource in a creative and innovative manner. Sometimes these institutions are folkloric in origin. Sometimes these are traditional institutions governed by historically evolved norms. There are also cases as described above where traditional institutions broke down and through some social entrepreneur new institutions have emerged. The most important aspect of institutional regeneration or renewal is the long term sustainability of a natural resource. The lack of institution building has been the single most important factor responsible for decline of watershed projects as well as other community afforestation projects after a period of time.

Essentially, there are four dimensions of institutions building. First the issue of boundary management, i.e., consensus about who is in, who is out. We could also call them as access rules. The second aspect is the allocation of resources in terms of who gets what, when, how, etc. This requires some norms for regulating inter-personal access and also for maintaining the resource use. In the absence of regeneration and value addition, mere distribution of what exists is unlikely to provide incentives of long term cooperation. The third aspect is conflict resolution. How do we resolve the conflicts about access, resource allocation, value addition, etc. The conflict resolution systems are unlikely to be sustainable if these don't build upon multi enterprise multi level involvement of households in different portfolios. The fourth dimension relates to leadership which has to assure the group about complaints of norms and generation of new agendas to keep up the motivation of the group. The ethical aspects of access, assurance and ability (skills) of people obviously impinge on the viability of institutions. The long term sustainability is unlikely if the interest of only human beings and that too of only present generation are taken into account. The responsibility towards next generation and towards the non-human sentient beings cannot be ignored in any institutional arrangement.

Linking formal and informal science for generating incentives for conservation

Peoples' knowledge systems about natural resources is not only multi functional but also multi dimensional. Tulsi is revered as a sacred plant, but it is also found useful for various purposes in human, livestock and agricultural systems. Sometimes, the scientific mind refuses to take note of the sacred and insists on dealing with only the secular aspect of knowledge systems. It may not pose any problem so long as the secular aspect of knowledge systems are considered independent of the sacred aspect. I have argued that just like a double helical structure of DNA, the secular and sacred are intertwined. The sacred space provides identity to nature in which eco centric views can override the anthropocentric view. The secular knowledge system generally provides for the priority for human preferences in dealing with nature.

The nature of research that people do in linking cause with effect may not necessarily meet all the conditions of formal science. However, the indigenous taxonomy, clouds, soil, winds, waves, etc., do provide us new ways of looking at survival options under environmental stress conditions which do not necessarily emanate from the formal scientific categories. Blending of formal and informal science is a natural corollary of blending secular and sacred though not always in a straightforward manner. For instance, one can develop a drug from a plant used by native people for a given purpose disregarding the sacred aspect of the knowledge system. However, the anthropocentric view cannot offer guarantee the fact the plant will always remain available. For continuance of biodiversity, blending of secular and sacred consciousness is inevitable.

In part two I summarize seven principles of sustainability including the six dimensions of bio-ethics to conclude how the Honey Bee approach provides a new perspective in this challenging task.

Part two: Seven steps to sustainability: learning from indigenous ecological thought

Sustainability of Spirit is the key

Even if we have technologies which can help in use of resources within sustainable limits, will appropriate institutions emerge if the spirit is absent? Such was the question posed once in an Indian epic, Ramayana. In this epic, Lord Rama symbolizes the Dharma (noble conduct) and Ravana (who otherwise was a very wise sage) the Adharma (bad conduct).

> Ram was very frustrated on knowing that his wife, Sita (abducted by Ravana) was just on the other side of a vast expanse of water and he didn't have wherewithal to cross the sea or build a bridge. His followers were equally restive. The task appeared impossible. Suddenly a ray of hope emerged.
> Ram observed that a squirrel was behaving in an odd fashion. She was wetting her tail in water, coming back to the shore, rolling in sand and going back to the sea and washing her tail. She was doing it repeatedly and almost furiously, as if in a great hurry. Ram called that squir rel and asked her the reason for her odd act. She replied that knowing the challenges before them, she was contribut ing her mite. She was trying to fill the sea by the sand attached to her tail so that a path could be built.
> The entire work force of Ram felt ashamed at their despondency. And soon, with their collective effort, the path was built.

The projects like the squirrel's efforts are seldom sustainable. But a non-sustainable act like this could inspire a sustainable process. The trick thus is to unfold the locked up entrepreneurial energy of all those around. The momentum so generated may eventually solve the problem or generate the ripple which unsettles those believing in maintenance of status quo. The spirit of sustainability is prior, the substance is subsequent.

Sustainability requires acknowledgement of rights of the 'Others' - the sentient beings (birds, beasts and unborn human and non human life)

In most societies and cultures, strands of philosophy are found which justify the rights of the 'perfect strangers' like the unborn and other living forms which provide the much needed biodiversity. It is necessary for us to understand the process through which such a consciousness is ingrained in the day to day use of resources and observance of boundaries. A folk song I heard (as a part of our discussions in an action research project on water-

shed management in Shimoga district of Karnataka state in south India)
suggests how societies have kept the germ of this consciousness alive.

Paradox of Parrot:

> In a drought year, the crop has suffered very badly. A woman is com-
> ing back from the field after picking up whatever grains she could. On
> the way she meets a parrot. The parrot starts staring at her. She asks
> the parrot as to why was he looking at her so intently. The parrot re-
> plies that he was actually confused after looking at the woman's
> necklace. The necklace had a green agate stone. He mistook it to be a
> grain. Only when the woman came closer, he realizes it was just a
> stone. The woman asks him, whether he had not got anything for eat-
> ing. The parrot replies that hadn't she brought all the grains from the
> field even the ones which had fallen on ground. The women realizes
> that the parrot was hungry, and she also needed the grains very badly
> for her children. She asks the parrot to come home with her and share
> whatever she gives to her children. But the parrot flies away leaving
> the woman dumb founded.

Why did the parrot fly away? Did the parrot realize that if he delayed
the search for grains other people would also pick up whatever grains were
left in the fields? He remembered his young ones who were waiting to be
fed. Did he think that poor humans were so meek and weak that they
could search for grains only in a limited space whereas he could fly over
long distances? He should thus leave the grains for the poor woman.
Maybe he thought, he had a right over the grains so long as these were in
the field. Once these were in the hands of a human being, she had the right
over it (an instance of an ethic superior to the one we humans use!).

There could be many other interpretations!

The song speaks about a cultural system in which the right of birds are
being debated vis-a-vis the right of human beings even in a drought year.

How does one interpret this song would also depend upon how one
conceptualizes the right of different claimants over natural resources. If
birds were also considered as legitimate stake holders in the natural resources, then the viability, sustainability and effectiveness of any institution would have to be interpreted very differently. Many times, physical
resource scientists have taken a very limited view of human nature a view
which excludes the rights of other natural beings. The modern conservation ethic anchored on such a view can seldom produce sustainable outcomes.

Sustainability through creative culture bound indigenous institutions of management of Common Property Resources(CPRs)

Most of the sustainable arrangements for natural resource management
require group action through some kind of CPR institutions. While many
of the available frameworks of analyzing such institutional arrangements

have emphasized either game theoretic or utilitarian perspectives, I stress the need for giving importance to the process of rule making as much if not more than the rules per se. Further, I also feel that there is an admixture or what I may call double-helical intertwining of explicit and implicit, secular and sacred and `this' and the `other' worldly consciousness in these indigenous institutions.

Feeding the birds for poaching the trees:

A village panchayat (an assembly of eldermen) in Rajasthan devised a unique way of punishing persons who cut some branches of trees from common land where such poaching was prohibited. The offense was discussed by the village assembly of elders. Hours of discussion about various issues followed such as: when did such an offense take place last time; what were the choices considered then; did the culprit commit any or similar offense earlier; etc. The punishment given was to ask the culprit to stand barefooted under open sun in the hot summer and feed the birds two and a half kilograms of grains from morning to evening (Agar wal, 1990).

It may be difficult to establish a relationship between the cutting of tree branches, reduction of bird arrival, increases in the pest attack or decrease in the biodiversity because of lack of seeds brought by the birds and the feeding of the birds. This relationship is entirely my speculation. It is quite possible that this punishment would have been interpreted differently by different people in the village with some common meaning, but some uncommon meanings too. On the one hand the culprit was punished and on the other, he was supposed to have been blessed by the Gods for having fed the birds in such an hot environment standing barefoot.

An element of ambiguity characterizing such judgments provides a creative ground for exploration and speculation. Institutions are seen to be embedded in the socio cultural and religious world view of the people. It is quite possible that access of various social groups or classes to the same common lands may not have been equitable for all the resources. However, to infer from inequity in availability of one resource, say, wild berries from common lands that inequity or indifference should exist in the institutions for other resources, be they of aesthetic or material nature would be a mistake. In this case the deliberations were guided not just by keeping the interest of human claimants on the natural resources in view.

The global concern for sustainable development and conservation of bio-diversity is dominated by the strategies and styles suitable for essentially the degraded environments. Since degradation in environment inevitably is accompanied with the degradation of the institutions, these policies take absence of institutions as given. Much greater reliance is placed on public interventions which in turn mean bureaucratic interventions. The case given above questions such a bias.

Sustainability through multi-functional institutions of restraint, reci-procity and respect generating collective responsibility for nature

There is a custom that people go together to the forest for collection of shingle wood in Bhutan on a particular day. There are several implications of this practice.

a) While collecting wood on the steep slopes, if somebody falls down, there are people around to help in the emergency.

b) Everybody monitors everybody else's collection of wood.

c) Since collection of wood has to be done keeping in mind the age, health, and condition of the tree, corrective restraint helps in maintaining those conditions.

d) Some people are either too old, handicapped, weak or their require-ments are larger than they can manage on their own. Groups help in such cases and carry the extra burden.

e) There are sites which might have suffered some damage due to rain, landslide or otherwise. Since such sites are observed together, it enables mobilization of the collective will for corrective action more easily.

f) In addition to the utilitarian dimensions mentioned above, the group action is its own reward when there is music, fun and laughter around.

Thus, emphasis on only the economic part of a resource would not pro-vide sufficient information or insights for building institutions that can help in managing resources sustainably. Development is possible only through creative institutions which constrain individual choices to some extent and yet provide scope for entrepreneurship.

Sustainability through blending of holistic and reductionist perspective for regenerating resources

I intend to take help of a story from another epic of India viz: Mahabharat. There was a famous Guru (teacher) who had an ashram (college) situated in a forest specially meant for royal scions. His name was Dronacharya. Five brothers (sons of the king Pandu) were his choicest students. Since Droanacharya was the best known teacher of the art of Archery, his stu-dents were supposed to share this excellence too. Once he took all the five brothers for an examination to a nearby place. He hung a bird on the tree and asked each one of them one by one to take an aim at that bird and tell him what they saw. When the turn of the eldest brother (Yudhister) came, he said that he saw the entire cosmos of which the earth was a part, of which the tree was a part and finally he saw a bird on the branch of that tree. Dronacharya asked him to sit down. The next brother came. He said that he saw he earth, tree, branch and the bird. He was also asked to sit down. Then came the turn of his favourite student Arjun (the hero of the famous story of Gita). He could see only the eye of the bird. Undoubtedly, he became the best known archer of his time (surpassed only by a tribal student Eklavya who was denied admission by Dronacharya to his Ashram since he was a commoner).

The eye of the bird reflects the extreme reductionist attitude just as the whole cosmos shows the holistic perspective. My contention is that we need both the perspectives i.e. reductionist as well as holistic, and not just any one as many environmentalists claim. Any theory building process requires drawing a picture being studied as partial. On the other hand we need a holistic view so that interconnections of different parts of nature can be seen. Sustainability requires balancing the sea saw of these two ends of the same spectrum.

Bio-ethics for sustainability

The sustainability of a use of a resource requires development and demonstration of an ethics which guides decisions regarding current versus future consumption of resources. The conception of nature and relationship between human and non-human, animate and in-animate, born and unborn etc., are defined if not determined by this ethics. The bio-ethics can raise the following choices:

a) Do I draw natural resources at a rate that the resources renew themselves within a short cycle.

b) Do I draw as much as I can as long as it is available and once exhausted, I shift or change the resource base.

c) Do I draw less than what can be used without impair ing the ability of the resource to renew itself.

d) Do I draw resources only as much as I need simulta neously ensuring that the genuine needs of others are also met and the resource is renewed before it drains down to its critical limits.

e) Do I draw as much as possible, hoard it if feasible and then market it at a very high price to ensure some kind of rationing of its use.

f) Do I develop an institution which through its inef ficiency (or coercion, or both) generates a constraint on the maximum sustainable yield.

These vectors of human choices confront every decision maker involved in resource restoration. To what extent these choices actually influence the design of organizations is a matter to be pursued further.

One of the persistent reasons why many externally induced interventions fail is because the local knowledge system is often discounted and if considered, is seen only in an utilitarian perspective (Gupta, 1980, 1981, 1987, 1989; Richards, 1985, 1989, 1992; Verma and Singh, 1969; Dharampal, 1971; Chambers, 1983;

Bebbington,1992; Periera,1991). This realization has dawned on the development planners now for some time, but the mechanisms chosen to build upon local knowledge are often worse than the problem. Various short cut methods popularly called rapid rural appraisal are used to get a handle on the local situation. We have critiqued these methods on ethical as well as efficiency grounds separately (Gupta and Patel,1992). It is necessary to note here that organizations of creative people whether in the form of networks or informal cooperatives or just loose associations would generate a very different pressure on society for sustainable development. The spirit of excellence, critical peer group appraisal, competitiveness and entrepreneurship, so vital for self reliant development may emerge, to our mind only in the networks of local `experts', innovators and experimenters. It is true that every farmer or artisan does do experiments. But not every one is equally creative and not in the same resource related fields. The transition of the developmental paradigm from the victim's perspective to that of the victor's and the organizational principles for a creative group are also likely to be different than for the rest.

The iterative, rotational and interactive leadership models are the only ones which sustain local community organizations. A study of Chenchu food gathering and hunting tribe in Andhra Pradesh (Gupta, 1983, 1987) revealed three principles of sustainable organizational sustenance:

1. The leader and follower can iterate. The leader in honey collection subgroup has no particular skill in hunting and becomes just a follower in that group.
2. The skills and not status determine the leadership. The person who knows the most critical functions in a task becomes the leader rather than the one who is chief of the tribe or his kin.
3. The pooling is independent of the redistribution. The honey, game, food, fish or fuel is shared in this tribe among all the members and not just the one who went on the expedition.

The original model of IB which emphasized on the inter-organizational changes is less useful now. The evolution of stakeholders' interest in the organization plays a vital role in the self reliance process.

The organizational principles which guide collective action in different regions would obviously have some common but many uncommon dimensions. The Institution Building process involves simultaneous intervention in eight dimensions of organizational change, viz: Leadership, Stake Building, Value Reinforcement, Clarifying Norms and Rule Making Processes, Capacity Building, Innovation and Creativity, Self-Renewal, and Networking. The theory of Institution Building (IB) has to be significantly remodelled for historical reasons. The IB processes evolved to increase the capacity of third world organisations to receive aid funds and use them efficiently and effectively. The problem was defined from an external perspective and resolved or sought to be resolved. They have provided only limited insights for strengthening the capacities of organizations which have emerged autonomously at local level.

Some values are brought by the members of any organization along with them, but some are acquired in the organizational life experience. It is these values which have to be so shaped that reliance on external instruments of control and supervision becomes less important.

The rule making process is the one of the most crucial aspect of the IB in any organization. The fine tuning of rules, norms and belief system in accordance with the strategic future directions is not a function of just the leader. The group has to collectively evolve the norms and changes therein so as to ensure that collective spirit is maintained.

One of the most inappropriate terms in the developmental jargon is that of `unskilled`labour. There is practically no person who has no skill whatsoever. The challenge is to provide space within and outside the organizations for each member to grow. The learning systems at individual and collective levels are to be strengthened in such a manner that the errors are not masked and corrections are not delayed too much.

The process of self renewal requires recalibrating the scales of measurement periodically. It is the ability to discriminate finer shades of the colour of life which in the normal course may be missed. The historical perspective helps just as does the urge to relate to larger social causes. One cannot discover the immense source of energy to pursue any specific goal till one finds the broader dimension of growth.

References

Agarwal Arun, 1990, Personal Communication

Bebbington Anthony, 1992, Farmer Knowledge, Agroecology and Institutional Arrangements for Sustainable Development: Comments from a Case Study, IIED, London

Chand,Vijay Sherry, Shailesh Shukla and Anil K Gupta, 1994, Incorporating Local Ecological Knowledge in Primary Education, Ahmedabad: Ravi J Matthai Centre For Educational Innovations, Indian Institute of Management, Ahmedabad, Mimeo.

Chambers, R. Rural Development: Putting the Last First, London: Longman, 1983.

Dharampal, Indian Science and Technology in the Eighth-ninth Century, Academy of Gandhian Studies, Hyderabad, 1983, 325 p.

Gupta, Anil K., 1980, Communicating with Farmers Cases in Agricultural Communication and Institutional Support Measures,IIPA, New Delhi, 1980, 92 p.

Gupta Anil K., 1981, A Note on Internal Resource Management in Arid Regions Small Farmers-Credit Constraints: A Paradigm, Agricultural Systems (UK), Vol.7 (4) 157-161

Gupta Anil K., 1987, Role of Women in Risk Adjustment in Drought Prone Regions with Special Reference to Credit Problems, October 1987, IIM Working Paper No. 704.

Gupta Anil K., 1989, Managing Ecological Diversity, Simultaneity, Complexity and Change: An Ecological Perspective. W.P. no 825. IIM Ahmedabad. P 115, 1989, Third survey on Public Administration, Indian Council of Social Science Research, New Delhi

Gupta Anil K., and Kirit K Patel, 1992, Survey of Innovations for Sustainable Development: Do Methods Matter?, Proceedings of International Conference on Indigenous Knowledge and Sustainable Development, IIRR, Silang, Cavite, Philippines during September 20-26, 1992

Gupta Anil K., Kirit K Patel, A.R.Pastakia and P.G.Vijaya Sherry Chand) Building Upon Local Creativity and Entrepreneurship in Vulnerable Environments, published in Empowerment for Sustainable Development: Towards Operational Strategies (Ed.Vangile Titi and Naresh Singh), Nova Scotia, Canada; Fernwood Publishing Limited; New Jersey, Zed Books Limited, 1995.

Periera, Winnin, 1991, Asking for the Earth. London:ITDG publications

Richards P, 1985, Indigenous Agricultural Revolution, London: Hutchinson Press

Richards P, 1989 Agriculture as a Performance, Farmers First: edited by Robert Chambers, Arnold Pacey and Lori Ann Thrupp, Farmer Innovation and Agricultural Research, London: Intermediate Technology Publication

Verma M.R. and Singh Y.P., 1969, A Plea for studies in Traditional Animal Husbandry, Allahabad Farmer, Vol. XL III (2), pp. 93-98.

Grameen's Programme for Research on Poverty Alleviation and Biodigester Experiments in Grameen

Dr. Syed M. Hashemi

*Director of the Programme for Research on Poverty Alleviation,
the Grameen Trust, Mirpur Two, Dhaka-1216, Bangladesh
E-mail: hashemi@citecho.net*

Summary

Presently research in Bangladesh suffers from being donor driven. In order to promote a self-reliant research agenda responding to the needs of the poor the Grameen Programme for Research on Poverty Alleviation has been set up. The Programme support research on a wide range of topics such as popular mobilizations of the poor against social injustices, the structural basis of gender inequality to appropriate technology for increasing productivity. As an example of the latter the paper includes a discussion of experiences with the plastic bag biogas digester.

Key words: Research, poverty alleviation, independent research agenda, integrated farming, biogas.

The historical background

Social Sciences research in Bangladesh has had strong foundations from the early 1960s. In fact the movement for the war of independence in 1971 was to a great extent motivated by the research of Bangladeshi social scientists that exposed the regional inequality and increasing poverty that the Harvard backed Pakistani development model was creating. Sadly enough, however, the independence of Bangladesh saw a massive flow of foreign funds in the name of humanitarian and development aid. This foreign aid sponsored the growth of a small group of indentors, consultants, and contractors, who along with members in the highest echelons of the civil and military bureaucracy, dominated the economic and political life of the country. The flow of foreign funds also trickled into the arena of academic research. Compared to what was available locally, this foreign funding was so massive that it greatly distorted the research market. While the salary for an university professor is $250, a good consultant can easily command fees of over $2,500.

Development research in Bangladesh has thus become donor driven. A direct consequence of this is the fact that a growing share of the research underway today in Bangladesh originates from priorities set by various donors who fund such studies and not from our priorities here in Bangladesh. While some research may indeed be important and relevant the fact that they are initiated externally implies that they are not used by our policy makers or the people in whose name much of this is initiated. Indeed most of this externally funded research is never disseminated locally and more importantly never tested for their academic merit. Hence a whole series of research projects are completed by high paid consultants who are not made accountable either for the analytical rigor of what they write or the policy prescriptions they make. This non accountable transfer of massive funds to academics have had a corrupting influence. It has undermined the quality of research here and made it difficult to initiate research stemming from Bangladesh's own needs because of the unavailability of provisions of high remunerations.

There is thus a desperate need for recapturing control over our own research agenda, which is itself part of the wider question of recapturing control over our own development priorities. This forms the raison d'etre of Grameen's Programme for Research on Poverty Alleviation. While we realize that the Programme in itself can only make a modest contribution towards achieving this goal, it nevertheless intends to make that all important first step.

Programme objectives

Taking the first few steps towards recapturing control over our research agendas forms the primary objective of the Programme. It is hoped that through the Programme Bangladeshi researchers can be asked to think out their research goals and needs, independent of the needs of donors or the demands of the consultancy market. It is hoped that the Programme can interest Bangladeshi researchers to undertake studies that would be of academic interest as well as being important for the needs of the country.

The second goal of this Programme is to make such research, with a developmental focus, sensitive to the needs and concerns of its users. While academic research is important in its own right, it gains in merit by serving a public purpose whereby its outcomes can serve to change policy, or improve the conditions of life of the people.

The focus given to this Programme on promoting research which serves to generate ideas designed to alleviate poverty, makes it particularly important that such research be serviceable to the concerns of the poor majority of Bangladesh. All research being funded under this Programme is to focus on the use value of the research on poverty. Attempts to design the project through a process of interaction with prospective beneficiaries is emphasized. The research will have to prove itself not just by its publishability but by its ability to serve the target groups. To this end mechanisms for disseminating the research to policy makers and end users will be emphasized and mechanisms put in place to mediate this linkage between producers and consumers of research.

The Programme also aims to bring into the research arena a whole new group of young researchers, with interesting ideas, who have hitherto been frustrated with minimal access to resources, so that their potential creativity has remained under utilized.

While these objectives remain ambitious they do set the stage for a rethinking on development and research priorities in Bangladesh. The Programme will have been a success if it can force the research community to start setting some of their own agendas based on the needs of the poor, and if it can assist in the development of even a small group of young researchers.

Research work undertaken

While research in diverse areas are being funded, the current emphasis, based on actual imperatives, have been in the following areas:

1. Popular mobilizations of the poor against injustices
   - the struggle against shrimp cultivation which is environmentally disastrous and which destroys the livelihood of the poor
   - state violence and the issue of human rights
   - restrictions in access of the poor to courts and legal assistance

2. The structural basis of gender inequality
   - purdah and restrictions in women's mobility
   - fundamentalist opposition to women's mobilizations
   - reproductive health

3. Appropriate technology for increasing productivity
   - low cost housing for the poor
   - technological innovations in weaving
   - use of animal waste for fertilizer, fish feed and biogas

Grameen's venture into integrated farming:

Though Grameen Bank itself is a bank (for the poor) the basic philosophy is one of placing the poor at the center of the development paradigm. This essentially implies that development can occur, poverty alleviation can take place, only when strategies are taken that directly include the poor as the principal actors. This implies that the poor can themselves take care of their own destinies; what is required is only ensuring that the space for this is opened up and made available. With Grameen Bank therefore the focus was on credit. The Bank ensured that the poor had access to credit. The poor themselves ensured that the credit was utilized, repayments made, and income increases, asset acquisitions and economic well-being achieved. Grameen subsequently turned its attention to agriculture and fishery.

Dependence on foreign aid and donor technology in Bangladesh has led to a situation where the use of chemical fertilizer and pesticide has increased dramatically. While this provides major profits to multinational chemical companies and their local distributors as well as government functionaries, for the vast majority of the peasants this implies dependence on products over which they have no market control. This has meant that with price increases, often due to monopoly market control, peasants are stuck having to make high payments and end up with huge losses. In fact the major fertilizer crisis last year as well as this year has meant that even with double the price, poor peasants are left with no access to fertilizers. The environmental and health hazards due to chemical fertilizers and pesticides are far more important but concerning which very little information exists.

Grameen's attempt at reducing this dependence as well as ensuring sustainable agricultural development has led it to establish two institutions -- the Grameen Krishi (Agriculture) Foundation and the Grameen Motso (Fishery) Foundation. It is through the Motso Foundation that experiments have been conducted in integrating farming.

The Mostso Foundation originated from the leasing of ponds from the Government of Bangladesh with the intent of increasing fish production and involving the poor in this process. The Grameen groups of landless were organized to look after the ponds under Grameen management and technical supervision. Initially chemical fertilizers were predominantly used. It was only later that Grameen learnt about organic fertilizers and started using animal waste (cow dung). Between 1987-88 and 1993-1994 the use of

Table 1.   *Use of chemical fertilizer and organic fertilizer in Joyshagor Fish Farm*

Year	Area of Waterbody (Hectare)	Chemical Fertilizer (Metric Ton)	Cow dung (Metric Ton)	Compost (Metric Ton)
1986-87	225.0	25.32	02.77	-
1987-88	335.0	98.25	390.00	15.00
1988-89	398.0	28.00	365.00	-
1989-90	389.0	4.00	1185.00	20.00
1990-91	405.0	22.50	2636.00	0.50
1991-92	382.5	23.87	3106.00	1612.50
1992-93	382.5	18.20	3609.00	2891.00
1993-94	385.5	15.25	5233.00	3163.00

chemical fertilizers declined from 98 metric tons to 15 metric tons; cow dung use increased from 390 metric tons to 5233 metric tons, and compost from 15 metric tons to 3163 metric tons. This has also meant a drastic reduction in costs by 50 %.

In 1994 Grameen took a major initiative in integrated farming by introducing biodigesters. Dr T.R. Preston of the University of Agriculture and Forestry in Ho Chi Minh City, Vietnam visited Bangladesh and constructed a polytene biodigester model for only US$ 30. In the second half of 1994 Professor George Chan's visit to Grameen and his seminar on "Integrated Farming" excited everyone and provided further encouragement to the use of biodigesters.

Research on biodigesters

A biodigester is a "chamber" in which natural waste such as cow and poultry manure, algae, human waste, water hyacinth and other forms of biomass are placed for bacterial fermentation. The process results in the production of methane gas and high quality fertilizer/effluent from the natural waste. Research has shown that when natural waste is processed inside the biodigester, the majority of its contents are converted into minerals (NO_2 and NO_3). This is why the resulting waste product is such a rich effluent. It is a vital ingredient for maintaining the state of balance of the integrated farm and is used in the cultivation of fish, crops, plants, vegetables and trees. Comparative experiments conducted by the Fishery and Agricultural Foundations resulted in faster growth of trees and fish for those plants and ponds that were treated with biodigester processed cow dung as opposed to regular cow dung or decomposed cow manure treated with TSP and Urea.

The use of biodigesters seem ideal for the poor. It has been estimated that an average rural household of five with two cows could produce enough cow dung that, when processed through a biodigester, would generate enough gas to meet all their cooking needs. In addition the effluent could be used in ponds or in agriculture to meet their own needs or be marketed. Currently there is a great demand for such effluent through the Grameen Krishi and Motso foundations. In fact since Grameen Bank provides credit to buy cows, a Grameen family could very easily generate enough incomes to meet repayments through the sale of such effluent and use the gas as extra. The critical issue however is in producing household unit biodigesters that are inexpensive as well as long lasting to make it feasible for the poor. It has been calculated that this would imply making biodigesters available at less than US$ 45 and lasting for four to five years. It is here that the Program for Research on Poverty Alleviation has stepped in and supported research to make this feasible.

Till date however experiments on biodigesters have not been completely successful. The plastic sheets initially being used to build biodigesters were brittle and susceptible to cracking, especially in the summer heat. The heavy duty plastic sheets later used were too expensive. Further, given that people and animals were stepping on them they too were cracking. A lot of experiments were conducted with synthetic textile material and chemical spraying on them. However there seemed to be gas seepage. Grameen does not want to introduce concrete biodigesters since that too would be too expensive and make biodigesters out of the reach of the poor. The struggle is on now to find alternative building material; the struggle is on to ensure access of biotechnology for the poorest.

Conclusions

The Programme for Research on Poverty Alleviation has been undertaken as a challenge. It is not merely another foreign funded project. In fact it is precisely to limit foreign intervention in our research, in setting our development priorities, that this Programme was started. In this the Programme challenges current norms and values. There have been moments in our own past however when Bangladeshis have shown grate capacity for self reliance, for charting their own future. The Programme is intended to set into motion this attitude of independence in our research. It will have been a success if the first steps towards this reorientation is taken during the course of the Programme.

Conclusions and Recommendations

Integrated Farming in Human Development

Introduction

Specialists with experience from developing countries in Asia, Africa and Latin America, employed by NGOs in the North and the South, major Consultancy Companies, International Aid Agencies, and Universities, met in Denmark from 25 to 29 March 1996 to conduct a workshop on the role of integrated farming in human development.

Conclusions

Integrated farming

General issues

The role played by livestock in farming systems for poor farmers is multi-faceted and synergistic and must be seen not as a primary form of production, but rather in terms of their overall contribution to the total farming system and to the immediate needs of the family.

The above concepts are not well understood by decision makers, at both Government and Agency level. As a result policy at all levels - in Ministries of Agriculture, at Universities and National Research Institutions - is still mainly sectorial along lines of scientific discipline rather than being focused on a holistic and multi-disciplinary approach.

The solution to this problem is long term and requires changes in attitudes by decision makers and in curricula at institutions concerned with human resource development. Agriculture must increasingly be taught from a biological standpoint, as regards theory; and from a systems approach as regards practical application.

Integrated farming offers unique opportunities for maintaining and extending biodiversity. The emphasis in such systems is on optimising resource utilisation rather than maximisation of individual elements in the system. Animals and plants of high genetic potential, and specialised output, require high quality inputs; by contrast, the resources available in integrated farming systems are diverse, available often in large quantities, but usually are of lower quality. Local breeds of livestock, and local varieties of crops, are often better adapted to utilize such resources. The buffalo is a good example of the former; the sugar palm tree is an example of the latter.

Local breeds

Local breeds are better adapted to local environments and local resources. In order to facilitate greater use of local breeds research is needed in simple breeding systems which optimize adaptation to the environment on the one hand and improved production capacity on the other. This requires identification of the merits of local breeds in the context of integrated farming systems, and ways of increasing their participation (and hence their numbers) in breeding systems. The Mong Cai breed of pig in Vietnam is favoured by farmers for reproduction but is crossed with exotic geno-types for fattening. In this way [local breeds with good characteristics for reproduction can be crossed with exotic breeds] an economic role (for re-production) is created for the local breed while the need for more efficient meat production is satisfied by crossbreeding.

At a practical level the need is to increase the availability of semen, and sires, of local breeds and their FI crosses at village level. Such activities will be better administered by farmers than by breeding centres managed by the State or commercial companies.

Local feed resources

The advantage of integrated farming systems is the opportunity of using residues, by-products and wastes generated in the different activities which are the basis of the integrated approach. Selection of component sub-systems should be predicated on their suitability to maximise solar energy capture (eg: multi-strata and associated crops), generate products with multiple uses and which are friendly to the environment. Crops which produce products that can be consumed by the family or sold and residues that can be fed to livestock should have priority. Thus specialised crops for livestock (eg; sown pastures or forages) are counter indicated. By contrast, there are opportunities for new sources of biomass in the form of trees which will contribute positively to the environment as well as pro-viding a varied range of products for use by the family.

Biodigesters

Biodigesters play a pivotal role in integrated farming systems by facili-tating control of pollution and at the same time adding value to livestock excreta through production of biogas and improved nutrient status of the effluent as fertilizer for ponds and crop land.

Impact of biodigesters has been limited by the high capital cost of tradi-tional designs (floating canopy and fixed dome cost upwards of US$400). Low-cost plastic tube biodigesters offer a solution to the problem of initial cost (US$30-70). But results have been variable ranging from excellent (90% success rate reported in South Vietnam) to moderate (60% success rate in Tanzania). Adoption has also been poor in Bangladesh. Experience indi-cates that the problem is not so much technical (eg: quality of the plastic) as the location (eg: access to other fuels) and the way in which the technology is introduced, supported, adapted and improved according to local conditions. There is need for on-farm research in different socio-ecological situations to evaluate management factors that influence digester performance and reliability. Closer cooperation and exchange of experiences among the various groups interested in the low-cost plastic tube digester technology will facilitate this process.

Integrated farming for human development

What is integrated farming? While it was appreciated that a single definition of integrated farming may not be forthcoming, it was discussed that the following are elements of integrated farming:

- It involves the utilization of locally available resources. These resources may include feeds, wastes and other outputs from the subsystems within.

- There is a high degree of nutrient recycling and hence reducing energy subsidy and cost.

- The total farming system is enhanced through reduction of waste; creating interdependence and overall economic efficiency.

- The system is made more sustainable ecologically, economically and socially.

What is the potential for integrated farming within human development?

Integrated farming can have a great impact on human development both in material and social terms.

- Because of the different systems involved there is substantial generation of knowledge and innovations. It can make farmers more self sufficient as well as self-reliant.

- Poor farmers or landless farmers can be assisted to create income and hence participate in the development process.

- External inputs to the system are very much reduced.

- Integrated farming has a great potential in reaching all levels of income. While the needs of the poor can be adequately addressed there is also the possibility of meeting the aspirations of the farmers with higher incomes.

- It was shown the BRAC poultry model in Bangladesh can reach the poorest of the poor of women and the model deserves the greatest attention of governments and donor agencies. However, it may be equally desirable to prevent others from sliding into poverty. This can be achieved by integrated farming as it responds to all facets of crop and livestock production as well as other important concerns like energy.

- In areas where farmers are faced with problems of land fragmentation, integrated farming will go a long way to address this problem by giving more opportunities for productive activities.
- By appreciating the fact that women contribute far more to the household security, integrated farming can improve the welfare of women.
- It provides for better utilization and distribution of labour.
- Costs are generally reduced and productivity of labour is increased.

There is therefore a general improvement in the sustainability of the system, creating more wealth on a more equitable basis and on a more environmentally friendly basis.

The principles of integrated farming are based on the fact that a number of resources are of finite nature and must be utilized judiciously to bring about positive change in the economic and social wellbeing of people.

Integrated farming must facilitate the application of on farm studies to improve the productivity and sustainability of the land.

The main concept is that the farm consists of subsystems which all work together in a synergistic manner, one subsystem creating inputs to the others and eventually ending up in a more closed cycle with least external input.

Recommendations

- Research and extension efforts must address the needs of the poor farmer in a more holistic manner as well as other farmers.
- Credit and the BRAC rural poultry model can be used as important tools of intervention for human development while generating income and preventing environmental degradation.
- More research should take its origin in problems perceived by farmers.
- More research is needed on problems found important by the farmers. The farmers should be consulted on possible known, new and potential solutions at all stages. On-farm research is important, but farmers can also contribute to a realistic planning and evaluation of "on -station" research.

Impact assessment

1. Definition
Impact can be defined as "the changes (positive or negative, direct or indirect) that occur as a result of project activities". Impact should relate to the

project's target groups and also neighbouring communities and should include the project and post-project periods.

2. Issues

Any impact assessment should start with selection of the main issues, but before deciding on these the project objectives should be clearly stated and the issues related to them. The assessment should include the target community before (base line data collection), during and after the project phase. External factors (eg government political and economic decisions) in addition to project activities will influence the situation of the target groups and could influence their decisions and actions vis-a-vis the project. It is important therefore to include other communities in the assessment - these should be as similar as possible to the target community, but be sufficiently far removed physically so as to be unaffected by project activities. Examples of key issues that will be considered for inclusion in any impact assessment exercise could be:

- Changes in *economic status* of the members of the community
- *Equity* issues, in particular *gender* and *age-ism* aspects
- *Environmental* changes
- *Socioeconomic* changes
- Changes in *nutritional status* of vulnerable groups - young children, lactating women, the old
- *Performance parameters* (eg. of livestock)
- *Sustainability* aspects (requires post-project assessment, but more work is needed to decide for how long after end of project)
- *Customs, taboos*
- *Practices and traditions*

For each of the issues a number of assessment techniques can be applied, which obviously will vary according to the particular issue under consideration, but will always include a set of indicators. Indicators should be selected in advance (before the start of the project) and ideally should include inputs from the target community. Flexibility in the choice of indicators throughout the project should be retained as important issues may appear through the influence of external factors.

3. Assessment techniques

Methods for assessment include the measurement of indicators, their summary either formal or informal, and their interpretation.

A selection of typical indicators for each of the above issues is :

Economic changes
Measures of input and output costs and expenditure; GNP; growth rate; access to credit; savings; income; purchasing power; taxation

Socio-economic/equity/customs/taboos
Planning policy; education rate; food supply; labour availability; social interaction; infrastructural change; medical assistance; housing and equipment; gender and age-ism issues; consumer choice and preferences; tenure; witchcraft; religious beliefs; life expectancy; child labour

Environment and sustainability
Soil erosion; overgrazing; deforestation; soil fertility; use of byproducts; pollution; sanitation; disease; dependence of project on external inputs; replication of the project; generation of internal funds; technological adaptations; capacity building of the staff

Performance and farming practices
Production per head; yields; mortality; fertility; growth rates; consumption; disease; types of farming; varieties; land use systems.
All of these may be measured at different times of the project and beyond and also within neighbouring communities.

4. *Other factors*

External factors
Weather; political changes; government policy; epidemics and pests; other research projects; external fund supply.

Project activities
Withdrawal or change of direction of the project and change of staff.

Assessment activities
- internal assessment
  (i)
     * management systems
     * accountability of the project
     * staff participation
  (ii)
     * community assessment
- external assessment
     * donors and consultants
     * neighbouring farmers

Measurement
- need for primary and secondary indicators possibly in groups; need for flexibility.

Data
Early collection of systematic databases is essential with appropriate scales of measurement for baselines and comparisons over time.

The way forward: recommendations

Pre-project assessments of probable impacts are in theory desirable, but in practice will delay the start of the project, and presuppose a degree of flexibility regarding project direction and activities. If they are carried out they should be done as quickly as possible.

Training and communication

- The M.Sc. course on 'Integrated livestock production based on local renewable feed resources' conducted at SLU, Uppsala is COMMENDED. Such courses should be encouraged because they are practical in nature and tailored to address farmer needs.

- It has clearly been observed that too few professionals trained in both developed and developing countries do not want to work with poor rural farmers. In order to alleviate the situation it is recommended:

  (a) to revise the training curricula for agriculture and development in developing countries to put more emphasis on the integrated farming perspective and farmer participation in research.

  (b) to re-orient attitudes of professionals (researchers, academicians and extension agents) towards more collaborative working with farmers for mutual learning.

- It is realised that in some situations professionals willing to go and work with farmers can not do so because of financial and logistical reasons. It is recommended that all development agencies facilitate on-farm research work.

- Appreciating the work done by the professionals of BRAC and Grameen Bank in Bangladesh, which involves collaborative work with farmers for mutual learning, it is recommended similar practices should be adapted to other developing countries. Farmer training sessions should be conducted to reflect the two - way flow of information.

- In order to retrain professionals to have more on-farm work with farmers, it is recommended that funding agencies facilitate researchers and professionals to exchange information using the electronic mail system which is a valuable tool for communication as demonstrated in this workshop. More use of computers to store and transfer information should be encouraged to minimise costs.

- While the unique features of the Tune courses are acknowledged including their roles in informing decision makers (consultants, scientists, administrators) it is RECOMMENDED also to hold future Development

Workers Courses in developing countries where most on-farm research activities take place.

- It is recommended that all development agencies facilitate on-farm research work by providing financial and logistical support.

Participants

Mr Bui Xuan An
Univ.Agriculture and Forestry
Thu Duc Ho Chi Minh, Vietnam
Tel: +84 8 961472
Fax: +84 8 960713
E-mail: an%sarec%ifs.plants@ox.ac.uk

Mrs Tania Beteta
Universidad Nacional Agrarid
P.O.Box 453, Monogua, Nicaragua
Tlf. 33 19 50
E-mail: vblandon@ibw.com.ni

Professor George Chan
Consultant to United Nations University's
Zero emission research initiative
Skibuya-Ku, Tokyo 150, Japan
E-mail: 100075.3511@Compuserve.Com
Mail: 14 Poivre St. Beau Bassin, Mauritius

Ms Lotte Cortsen
MA student of political science
University of Aarhus, Denmark
Gøteborgalle 14, 12B
8200 Århus N.
Denmark
E-mail: ifsk902004@ecostat.aau.dk

Mr Fekadu Dereje
Alemaya University of Agriculture,
P.O.Box 138, Dire Dawa, Ethiopia.
E-mail: Tesfaye_Salilew@gnfido.fidonet.org

Mr Tadelle Dessie
P.O.Box 32
Debre Zeit
Ethiopia
Phone: + 251 1 338555
Fax: +251 1 338061
E-mail: dzarc.ncic@padis.gn.apc.org

Mr Frands Dolberg
Dept of Pol Science
Univ of Aarhus
Phone and fax: +45 86 152704
Fax: + 45 86 139839
E-mail: frands@po.ia.dk

Mrs Nguyen Nhut Xuan Dung
Agricultural Faculty
Cantho University
Cantho province
Vietnam
E-mail: xdung%Cantho%sarec%ifs.plants@ox.ac.uk

Mrs Pernille Fenger
Chief Programme Coordinator
DANCED (Danish Cooperation for Environment and Development)
Ministry of Environment and Energy
Danish Environmental Protection Agency
Strandgade 29, DK-1401 Copenhagen K
Denmark
Phone: +45 32 660100
Fax: +45 32 660479
E-mail: fenger@mst.mst.min.Dk

Mr Erik Fiil
Head of Environment Section
Danida, Danish Ministry of Foreign Affairs,
Asiatisk Plads 2, 1448 Copenhagen K,
Denmark.
Phone: +45 33 920000
Fax: +45 33 920493
E-mail: um@um.dk

Mr Ezekiel Goromela
ZRTC, Mpwapwa
P.O.Box 202, Mepwapwa
Tanzania.
Phone/Fax: +255 61 24526

Dr Abdi Osman Haji-Abdi
Værebrovej 58, 6.1
2880 Bagsværd

Professor Syed M. Hashemi
Programme for Research on Poverty Alleviation
Grameen Trust
Grameen Bank Bhaban
Mirpur.Two, Dhaka 1216
Bangladesh
Phone: +880 2 805422
Fax: +880 2 803559 or 806319
E-mail: hashemi@citechco.net

Mr Olaf Havsteen
Agronomist and project adviser
Danchurchaid
Nørregade 13
1165 Copenhagen K
Denmark
Phone: +45 33 152800
Fax: +45 33 911305
E-mail: dca@dca.dk

Dr Bente Ilsøe
Project administrator
Dept of Evaluation, Research and Documentation
Sts. 4, Danida,
Ministry of Foreign Affairs,
Asiatisk Plads 2, Denmark
Phone: +45 33 920942
Fax: +45 33 920493
E-mail: um@um.dk

Mr Hans Askov Jensen
Agronomist, poultry specialist
Strindbergsvej 104
2500 Valby
Phone: +45 31 161456
Fax: +45 31 161406
E-mail: askov@ibm.net

Jens Duedahl Jensen
MSc, agriculture (Lecturer)
Nordisk Landboskole
Rugårdsvej 286
5210 Odense NV
Phone: +45 66 16 18 90
Fax: +45 66 165690
E-mail: noragric@inet.uni-c.dk

Mr Borin Khieu
Dept. Animal Health and Production
Monivong Street Phnom.Penh.
Cambodia
E-mail: borin@forum.org.kh

Dr Aichi Kitalyi
ZRTC, Mpwapwa
P.O.Box 202, Mepwapwa
Tanzania.
Phone/Fax: +255 61 24526
Temporary E.Mail till July 97:
Aichi.Kitalyi@Fao.org

Mr Carl E.S. Larsen
Associate Scientist
ILRI
P.O.Box 5689
Addis Ababa
Ethiopia.
Phone: + 251 161 3215
Fax: + 251 161 1892
E-mail: c.larsen@cgnet.com

Dr Niels Kyvsgaard
Center for Experimental Parasitology
Danish Agric Univ
Bulowsvej 13
1870 Frederiksberg C, Denmark.
Phone: +45 35 282785
Fax: +45 35 282774
E-mail: niels.c.kyvsgaard@vetmi.kvl.dk

**Ass. Professor F.P.Lekule and
Dr Georg C. Kifaro**
Department of Animal Science & Production
Sokoine University of Agric.
P.O. Box 3004, Morogoro
Tanzania
Fax: +255 56 4562
Telex. 55308 UNIVMO G T Z
E-mail: SVSarwatt@hnettan.gn.apc.org.
Tel: +255 56 4617 or +255 56 4562

Karsten Lundsby
Course leader
Nordisk Landboskole
Rugårdsvej 286
5210 Odense NV
Phone: +45 66 161890
Fax: +45 66 165690
E-mail: noragric@inet.uni-c.dk

Mrs Joyce Mandal
Gl. Kalkbrænderivej 36, 3.tv.
2100 København Ø
Tlf. 35262100

Torsten Mandal
Agronomist, Ph.d.student
Danish Agric Univ
Thorvaldsensvej 40, 6th floor
1871 Frederiksberg C.
Phone: +45 35 283497
Fax: +45 35 283384

Dr Michala Marstrand
Nordisk Landboskole
Rugårdsvej 286
5210 Odense N
Phone: +45 66 16 18 90
Fax: +45 66 16 56 90
E-mail: noragric@inet.uni-c.dk

Professor John Martinussen
International Development Studies
Roskilde University, P.O. Box 260
Denmark
E-mail: johnm@iu.ruc.dk

Miss Malene Lassen
MA student (political science)
Absalonsgade 43, st.tv.
8000 Aarhus C.
Denmark
Phone: +45 86 20 10 85
E-mail: 891355@ps.aau.dk

Mrs Nguyen Thi Loc
Department of Animal Nutrition,
Hue Agricultural and Forestry University,
Vietnam.
E-mail: loc%hue%ifs.plants@ox.ac.uk

Mr Bui Xuan Men
Faculty of Agriculture, Cantho University,
Cantho, Vietnam.
Phone: + 84 71 831348
E-mail: men%cantho2%cantho%sarec%ifs.plants@ox.ac.uk

Mrs Nguyen Thi Mui
Goat and Rabbit Research Centre,
Sontay, Hatay, Vietnam.
E-mail: mui%bavi%hue%ifs.plants@ox.ac.uk

Hanne Nielsen
M.A. Social science
DARUDEC
Stamholmen 112
2650 Hvidovre
Denmark
Phone: +45 36 39 07 27
Fax: +45 36 77 28 29
E-mail: darudec@inet.uni-c.dk

Helle Katrine Nielsen
M.Sc. student (biology)
University of Aarhus
Trøjborgvej 36, 3
8200 Århus N.
Denmark
Phone: +45 86 10 17 14
E-mail: katrine@stat.bio.aau.dk

Dr Brian Ogle
Sveriges Lantbruksuniversitet
Institutionen för husdjurens utfodring och vård
Box 7024
S-750 07 Uppsala
Fax +46 18 672995
Phone +46 18 672061
E-mail: Brian.Ogle@huv.slu.se

Dr Anders Permin
Dept of Microbiology
Danish Agric Univ
Bülowsvej 13
1871 Frederiksberg C
E-mail: anders.permin@vetmi.kvl.dk

Docent Poul Henning Petersen
Dept of Animal Science & Animal Health
Danish Agric Univ
Bülowsvej 13
1870 Frederiksberg C.
Phone: +45 35 283093
Fax: +45 35 283087
E-mail: php@kvl.dk

Jens Ole Poulsen
Agricultural teacher
Voldtoftevej 36 A
5620 Glamsbjerg
Fax: +45 64 731857
Phone: +45 64 73185700
Nordic Agric. College
Fax +45 66 165690
E-mail: noragric@inet.uni-c.dk

PhD, DSc Thomas R. Preston
University of Acriculture and Forestry
Thu Duc, Ho Chi Minh City
Vietnam
Tel: +84 8 961472
Fax: +84 8 960713
E-mail: thomas%preston%sarec%ifs.plants@ox.ac.uk

Dr Jules Pretty
Sustainable Agriculture Programme
International Institute for Environment and Development
3 Endsleigh Street, London WC1H ODD
Phone: +44 171 388 2117
Fax: + 44 171 388 2826
E-mail: iiedagri@gn.apc.org

Dr Janet Riley
IACR-Rothamsted
Harpenden
Hertfordshire AL5 2JQ
England
Phone: +44 1582 763133 ext 2370
Fax: +44 1582 467116/760981
E-mail: janet.Riley@bbsrc.ac.uk

Miss Lylian Rodriquez
University of Acriculture and Forestry
Thu Duc, Ho Chi Minh City
Vietnam
Tel: +84 8 961472
Fax: +84 8 960713
E-mail: lylian%sareclr%sarec%ifs.plants@ox.ac.uk

Professor M.Saadullah
Department of General Animal Science
Bangladesh Agricultura University
Mymensingh-2202, Bangladesh
Phone (office) +880 91 5695-7/289
(private) +880 91 5815
E.mail: saad@drik.bgd.toolnet.org

Dr Md. A. Saleque
Manager, EIG, RDP
BRAC, 66 Mohakhali
Dhaka 1212, Bangladesh
Phone: +880 2 884180-7
Fax: +880 2 883542
E-mail: brac@drik.bgd.toolnet.org

Mr Karsten C. Schleiss
M.Sc. (Agric.)
Senior Consultant
Danagro Adviser a/s
Granskoven 8
2600 Glostrup
Denmark
Tel: +45 43 434590
Fax +45 43 434049
E-mail: kcsdan@carlbro.dk

Mr Olav Sejerøe
Agricultural teacher
Lyngby Landbrugsskole
Ledreborg Alle 50
4000 Roskilde
Phone/Fax: +45 46 480532

Mrs Vibeke Tuxen
The Rainforest Group Nepenthes
Odensegade 4 B
P.O.Box 601
8100 Aarhus C
Tel: +45 86 13 52 32
Fax: +45 86 12 51 49
E-mail: nep.dk@inet.uni-c.dk

Dr E.R. Ørskov
The Rowett Research Institute
Bucksburn
Aberdeen AB2 9SB
Scotland
Phone: +44 1224 712751
Phone: +44 1224 716614 Direct
Fax: +44 1224 716687
Home: +44 1651 862351
E-mail: ero@rri.sari.ac.uk